Learning Differential Equations through DERIVE

John Berry
Centre for Teaching Mathematics
University of Plymouth

Brian Lowe
Division Mathematics
School of Accounting and Mathematics
University of Glamorgan

Studentlitteratur

Other books in the "Learning through Computer Algebra" Series
Published by Chartwell-Bratt and Studentlitteratur. Series editor: John Berry.

"Learning Mathematics through DERIVE", J. S. Berry, E. Graham, A. J. P. Watkins
"Learning Numerical Analysis through DERIVE", T. Etchells, J. Berry
"Learning Linear Algebra through DERIVE", B. Denton
"Learning Differential Equations through DERIVE" by B. Lowe, J. Berry.
"Learning Modelling through DERIVE", S. Townend, D. Pountney.

British Library Cataloguing in Publication Data
A catalogue record for this book is available from the British Library

http://www.studentlitteratur.se

Printed in Sweden
Studentlitteratur, Lund
Art.no. 6694
ISBN 91-44-00864-3

Printing/Year 1 2 3 4 5 6 7 8 9 10 | 03 02 01 2000 99 98

Contents

Preface

Everywhere around us we can observe things that are continually changing. On a large scale, the position of the earth changes as it revolves around the sun while at a more everyday level the speed of a car changes when we brake or accelerate and the temperature of a freshly made cup of tea drops while we wait for it to reach a drinkable temperature.

The mathematical subject area that describes how quantities change is calculus; this was first developed in the latter half of the seventeenth century by Newton and Leibnitz (independently of each other). One of Newton's main interests was investigating and describing the motion of the planets about the sun. At this time Kepler had already predicted, empirically using astronomical data on the positions of the planets, that each planet moves in an elliptical orbit about the sun. Using calculus, Newton was able to formulate his laws of motion and gravitational force and confirm by calculation Kepler's assertions.

The rate of change of one quantity (such as velocity) with respect to another (such as time) is called the derivative (of velocity with respect to time). When derivatives are combined with other quantities in an equation we have a differential equation, which is the mathematical description of the system being studied (or modelled). Solving the differential equation reorganises the information into a form which can predict how the values of the quantities will change relative to each other.

For example, Newton's Second Law of motion is

$$\text{mass} \times \text{acceleration} = \text{forces acting}$$

This can be expressed symbolically as $ma = F$ or, using the result that acceleration measures the rate of change of velocity, or $a = \frac{dv}{dt}$, we obtain the differential equation $m\frac{dv}{dt} = F$. From this equation we can, provided the force F can be quantified, predict the velocity v at any time t.

Many differential equations arise from the study of physical situations where the problem is modelled using mathematics. The steps in mathematical modelling are described in Section 1 of Chapter 1 and differential equations arising from the study of various real world situations are derived in Section 2.

Chapters 2, 3 and 4 deal with first order differential equations. In Chapter 2 we look at how these equations can be solved analytically, while Chapter 3 deals with their numerical solution. Chapter 4 looks in greater depth at some problems from the real world which can be modelled with first order differential equations.

Chapters 5 and 6 deal with second order differential equations. In Chapter 5 we discuss their analytical solution and Chapter 6 looks at some of their real world applications.

This text contains the 'standard theory' of first and second order differential equations, worked examples, exercises, modelling problems and DERIVE Activities. The aim of each DERIVE Activity is to show how symbolic algebra systems such as DERIVE can be used in the solution phase of a modelling task. This allows the problem solver to concentrate on the formulation of the model and interpretation of the solution. The DERIVE Activities could be used as a resource for workshop sessions with large groups and for demonstrating solutions of differential equations and their graphical interpretations in lectures and tutorials.

What makes this text different from the many others on the market at this level is the use of the computer algebra package DERIVE. This powerful tool does for algebraic manipulation what the calculator does for arithmetic. However a word of warning! The authors believe that the introduction of DERIVE into the learning classroom *does not replace* the need to learn and understand algebraic manipulation and the basic methods of solving differential equations. Our experience suggests that the *best* users of DERIVE are those students who are good at algebraic manipulation and understand the concepts behind algebra and calculus. The text attempts to blend the traditional skills with the power of DERIVE. To encourage good practice many of the exercises should be done 'by hand' first and then the results checked using DERIVE.

We have assumed that the reader has had an introduction to DERIVE before coming to this text. The commands that you should be familiar with are **Author**, **Plot**, **Substitute**, **Expand**, **Factor**, **Remove**, **Simplify**, **Approximate**. The instructions in the DERIVE Activities, screen dumps and many of the figures have been produced using DERIVE for Windows. This should provide no problems for DOS users of DERIVE as the same command words are used in each version.

We are extremely grateful to Sharon Ward for her skill and meticulous care in producing the camera ready copy for this text. We appreciate her patience and advice in the many changes to the page design and layout. We also thank Pauline Framingham for producing the artwork and Mike Herring for his helpful comments.

We would welcome comments from teachers and students for improving this text. Since this is one of the first textbooks to fully integrate a computer algebra package into the learning of mathematics it is inevitable that the mix of DERIVE activities and standard theory could be improved. If you have constructive criticisms to help the writing of a second edition please write to the authors in Plymouth.

We hope that the teacher and learner will enjoy solving differential equations.

Brian Lowe
Division of Mathematics
School of Accounting and Mathematics
University of Glamorgan
South Wales
CF37 1DL
email: bclowe@glam.ac.uk

John Berry
Centre for Teaching Mathematics
University of Plymouth
Drake Circus, Plymouth
Devon PL4 8AA, UK
Telephone/FAX: 01752 232772
email: jberry@plymouth.ac.uk

Series Preface

This series is designed to encourage the teaching and learning of various courses in mathematics using symbolic algebra. Each text in the series will include the 'standard theory' appropriate to the subject, worked examples, symbolic algebra exercises and Activities so that students have a sound conceptual base on which to build. It is through these Activities that the learning of mathematics through investigations and discussion is encouraged. The aims of each Activity are to provide an investigation to introduce a new topic and to introduce the symbolic algebra commands for using as a manipulator in mathematical problem solving. Mathematics teaching and learning is changing as new technology becomes more readily available and these texts are written to support the new methods.

John Berry
Centre for Teaching Mathematics
University of Plymouth

1

A Modelling Approach to Differential Equations

1.1 MATHEMATICAL MODELLING

In order to investigate a problem mathematically it is necessary to describe it using symbols and equations, which may then be solved using standard mathematical techniques. We hope that the mathematical solution matches the real world solution and so can help explain past behaviour, predict what will happen in the future and analyse the effect of changing values or situations on the solution.

A model is a description of a problem using mathematics and could involve simultaneous or polynomial equations or a set of inequalities, but here we are interested in models that result in differential equations.

There are three main stages in the modelling process:

formulate the given problem as a mathematical problem
solve the mathematical problem. In this text this will be a differential equation
interpret the mathematical solution by relating back to the real world situation.

The activities involved in mathematical modelling can be summarised in the following diagram, which is based on that used in several Open University courses.

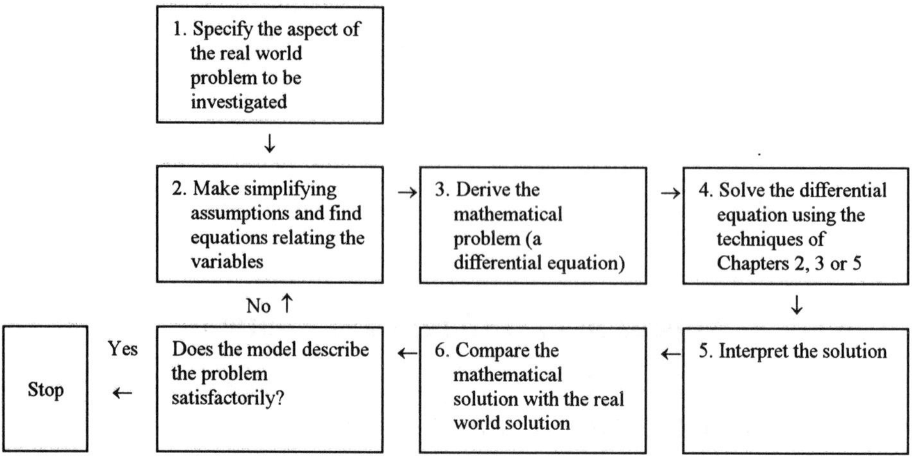

Box 1 A real problem will have many aspects. A model will, to start with at least, look at one of them in a simple way. Before using any symbols describe in words the problem precisely.

Box 2 List everything you can think of that is connected with the problem and decide which features of the problem are to be investigated. In order to simplify and obtain a straightforward initial model make some assumptions. This will eliminate some variables or the investigation of a feature that you think will have a marginal effect on the problem. Use any relevant physical laws to write down relationships between the variables.

Box 3 Interpret the specification in Box 1 in terms of mathematics. Is the problem solvable? If not return to Box 2 and simplify further.

Box 4 Solve the mathematical problem. In this text this will be a differential equation which can be solved using the techniques of Chapters 2, 3 or 5.

Box 5 Obtain data for the parameters appearing in your model and find out what the solution of the mathematics predicts. Draw suitable graphs. It can be useful to make qualitative checks on the solution; does the solution follow common sense, how sensitive is the solution to (small) changes in the parameters?

Box 6 Obtain data concerning the real world behaviour of the system you are modelling and compare it with the results of your mathematical model.

If your model does not adequately predict what really happens then it is necessary to go round the loop again. Return to Box 2, refine your model by incorporating an extra feature or relaxing a simplifying assumption and repeat boxes

3 to 6. This can be done as many times as is necessary to obtain a model that satisfactorily describes the initial problem.

1.2 SOME MATHEMATICAL MODELS

Four different situations are modelled in this section, and the differential equations obtained are solved in succeeding chapters.

Application 1 Emptying the bath

How long does it take for a bath to empty?
How much faster does the first half empty than the second half?

Make the following assumptions.

A1. The bath has a flat bottom and vertical sides.

A2. The bath is initially full.

A3. The taps are turned off.

A4. Toricelli's law applies (this states that for a non viscous fluid flowing through a small hole the speed of the fluid particles at the hole is the same as if they were released from rest from the fluid surface and allowed to fall under gravity with no air resistance).

The variables and parameters are:

Variable/ parameter	Represents	Units
h	initial height of water in the bath	m
A	area of the bottom of the bath	m^2
S	area of the plug hole	m^2
m	mass of a particle of water	kg
t	time since the plug was removed	s
x	height of water in the bath at time t	m
v	speed of water leaving the plug hole at time t	ms^{-1}

The next step is to formulate relationships between the variables. These will form the mathematical model.

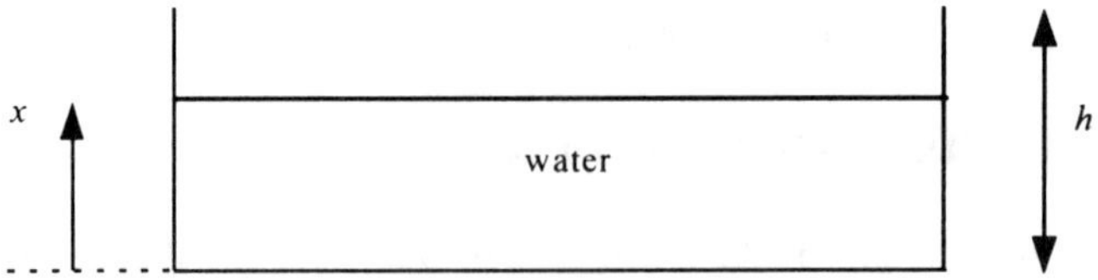

Consider a particle of water which falls from the surface of the water towards and through the plug hole

Loss of potential energy = $mg\,x$
Gain in kinetic energy = $\frac{1}{2}mv^2$

From A4 energy is conserved and so $\frac{1}{2}mv^2 = mg\,x$. Solving for v, the speed of water passing through the plug hole is $v = \sqrt{2gx}$.

The rate of flow of water through the plug hole is given by area × speed = $S\sqrt{2gx}$. This will be the same as the rate of decrease of water inside the bath = –rate of increase in volume $= -\frac{d}{dt}(Ax) = -A\frac{dx}{dt}$.

Hence we obtain

$$-A\frac{dx}{dt} = S\sqrt{2gx}$$

$$\frac{dx}{dt} = -\frac{S\sqrt{2g}}{A}\sqrt{x}$$

or $$\frac{dx}{dt} = -k\sqrt{x} \qquad \left(\text{where } k = \frac{S\sqrt{2g}}{A}\right)$$

The mathematical problem is to solve the equation

$$\frac{dx}{dt} = -k\sqrt{x} \qquad \text{given } x = h \text{ when } t = 0$$

and

(i) find the time T for which $x = 0$;
(ii) compare the time from $x = 0$ to $x = \frac{1}{2}h$ with that for $x = \frac{1}{2}h$ to $x = h$.

This equation is solved in Section 4.1 of Chapter 4.

□

Application 2 Population Models

Estimating the future growth or decline of populations is an important problem. In Britain the birth rate has been in decline and people are living longer. A good model of the future population is needed so that resources, which will come from the decreasing proportion of the population in work, can be allocated where most needed.

Some animal populations are exploited (and sometimes over exploited) for food. The whale population, for example, has been drastically overfished resulting in its near extinction.

Throughout the world many fisheries are under threat due to the conflicting demands of the consumer and conservation. Quotas suggested by scientists are regularly exceeded and the total catch has been decreasing. Unless the trend is reversed some fish stocks could collapse, but this is difficult to achieve as it could mean unemployment among fishermen and increased fish prices. Politicians are seldom able to take the long term view.

A mathematical model would help to investigate whether there is a sensible compromise, whereby no more are taken for food than can be replenished by reproduction.

A population increases due to births and immigration.

A population decreases due to deaths by natural causes, deaths by unnatural causes (such as wars in humans or exploitation in animals) or by emigration.

Then

Increase in population = (births + immigration) – (deaths + emigration)

We make the following assumptions:

A1. There is no migration (immigration or emigration).

A2. All deaths are by natural causes.

We then obtain

Increase in population = births – deaths (1.1)

The variables and parameters are:

Variable/ parameter	Represents	Units
t	time since population was studied	days/months/years
P	population size at time t	no units
δt	small interval of time	days/months/years
δP	small change in population in time δt	no units
B	birth rate	$\frac{1}{\text{time}}$
D	death rate	$\frac{1}{\text{time}}$
P_0	population at time $t = 0$	no units
M	equilibrium population	no units

Suppose the population at time t is $P(t)$. Then the increase δP in the population in a small time interval δt is given by

$$\delta P = P(t + \delta t) - P(t)$$

Application 2A The Malthusian, or exponential, model

We make assumptions about the birth and death rates.

A3a. The number of births is proportional to the time interval δt and the population at the start of the time interval. Then,

$$\text{number of births} = BP(t)\delta t \qquad (B \text{ constant}).$$

A4a. The number of deaths is also proportional to P and δt

$$\text{number of deaths} = DP(t)\delta t \qquad (D \text{ constant})$$

Expressing (1.1) symbolically:

$$\delta P = BP\delta t - DP\delta t$$

$$\frac{\delta P}{\delta t} = (B - D)P$$

In the limit as $\delta t \to 0$ this becomes

$$\frac{dP}{dt} = (B - D)P \tag{1.2}$$

or $$\frac{dP}{dt} = \alpha P \qquad \text{where } \alpha = B - D.$$

The mathematical problem is to solve $\frac{dP}{dt} = \alpha P$ given the initial population $P = P_0$ at $t = 0$.

This equation is solved in Section 4.3 of Chapter 4.

Application 2B The Logistic model

From equation (1.2), $\frac{1}{P}\frac{dP}{dt} = \alpha$ and this means that the proportionate growth rate $\frac{1}{P}\frac{dP}{dt}$ is constant.

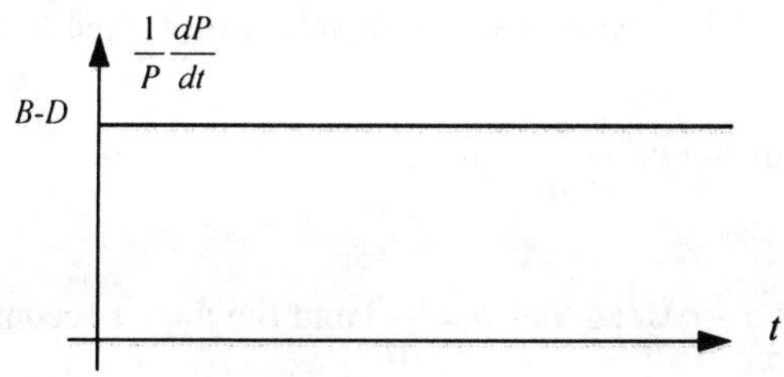

Figure 1.1 For the exponential model, $B - D$ is constant.

The logistic model assumes that this is not the case. At low population levels there might be a high growth rate due to the abundance of living space and food, whereas when the population is large lack of food and cramped living space might reduce the birth rate and more disease increase the death rate. That is, the proportionate growth rate will decrease as the population increases. We make the assumption

A3b. The proportionate growth rate $\frac{1}{P}\frac{dP}{dt}$ is a decreasing linear function of P.

Then $\frac{1}{P}\frac{dP}{dt} = a - bP$, where a and b are positive constant.

We further assume that there is an equilibrium population level where there is just enough space and food to sustain the population and then the population will neither grow nor decline. The births will then exactly cancel the deaths.

A4b. The proportionate growth rate is zero at an equilibrium population level M. That is $\frac{1}{P}\frac{dP}{dt}=0$ when $P = M$. Figure 1.2 shows a graph of proportionate growth rate against population.

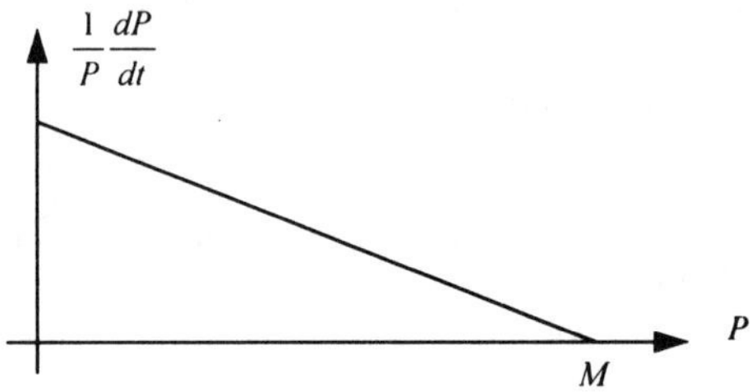

Figure 1.2 Proportionate growth rate against population

From A3b our model is $\frac{1}{P}\frac{dP}{dt}=a-bP$.

Using A4b: $0 = a - bM$, so that $b=\frac{a}{M}$ and the model becomes

$$\frac{1}{P}\frac{dP}{dt}=a-\frac{aP}{M}$$

or
$$\frac{dP}{dt}=aP\left(1-\frac{P}{M}\right) \tag{1.3}$$

This is the logistic model and is solved in Section 4.3 of Chapter 4.

□

Application 3 Throwing a ball in the air

Suppose a ball is thrown vertically upwards. There are a number of questions we might ask. How high does it go? How long does it take to get there? Does the size of the ball make a difference? What about its weight? What happens on the downward flight - does it take the same time to come down as to go up and with what speed does it hit the ground?

We will follow the modelling diagram.

Box 1 The problem is defined as follows:
Given an object of known mass projected vertically upwards with known speed, predict the height the object will reach and the time taken to reach this height.

Box 2 Simplifying assumptions:

A1. The object is treated as a particle. (This means that its size is negligible, has no internal structure and is not subject to effects such as those due to spin.)

A2. Newton's laws of motion apply.

A3. The ball is thrown from ground level.

A4. The magnitude of the air resistance force is proportional to the speed of the ball. (A simpler model might choose to ignore the effects of air resistance.)

We can now define the variables and parameters:

Variables/ Parameters	Represents	Units
t	time from projection	seconds s
x	height above ground at time t	metre m
v	speed at time t	metres per second ms^{-1}
a	acceleration at time t	ms^{-2}
m	mass of ball	kg
g	acceleration due to gravity	ms^{-2}
u	initial speed of ball	ms^{-1}
F	weight of ball	newtons N
R	magnitude of air resistance force	N
k	constant of proportionality (air resistance)	kgs^{-1}

Figure 1.3 shows the forces on the ball during its motion.

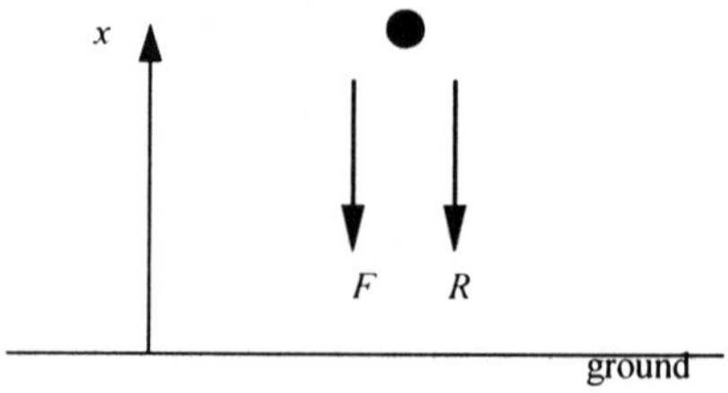

Figure 1.3 Force on the ball

The magnitudes of the forces are $F = mg$ and, from A4, $R = kv$. Applying Newton's second law of motion (mass × acceleration = forces acting)

$$ma = -F - R$$
$$ma = -kv - mg$$

As acceleration is the rate of change of velocity it follows that $a = \dfrac{dv}{dt}$ and the equation becomes

$$m\frac{dv}{dt} = -kv - mg$$

Also, velocity is the rate of change of distance, so that $v = \dfrac{dx}{dt}$.

Box 3 The mathematical problem is

(i) solve $m\dfrac{dv}{dt} = -kv - mg$ given $v = u$ at $t = 0$ to find the time T when $v = 0$;

(ii) solve $\dfrac{dx}{dt} = v$ given $x = 0$ at $t = 0$ to find x when $t = T$.

This problem is discussed further in Example 4J, Chapter 4.

□

Application 4 Car Suspensions

The vibrations caused by a car moving over a bumpy road are transmitted through the tyres and car body to the passengers. Manufacturers try to design suspension systems so as to make the ride for the passengers as comfortable as possible. If a car hits a bump or pothole the spring suspension will take the brunt of the ensuing vibration and the dampers will ensure that resulting oscillations die out quickly.

A simple model of a car suspension system is developed below, using the following assumptions.

A1. The car is a particle.

A2. The car suspension, shock absorbers and tyres are modelled by a spring and dashpot (the damper) in parallel. (A more detailed discussion on springs and dashpots is given in Section 6.1 of Chapter 6.)

A3. Hooke's law applies to the spring. That is, the force in the spring is proportional to its extension.

A4. The dashpot law applies to the damper. That is, the force always opposes motion and its magnitude is proportional to the speed of the plunger in the dashpot (or to the rate of change of length of the dashpot).

A5. The road surface is sinusoidal in shape.

Figure 1.4 shows a simplified spring/mass/dashpot system representing the car and its suspension system.

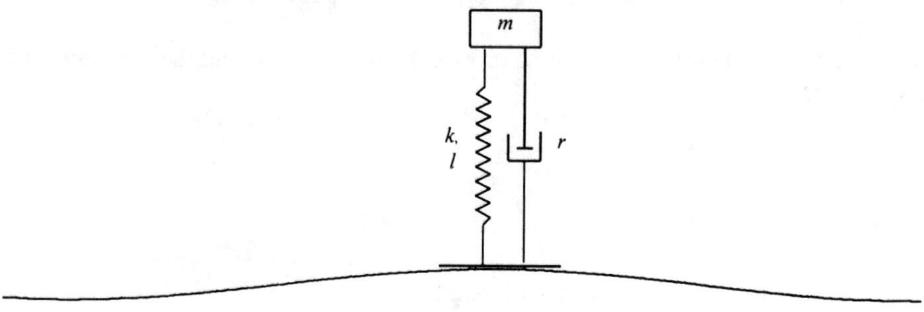

Figure 1.4 Model of car and suspension

The spring has natural length l and stiffness k. The dashpot has constant r. The mass of the car is m.

We are interested in the vertical motion of the car and it may be helpful to think of the car as being stationary horizontally, with the road moving past beneath it.

Define x as the height of the car above the mean road level. Define y as the height of the indulating road above the mean road level. In order to derive the equation of motion we assume that the spring is extended and the dashpot lengthening. The forces acting on the car are as shown in Figure 1.5.

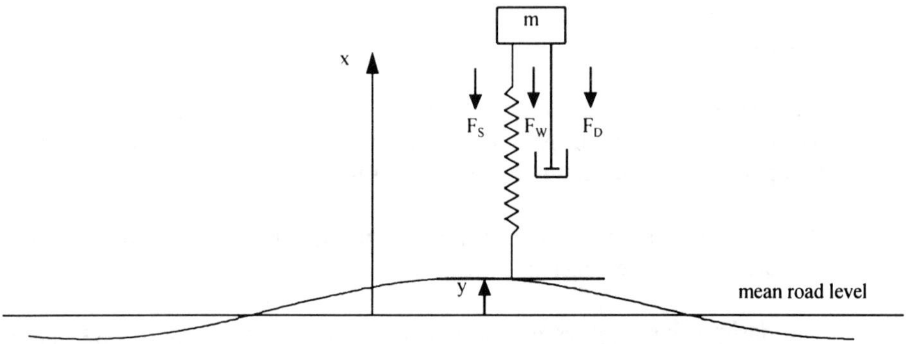

Figure 1.5 Forces on the car

The weight of the car is $F_w = mg$. Hooke's law for a spring states that

Tension $\propto$ extension

or Tension = $k \times$ extension (k, constant, is the stiffness)

Here the length of the spring is $x - y$, so its extension is $x - y - l$ and the magnitude of the spring force is

$$F_s = k(x - y - l)$$

The dashpot law states that

Dashpot force $\propto$ rate of change of length

or Dashpot force = $r \times$ rate of change of length (r is the dashpot constant)

Here the length of the dashpot is also $x - y$ and so the magnitude of the dashpot force is

$$F_D = r\frac{d}{dt}(x-y) = r(\dot{x}-\dot{y}) \qquad \text{(where } \dot{} \text{ stands for } \frac{d}{dt}\text{)}$$

Applying Newton's second law to the car:

$$m\ddot{x} = -F_s - F_w - F_D \qquad \text{(using } \ddot{x} = \frac{d^2x}{dt^2} \text{ for acceleration)}$$

Inserting the above expressions for the forces this becomes

$$m\ddot{x} = -k(x-y-l) - mg - r(\dot{x}-\dot{y})$$
$$m\ddot{x} + r\dot{x} + kx = ky + kl - mg + r\dot{y}$$

To put the equation in a form ready for solution we need an expression for y (and $\dot{y}$) in terms of t. From A5, and noting that the origin for y is the mean road level, the equation for y will be of the form $y = a \cos \omega t$. This has amplitude a and period $T = \frac{2\pi}{\omega}$.

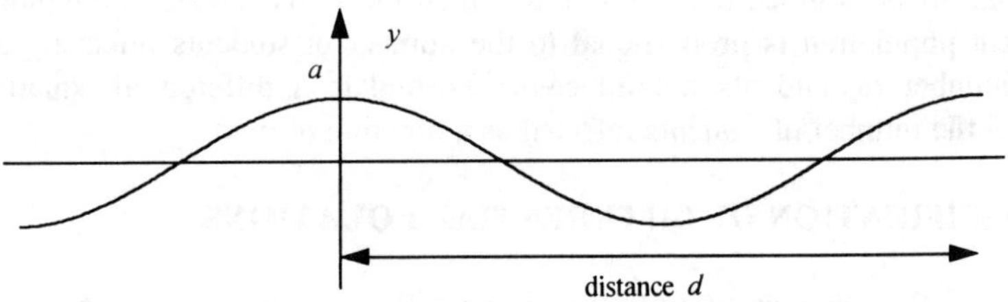

Figure 1.6 Model for the shape of the road

If the car is travelling at constant speed u then the time for it to cover the distance d between two successive maxima is $T = \frac{d}{u}$. Comparing the two expressions for T, $\frac{2\pi}{\omega} = \frac{d}{u}$, or $\omega = \frac{2\pi u}{d}$. Hence the road surface has equation $y = a\cos\left(\frac{2\pi ut}{d}\right)$ and so $\dot{y} = -\frac{2\pi au}{d}\sin\left(\frac{2\pi ut}{d}\right)$. The equation of motion of the car then becomes

$$m\ddot{x} + r\dot{x} + kx = kl - mg + ka\cos\left(\frac{2\pi ut}{d}\right) - \frac{2\pi aur}{d}\sin\left(\frac{2\pi ut}{d}\right)$$

The solution of this equation is discussed in Example 6N, Chapter 6.

□

Exercise 1A

1. What other assumptions should we make in formulating the model in Application 3?

2. List the variables/parameters used in Application 4. Explain what they represent and give their units.

3. During the decay of a radioactive substance, the rate at which mass is lost is proportional to the mass present at that instant. Formulate a differential equation model to describe the process. List your assumptions and define the variables/parameters used.

4. After a drug has been injected into the bloodstream, its concentration gradually decreases and the drug loses its effectiveness. Formulate a differential equation model to describe this situation. List your assumptions carefully. Define the variables/parameters that you have used.

5. An individual in a population of students in a University becomes infected with a virus. It is assumed that the rate at which the virus spreads throughout the student population is proportional to the number of students infected, and to the number of students not infected. Formulate a differential equation to model the number of students infected as a function of time.

1.3 CLASSIFICATION OF DIFFERENTIAL EQUATIONS

In section 2 of this chapter we introduced the following differential equations as models of physical situations.

Application 1: $$\frac{dx}{dt} = -k\sqrt{x} \qquad \text{with } x = h \text{ when } t = 0 \qquad (1.4)$$

Application 2: $$\frac{dP}{dt} = \alpha P \qquad \text{with } P = P_0 \text{ when } t = 0 \qquad (1.5)$$

$$\frac{dP}{dt} = aP\left(1 - \frac{P}{M}\right) \qquad \text{with } P = P_0 \text{ when } t = 0 \qquad (1.6)$$

Application 3: $$m\frac{dv}{dt} = -kv - mg \qquad \text{with } v = u \text{ when } t = 0 \qquad (1.7)$$

$$\frac{dx}{dt} = v \qquad \text{with } x = 0 \text{ when } t = 0 \qquad (1.8)$$

Application 4: $m\ddot{x}+r\dot{x}+kx = kl-mg+ka\cos\left(\frac{2\pi ut}{d}\right)-\frac{2\pi aur}{d}\sin\left(\frac{2\pi ut}{d}\right)$ (1.9)

Your initial reaction may be that differential equations come in all shapes and sizes. Is there a common thread that runs through them?

The first five equations contain the first order derivative only whereas equation (1.9) has first and second order derivatives. This feature of the highest order derivative will be an important part of the classification of differential equations that guides us to a method of solution.

Another important feature in classifying differential equations is illustrated if we compare equations (1.5) and (1.6). For equation (1.5) the right hand side is a linear function of P (i.e. it has power 1) whereas in equation (1.6) the right hand side is a non-linear function of P (in fact it is a quadratic function).

Now we put these observations together to provide a method of classifying differential equations which will lead to different methods of solution.

The order of a differential equation

The order of the highest derivative in a differential equation is called the **order** of a differential equation.

For example,

$$\frac{dy}{dx} = y^2 + x^2$$

is a first order differential equation with dependent variable y and independent variable x. The general form of a first order differential equation is

$$\frac{dy}{dx} = f(x,y)$$

where the function f does not contain any derivatives of y with respect to x.

$$\frac{d^2x}{dt^2}+4x = 3\cos t$$

is a second order differential equation with dependent variable x and independent variable t. The general form of a second order differential equation is

$$\frac{d^2y}{dx^2} = f\left(x,y,\frac{dy}{dx}\right)$$

The function f may contain first order derivatives but not those of second order or higher.

In the two general forms we have used variables x (independent) and y (dependent). We can of course use any symbols as long as we identify the independent and dependent variables. For example, if we write

$$\frac{dP}{dt} = f(t, P)$$

then t is the independent variable and P is the dependent variable.

In this book we investigate the solution of first and second order equations only.

Linearity of a differential equation

A differential equation is said to be **linear** if the dependent variable and its derivatives are linear i.e. they are all to the power 1. Otherwise the differential equation is non-linear. For example,

$$\frac{dy}{dx} = y + x^2$$ is a linear first order differential equation

$$\frac{dy}{dx} = y^2 + x^2$$ is a non-linear because of the y^2 term.

The general form of a linear first order equation in variables x and y is

$$\frac{dy}{dx} = a(x)y + f(x) \tag{1.10}$$

where $a(x)$ and $f(x)$ may be functions of x only.

The general form of a linear second order differential equation is

$$a(x)\frac{d^2y}{dx^2} + b(x)\frac{dy}{dx} + c(x)y = f(x) \tag{1.11}$$

where a, b, c and f may be functions of the independent variable x only.

Homogeneity of a differential equation

If the function $f(x) = 0$ in the general forms (1.10) or (1.11) then the differential equation is said to be homogeneous. Otherwise it is non-homogeneous (or inhomogeneous).

The following table shows examples of differential equations and their classification.

Table 1.1 Classifying differential equations

differential equation	order	linearity	homogeneity
$\dfrac{d^2x}{dt^2}+3x^2=4t\sin t$	second	nonlinear	non-homogeneous
$\dfrac{dy}{dx}+5y=x$	first	linear	non-homogeneous
$\dfrac{dy}{dt}=t+\sin y$	first	nonlinear	non-homogeneous
$m\dfrac{dv}{dt}=-kv-mg$	first	linear	non-homogeneous
$\dfrac{dP}{dt}=\alpha P$	first	linear	homogeneous
$\dfrac{dP}{dt}=aP\left(1-\dfrac{P}{M}\right)$	first	nonlinear	homogenous
$\dfrac{dx}{dt}=-k\sqrt{x}$	first	nonlinear	homogeneous
$\dfrac{d^2y}{dx^2}+4\dfrac{dy}{dx}-3y=0$	second	linear	homogeneous
$\dfrac{d^3x}{dt^3}-4\left(\dfrac{dx}{dt}\right)^2=3x$	third	nonlinear	homogeneous
$m\ddot{x}+r\dot{x}+kx=a\sin(\omega t)$	second	linear	non-homogeneous

Exercise 1B

Classify the following differential equations:

1. $\dfrac{dx}{dt} - 4x = 0$

2. $\dfrac{dy}{dx} + 3y^2 = x$

3. $y\dfrac{dy}{dx} = 4$

4. $\dfrac{d^2y}{dx^2} + 7\dfrac{dy}{dx} + 2y = 0$

5. $\dfrac{d^2x}{dt^2} + 8\dfrac{dx}{dt} + 6x = t\sin(4t)$

6. $\dfrac{d^4y}{dx^2} + 3\dfrac{d^2y}{dx^2} + 2y = e^{4x}$

7. $\ddot{x} - 3\dot{x} = \cos(2t)$

8. $\dfrac{dx}{dt} + x^2 = 4t - 3$

2

Analytical Solution of First Order Differential Equations

In Chapter 1 we saw how differential equations arise from the analysis of physical problems. For example we saw that the logistic model for population growth led to the equation $\frac{dP}{dt} = aP\left(1 - \frac{P}{M}\right)$. In this chapter we shall show how this, and other, equations can be solved analytically so that an equation can be found for P in terms of t.

In general we wish to find the analytical solution of first order equations of the form $\frac{dy}{dx} = f(x, y)$, but first we shall see how we can predict, with the help of DERIVE, the qualitative behaviour of the solution through the construction of a direction field.

2.1 DIRECTION FIELDS

The value of the derivative $\frac{dy}{dx}$ gives information about the graph of y. We know that $\frac{dy}{dx} = 0$ identifies stationary points and that the sign of the derivative indicates whether y is increasing or decreasing. In general the value of $\frac{dy}{dx}$ at any point gives the gradient of the tangent to the curve at that point.

For the equation $\frac{dy}{dx} = f(x, y)$ we calculate the gradient $\frac{dy}{dx}$ of the solution curve at various points in the (x,y) plane and draw a short line of the correct slope at each point. The set of all such lines is called the **direction field**.

As an example consider the equation $\frac{dy}{dx} = 1 - y$. Here the right hand side, $1 - y$, depends on y only. Calculating $\frac{dy}{dx}$ for various y values gives:

y (any x)	-1	0	1	2	3	4
$\dfrac{dy}{dx}$	2	1	0	-1	-2	-3

The direction field is then

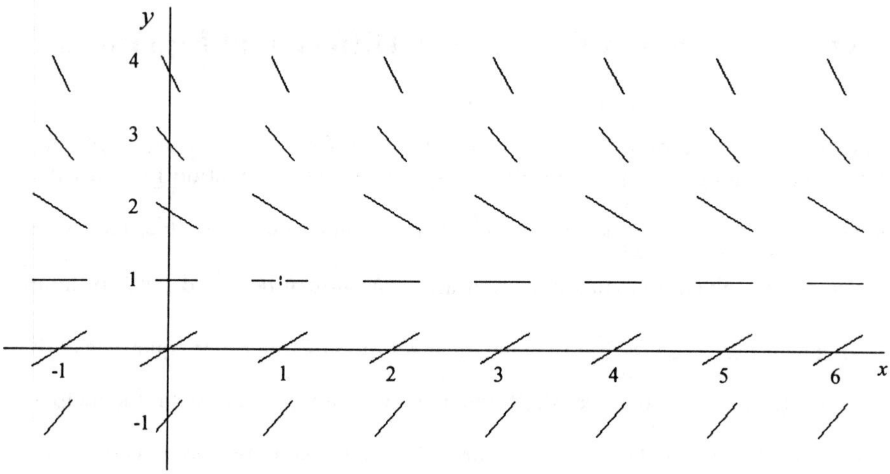

Figure 2.1 The direction field for $\dfrac{dy}{dx} = 1 - y$

Note that the gradient along any horizontal line is constant as the right hand side of the equation, $1 - y$, is independent of x.

A **solution curve** or **trajectory** can now be drawn from any given point (called an initial condition) so that its gradient agrees with the direction field at each point. Figure 2.2 shows solution curves for various initial conditions.

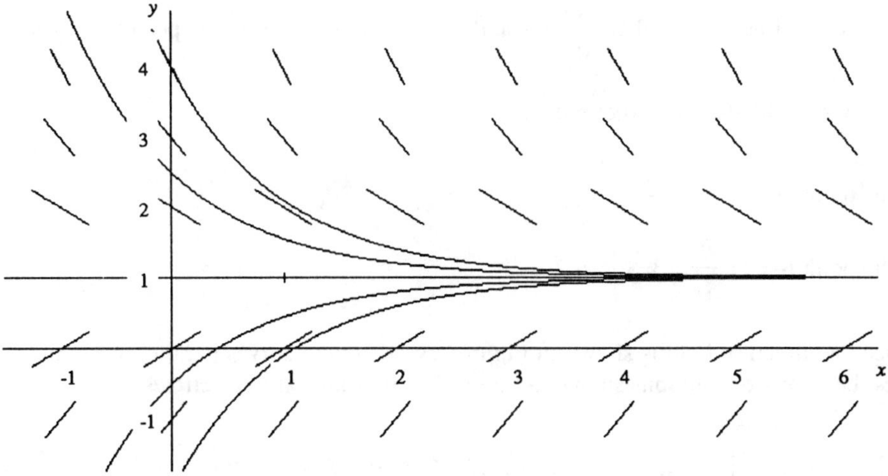

Figure 2.2 Solution curves for $\frac{dy}{dx} = 1 - y$

Example 2A

Draw a direction field for $\frac{dy}{dx} = -\frac{x}{y}$ and sketch a few solution curves. What shape are they? Check that $y = \sqrt{a^2 - x^2}$, where a is a constant, is a solution of the equation.

Solution

Rather than tabulate $\frac{dy}{dx}$ for a large number of x and y values we look for points where $\frac{dy}{dx}$ has the same value.

If $x = 0$ (the y axis) then $\frac{dy}{dx} = 0$ and the tangent is horizontal.

If $y = 0$ (the x axis) then $\frac{dy}{dx}$ is infinite and the tangent is vertical.

If $x = y$ (the line $y = x$) then $\dfrac{dy}{dx} = -1$ and the tangent is at 135° to the positive x axis.

If $x = -y$ (the line $y = -x$) then $\dfrac{dy}{dx} = 1$ and the tangent is at 45° to the positive x axis.

We could also plot other points:

At the point (1,2) $\dfrac{dy}{dx} = -\dfrac{1}{2}$

At the point (−3,1) $\dfrac{dy}{dx} = 3$

Hence the direction field is shown in Figure 2.3. Drawing curves parallel to the field gives the family of solutions shown in Figure 2.4; they appear to be circles.

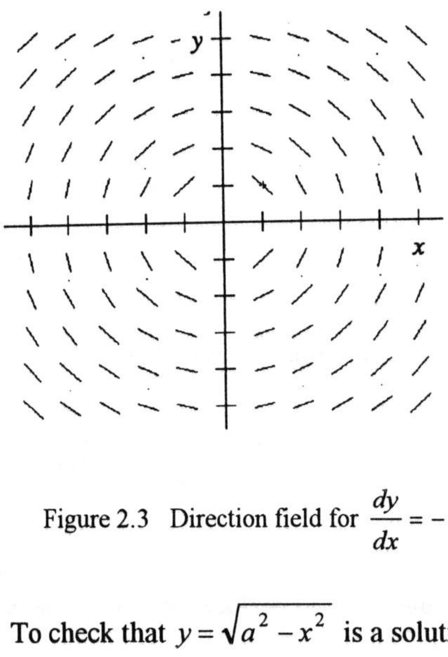

Figure 2.3 Direction field for $\dfrac{dy}{dx} = -\dfrac{x}{y}$

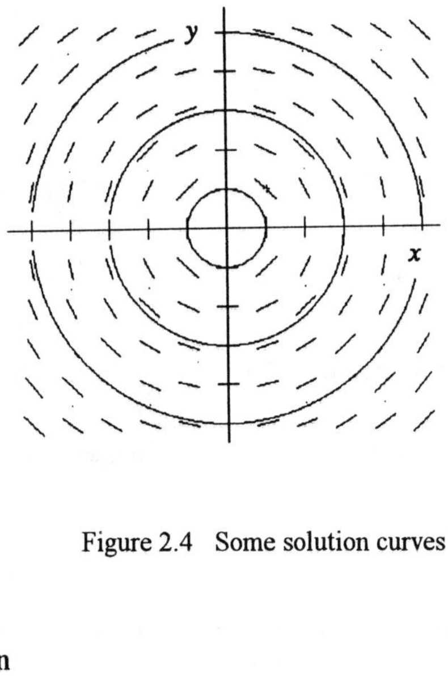

Figure 2.4 Some solution curves

To check that $y = \sqrt{a^2 - x^2}$ is a solution

$$\text{LHS} = \frac{dy}{dx} = \frac{1}{2}(a^2 - x^2)^{-\frac{1}{2}}(-2x) = -\frac{x}{\sqrt{a^2 - x^2}}$$

$$\text{RHS} = -\frac{x}{y} = -\frac{x}{\sqrt{a^2 - x^2}}$$

□

Example 2B

For the direction field shown in Figure 2.5 draw trajectories passing through the points A, B and C. Select the differential equation that is most likely to be satisfied by the trajectories.

(i) $y' = xy$ (ii) $y' = x + y$ (iii) $y' = \dfrac{x+1}{y}$

(iv) $y' = \dfrac{y+1}{x}$ (v) $y' = \dfrac{2-x-y}{xy}$

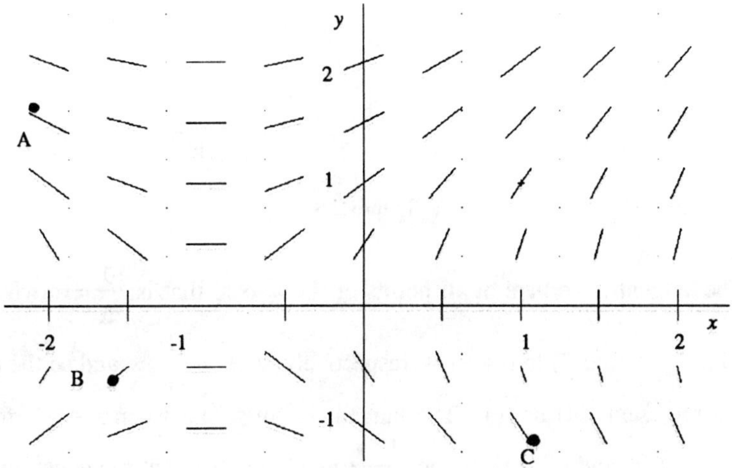

Figure 2.5

Solution

Trajectories through A, B and C are drawn so that they are always parallel to the direction field. Figure 2.6 shows these trajectories.

To determine which equation the field represents, choose suitable points to test each. Points where $\dfrac{dy}{dx} = 0$ or ∞ are especially useful.

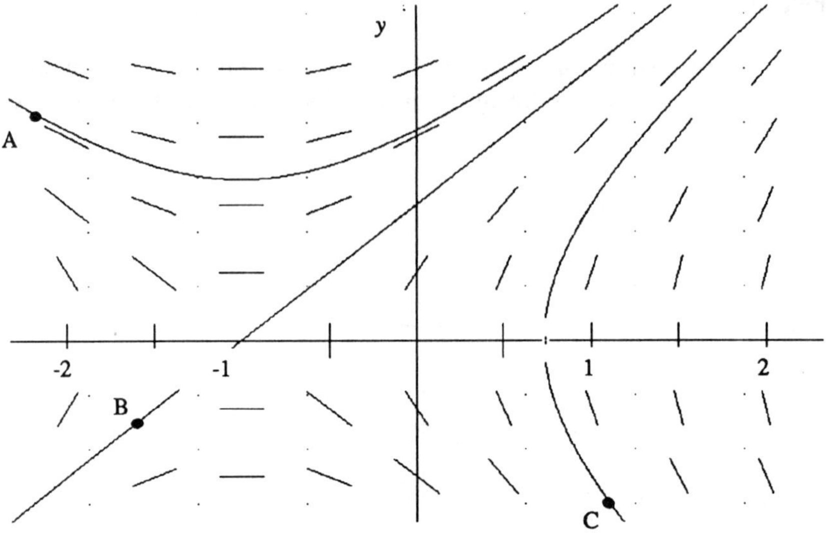

Figure 2.6

The tangent is vertical at all points of the x axis; that is $\frac{dy}{dx} = \infty$ when $y = 0$. Substituting $y = 0$ into (i) to (v) gives, respectively, 0, x, ∞, $\frac{1}{x}$, ∞ and so the direction field must represent (iii) or (v). The tangent is horizontal when $x = -1$, for any y. When $x = -1$, (iii) and (v) give 0 and $\frac{3-y}{-y}$ and it follows that the equation must be (iii), that is $\frac{dy}{dx} = \frac{x+1}{y}$. (Note that many other approaches are possible. For example, at (0,1), $\frac{dy}{dx} = 1$, but this is not satisfied by (i), (iv) or (v) and so they are eliminated.)

□

DERIVE ACTIVITY 2A

This activity will show you how to draw direction fields on DERIVE and use them to obtain solutions to some problems.

(A) (i) **File**, **Load** the **Utility** file ode_appr.mth.

(ii) Consider the differential equation

$$\frac{dy}{dx} = 5 - 2y$$

Author Expression DIRECTION_FIELD (5–2*y*, *x*, 0, 5, 10, *y*, 0, 5, 10) and click on OK.

If the DIRECTION_FIELD line is highlighted it can be transferred to the entry line by pressing F3. Then **Simplify Approximate** this expression choosing **Approximate** in the dialogue box, which will take some time. The **Simplify Approximate** expression you obtain can be plotted to form a direction field, by following the instructions in the next section.

(iii) **Window New 2D-plot** to draw the direction field. Define the format for the window by selecting the following:

Set Scale to **Horizontal** 0.75 and **Vertical** 0.75
Set Center Horizontal and **Vertical** positions to 2 and 2.5 respectively.
Options Points choose **Connected** and **Small.**
Plot!

You should now have a plot of the direction field shown in the screen dump in Figure 2.7.

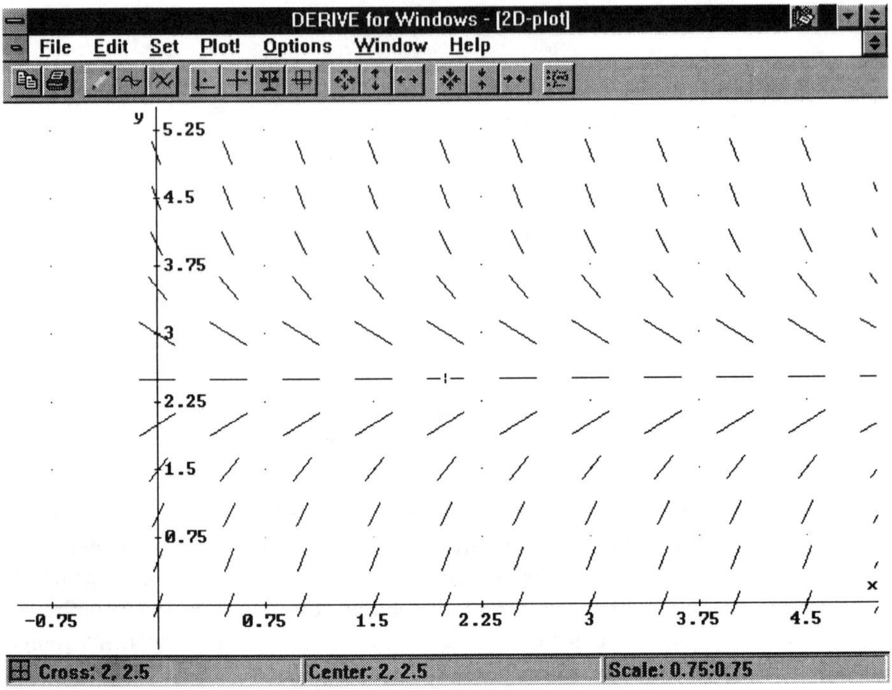

Figure 2.7 Direction field for $\dfrac{dy}{dx} = 5 - 2y$

(iv) What value does y approach as x increases?

(v) In the **Algebra** window **Author Expression** VECTOR(2.5–2ae^{-2x},a, –5,5). **Simplify Basic** and in the **2D-plot Window Plot!** the expression to produce a family of curves. The equation of this family of curves is

$$y = 2.5 - 2ae^{-2x}$$

where a can take any value. In the plot a takes values, –5, –4, ..., 5. This equation involving the unknown constant a is called the **general solution of the differential equation**. Each of the curves drawn that correspond to a particular value of a are known as **particular solutions**. Figure 2.8 shows the screen for this activity.

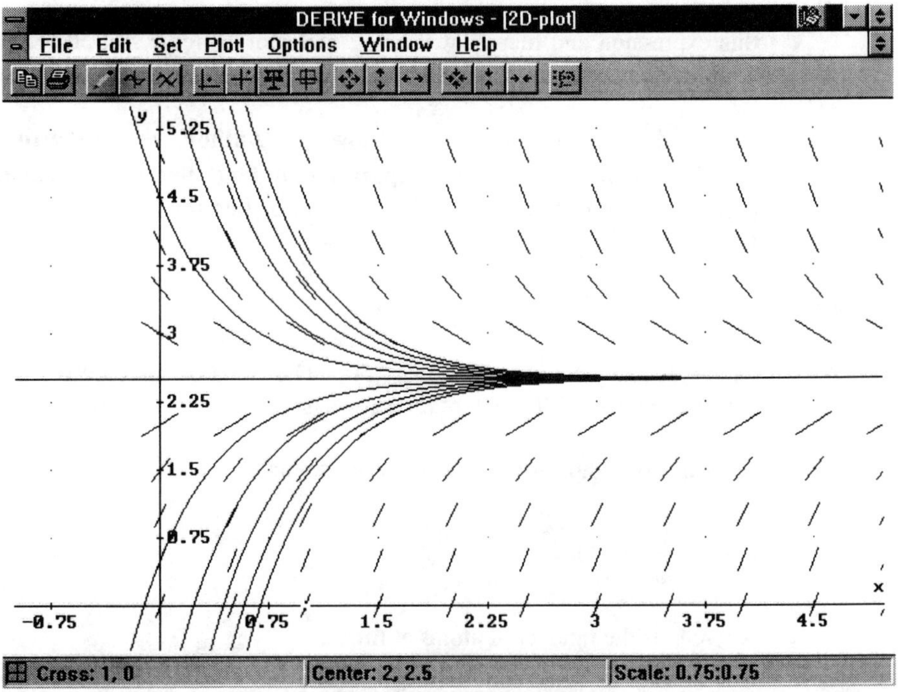

Figure 2.8 Solution curves for $\dfrac{dy}{dx} = 5 - 2y$

(v) Repeat the activity for the differential equation

$$\frac{dy}{dx} = \frac{y}{x}$$

What shape do you think the solution curves will have? Check your answer by simplifying and plotting VECTOR($\frac{a}{2}x$,a,–4,4).

(B) (i) A population grows so that

$$\frac{dP}{dt} = 0.05P$$

where P is the present population.

Author Expression DIRECTION_FIELD(0.05P,t,0,100,10,P,0,300,10). This will produce a direction field where t varies from 0 to 100 in 10 steps and P varies from 0 to 200 in 10 steps. **Simplify Approximate**

this expression and then **Plot** it using scales of 10 for x and 50 for y, with the **Centre** of the screen at (50,100).

(ii) In the **Algebra** window **Author Expression** VECTOR($10ae^{0.05t}$,a,0,6,0.3). **approX** and **Plot!** this expression to produce a family of curves. The equation of this family of curves is given by

$$P = Ae^{0.05t}$$

where A takes values 0, 5, 10, ..., 50. The equation above gives the general solution while each of the curves is a particular solution.

(C) (i) A sample of radioactive material decays so that

$$\frac{dm}{dt} = -0.1m,$$

where m is the number of atoms at time t.

Author Expression DIRECTION_FIELD($-0.1m$,t,0,20,10,m,0,10,10) then **Simplify Approximate** and **Plot** to produce a direction field where t varies from 0 to 20 in 10 steps and m varies from 0 to 10 in 10 steps. Choose your own **Scale** values and position for the **Centre** of the plot.

(ii) **Author** the **Expression** VECTOR($ae^{-0.1t}$,a,1,16,3) to produce a family of curves that you can **Plot!** over the direction field. The general solution of this differential equation is

$$m = Ae^{-0.1t}$$

(D) A parachutist falls so that

$$\frac{dv}{dt} = 10 - v$$

Plot a direction field for $0 \le t \le 10$ and $0 \le v \le 20$. What is the terminal speed of the parachutist?

(E) Plot the direction fields for the following differential equations, print out your fields and sketch some trajectories that represent particular solutions.

(i) $\dfrac{dy}{dx} = 1 - y$ (ii) $\dfrac{dy}{dx} = x^2 y$

(iii) $\dfrac{dy}{dx} = x^2 + y^2$ (iv) $\dfrac{dy}{dx} = y - x$

2.2 ANALYTICAL SOLUTIONS

We shall be describing different methods for solving first order differential equations of the form $\dfrac{dy}{dx} = f(x, y)$. Unfortunately there is no one method which solves all equations but a number of methods, each of which is used to solve a particular class of equations. There are also many equations for which no analytical method exists; these equations must then be solved numerically (see Chapter 3).

Here are six examples of differential equations arising from physical situations. In each case there is a statement, in italics, of the method to be used to solve the equation. It is important that when given an equation to solve you are able to correctly identify the method to be used. Notice how sometimes there is a choice of methods.

Example 1

The deflection y at distance x from the left hand end of a uniform horizontal cantilever satisfies the equation

$$EI\frac{d^2y}{dx^2} = \frac{w(l-x)^2}{2}$$

where E, I, w and l are constants.

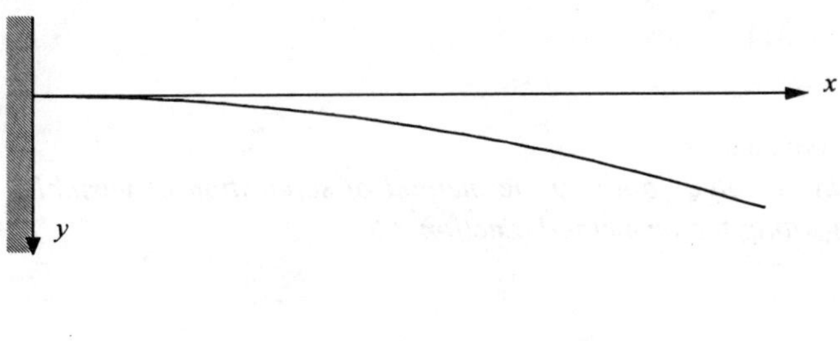

Figure 2.9 A cantilever

This equation can be solved using direct integration (Section 2.3).

□

Example 2

The equation of motion of a ball moving upwards under the action of the force of gravity and a resistive force proportional to the square of the speed of the ball is

$$\frac{dv}{dt} = -g - kv^2$$

where g and k are constants.
This equation can be solved using the method of separation of variables (Section 2.4).

□

Example 3

The current i at time t flowing in an electrical circuit consisting of a resistance R and inductance L connected in series to an alternating voltage $E_0 \sin \omega t$ satisfies the equation

$$L\frac{di}{dt} + Ri = E_0 \sin \omega t$$

This is a first order linear equation that can be solved using the integrating factor method (Section 2.5).

□

Example 4

Newton's law of cooling predicts that the temperature θ at time t of a body cooling in a room of temperature θ_0 satisfies the equation

$$\frac{d\theta}{dt} = -k(\theta - \theta_0)$$

where k is a positive constant.
This equation can be solved both by the method of separation of variables (Section 2.4) and the integrating factor method (Section 2.5).

□

Example 5

The logistic equation of population growth is of the form

$$\frac{dP}{dt} = aP\left(1 - \frac{P}{M}\right)$$

This equation is a Bernoulli equation (Section 2.6) and can also be solved by separating the variables.

□

Example 6

The equation of the reflector of a searchlight, designed so that the beams of light are parallel, is

$$\frac{dy}{dx} = \frac{y + \sqrt{x^2 + y^2}}{x}$$

This equation is homogeneous and is solved in Section 2.7.

□

2.3 EQUATIONS SOLVED BY DIRECT INTEGRATION

The general first order equation $\frac{dy}{dx} = f(x, y)$ can be integrated directly when the right hand side is a function of x only, and independent of y. The equation then becomes

$$\frac{dy}{dx} = f(x) \tag{2.1}$$

Integrating with respect to x

$$\int \frac{dy}{dx} dx = \int f(x) dx$$

or

$$y = \int f(x) dx + C \tag{2.2}$$

This is called the **general solution** (GS) of (2.1).

Any solution obtained by giving C a value is called a **particular solution** (PS). The value of C is usually found by applying an initial condition given in the question. This approach can also be used to solve higher order equations of the form

$$\frac{d^n y}{dx^n} = f(x) \tag{2.3}$$

y will be found after n integrations, and the general solution will contain n constants of integration.

□

Example 2C

Solve the equation $\frac{dy}{dx} = 3x^2$.

Solution

Integrating with respect to x

$$\int \frac{dy}{dx}\,dx = \int 3x^2\,dx$$

The general solution is

$$y = x^3 + C$$

The presence of the constant C indicates that there is an infinite number of functions which satisfy the differential equation, one for each value of C. Hence the general solution describes a whole family of solutions.

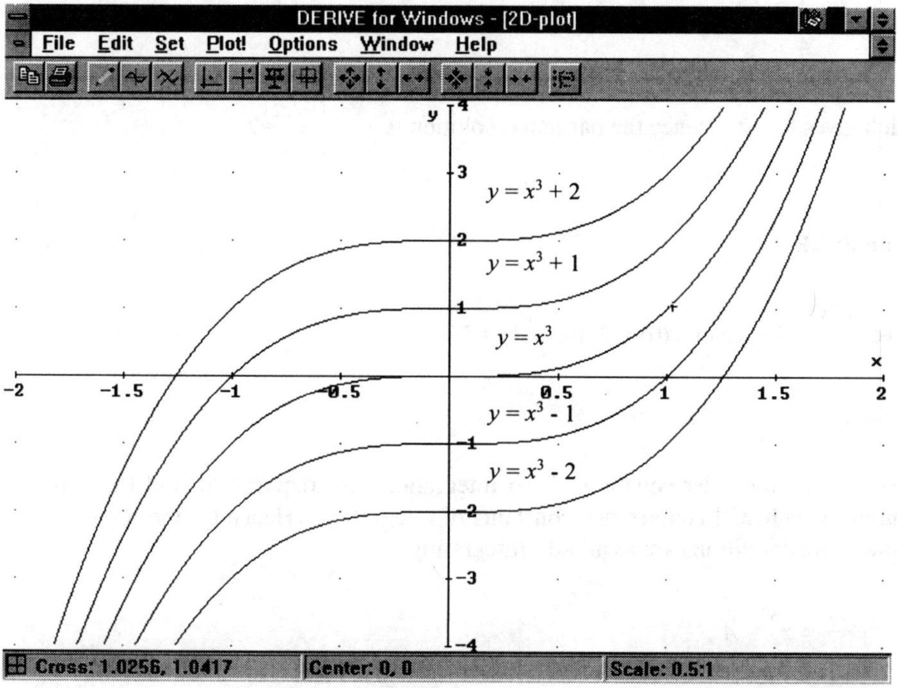

Figure 2.10

□

Example 2D

Solve $\dfrac{dy}{dx} = x^3$ given $y = 2$ when $x = 0$.

Solution

In this example we are given an initial condition as well as the equation. This condition will be used to find the constant of integration. Integrating

$$y = \int x^3 dx + C$$

$$y = \frac{1}{4}x^4 + C \qquad \text{(GS)}$$

Using the condition $y = 2$ when $x = 0$

$$2 = \frac{1}{4} \times 0^4 + C$$

which gives $C = 2$. Hence the particular solution is $y = \frac{1}{4}x^4 + 2$.

□

Example 2E

Solve $\frac{d^2y}{dx^2} = 2x$ given $y(0) = 2$ and $y(3) = 5$.

Solution

This is a second order equation. Two integrations are required to find the general solution, which will contain two constants of integration. Hence to find a particular solution two conditions are required. Integrating

$$\frac{dy}{dx} = x^2 + A$$

Integrating again

$$y = \frac{1}{3}x^3 + Ax + B \qquad \text{(GS)}$$

Using $y(0) = 2$

$$2 = 0 + 0 + B$$

giving $B = 2$. Using $y(3) = 5$

$$5 = 9 + 3A + B$$

from which $A = -2$. Hence the particular solution is

$$y = \frac{1}{3}x^3 - 2x + 2$$

□

Example 2F

Solve $x^2 \dfrac{d^2y}{dx^2} = 3x^4 - 1$ given $y = 0$ when $x = 1$ and $\dfrac{dy}{dx} = 2$ when $x = 1$.

Solution

We must first divide throughout by x^2 to put the equation in the form of equation (2.3). Then

$$\frac{d^2y}{dx^2} = 3x^2 - x^{-2}$$

Integrating

$$\frac{dy}{dx} = x^3 + x^{-1} + A$$

Integrating again

$$y = \frac{1}{4}x^4 + \ln x + Ax + B \qquad \text{(GS)}$$

Using $\dfrac{dy}{dx} = 2$ when $x = 1$

$$2 = 1 + 1 + A$$

from which $A = 0$. Using $y = 0$ when $x = 1$

$$0 = \frac{1}{4} + 0 + A + B$$

giving $B = -\dfrac{1}{4}$.

The required particular solution is

$$y = \frac{1}{4}x^4 + \ln x - \frac{1}{4}$$

□

Example 2G

A uniform horizontal cantilever is of length l and weight w per unit length. The deflection y at distance x from the left hand end satisfies the equation $EI\dfrac{d^2y}{dx^2}=\dfrac{w(l-x)^2}{2}$ (E, I constants). Find the maximum deflection.

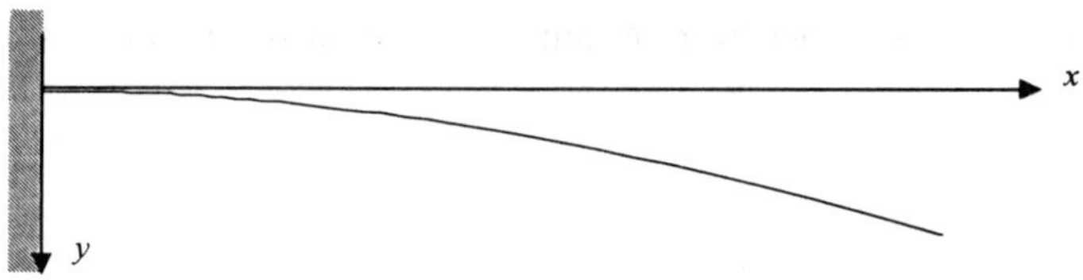

Figure 2.11

Solution

As E and I are constants it is not necessary to divide throughout by EI, the equation can be integrated as it stands. Integrating

$$EI\frac{dy}{dx}=-\frac{w(l-x)^3}{6}+A \tag{2.4}$$

Integrating

$$EIy=\frac{w(l-x)^4}{24}+Ax+B \tag{2.5}$$

The boundary conditions are, from the diagram, $y = 0$ at $x = 0$ and $\dfrac{dy}{dx}=0$ at $x = 0$ (the beam emerges from the wall horizontally).

Substitute $\dfrac{dy}{dx}=0$ at $x=0$ in (2.4)

$$0=-\frac{wl^3}{6}+A \quad\Rightarrow\quad A=\frac{wl^3}{6}$$

Substitute $y = 0$ at $x = 0$ in (2.5)

$$0 = \frac{wl^4}{24} + B$$

giving

$$B = -\frac{wl^4}{24}$$

The deflection at any point distance x from the left hand end is given by y, where

$$EIy = \frac{w(l-x)^4}{24} + \frac{wl^3 x}{6} - \frac{wl^4}{24}$$

The maximum deflection is at the free end ($x = l$), and this gives

$$y_{\max} = \frac{1}{EI}\left(0 + \frac{wl^4}{6} - \frac{wl^4}{24}\right) = \frac{wl^4}{8EI}$$

□

Exercise 2A

1. Which of the following equations can be solved by direct integration?

(i) $\frac{dy}{dx} = 2x$ (ii) $\frac{dy}{dx} = 2y$ (iii) $\frac{dy}{dx} = \frac{1}{x} - \sin x$

(iv) $x\frac{d^2y}{dx^2} = 2x - 1$ (v) $\frac{dy}{dx} = x + y$ (vi) $y\frac{dy}{dx} = \cos x$

2. Find the general solution of

(i) $\frac{dy}{dx} = 2x$ (ii) $\frac{dy}{dx} = \frac{1}{x} - \sin x$ (iii) $\frac{d^2y}{dx^2} = 24x^2 + 6x - 2$

(iv) $x\frac{dy}{dx} = 2x - 1$ (v) $\sqrt{1-x^2}\,\frac{dy}{dx} = 1$ (vi) $(x+1)(x-2)\frac{dy}{dx} = 3$

3. Solve

(i) $\dfrac{dy}{dx} = 3x^2 - \dfrac{1}{x}$ given $y(1) = 2$

(ii) $\dfrac{dy}{dx} + 2x = 6\cos 3x$ given $y(0) = 1$

(iii) $e^x \dfrac{dy}{dx} = x$ given $y(0) = 0$

(iv) $\dfrac{d^2x}{dt^2} = 6t$ given $x(0) = 1$ and $x(1) = 3$

4. The height h of a ball thrown vertically upwards from ground level satisfies the equation $\dfrac{d^2h}{dt^2} = -10$. If the initial speed of the ball $\left(\dfrac{dh}{dt}\right)$ is 20ms^{-1} find

(i) the time to reach the highest point,
(ii) how high the ball will go,
(iii) the total time of flight,
(iv) the speed with which the ball hits the ground.

5. A light beam of length l carries a load W at its mid point. The deflection y at distance x from the left hand end satisfies the equation $EI\dfrac{d^2y}{dx^2} = -\dfrac{Wx}{2}$, for $0 \le x \le l/2$. Solve the equation using the conditions $y = 0$ at $x = 0$ and $\dfrac{dy}{dx} = 0$ at $x = l/2$ and hence find the maximum deflection of the beam.

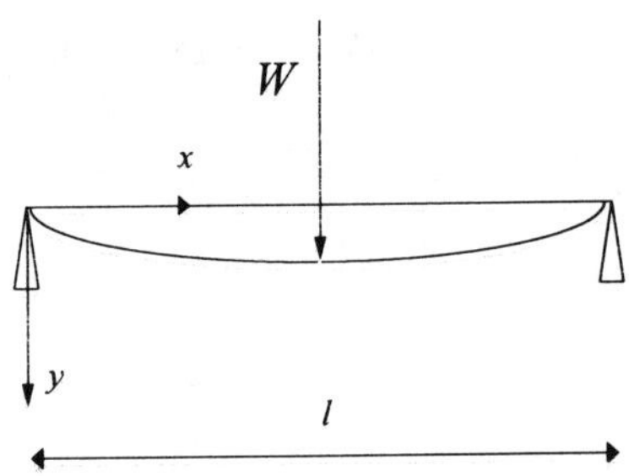

6. The intensity L of light at depth x metres below the surface of the sea satisfies the equation $\dfrac{dL}{dx} = -kx$. A diver knows from experience that diving to 4 metres in the Bristol Channel cuts the intensity of light in half. It is impossible to work without artificial light once the intensity of light falls below one tenth the surface value. What is the maximum depth a diver can work without artificial light?

7. The speed v of a rocket at time t satisfies the equation $\frac{dv}{dt} = \frac{cu}{M - ct} - g$, where c, u, M and g are constants. Given that $v(0) = 0$ find the speed of the rocket at time $t = \frac{M}{2c}$, when the fuel runs out, and the distance the rocket has travelled in this time.

8. Find the general solution of

(i) $\frac{dy}{dx} = x(1+x^2)^3$ (ii) $\frac{dy}{dx} = 4x\cos 2x$

(iii) $x^2\frac{dy}{dx} = x + 2x^3 e^{x^2}$ (iv) $(1+x^2)\frac{dy}{dx} = 1 + 4x + 3x^2$

9. (i) Solve $(1+x^2)\frac{d^2y}{dx^2} = 2$ given $y = 0$ and $\frac{dy}{dx} = 1$ at $x = 0$.

(ii) Solve $(4-x^2)^{\frac{3}{2}}\frac{d^2y}{dx^2} = 1$ given $y = 0$ at $x = 0$ and $\frac{dy}{dx} = 1$ at $x = 0$.

DERIVE ACTIVITY 2B

This activity uses the standard integration command of DERIVE to evaluate the integrals.

(A) (i) Consider the differential equation

$$\frac{dy}{dx} = x\sin x$$

Integrating

$$y = \int x\sin x\,dx + c$$

and the integral can be evaluated using DERIVE in the following way:
Author Expression $x \sin x$. Now select the commands **Integrate** (∫), author $x \sin x$, select **Indefinite** and **Simplify**.

According to DERIVE the general solution of the differential equation is

$$y = \sin x - x \cos x$$

Now amend the solution by adding a constant of integration c, to give

$$y = \sin x - x \cos x + c$$

In a **New 2D-plot Window**, use **Author Expression** VECTOR ($\sin x - x \cos x + c$,c,–2,2), **Simplify** and **Plot!** with scales 1.5 for both x and y. Figure 2.12 shows the graphs of these results.

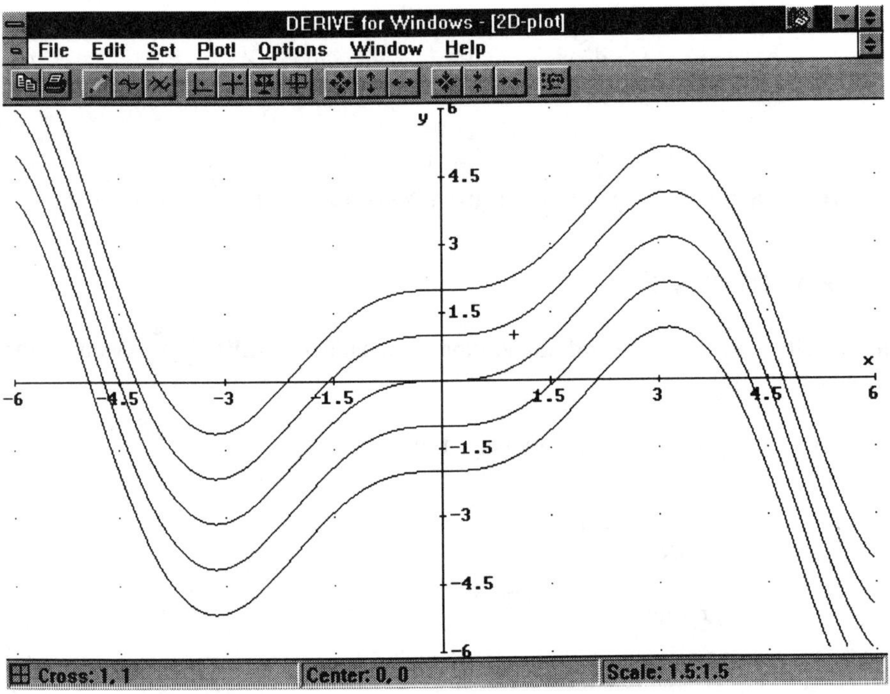

Figure 2.12

(ii) Describe the transformation between the graphs you have drawn.

(B) Using DERIVE to evaluate the integrals, find the general solution of the following differential equations:

(i) $\frac{dy}{dx} = x\cos 2x$ (ii) $\frac{dx}{dt} = te^{-t}$

(iii) $\frac{dP}{dt} = t^3 \ln t$ (iv) $\frac{dy}{dx} = \sin^3 x$

(v) $\frac{dy}{dx} = \frac{e^x}{e^x + 1}$ (vi) $\frac{dx}{dt} = e^{-2t} \sin 3t$

(C) By choosing appropriate limits in the **Calculus Integrate** commands DERIVE can be used to solve to find the particular solution for a given initial condition.
Consider the differential equation

$$\frac{dy}{dx} = \frac{30}{\sqrt{3x+4}}$$

given the condition $y(4) = 20$. Formally we can write the solution as

$$\int_{20}^{y} dy = \int_{4}^{x} \frac{30}{\sqrt{3x+4}}\, dx$$

Note that the lower limits in each integral are the given conditions. The left and right hand sides can be evaluated directly to give $y - 20$ and $20\sqrt{3x+4} - 80$ respectively.

Hence the particular solution of the problem is

$$y - 20 = 20\sqrt{3x+4} - 80$$

or

$$y = 20\sqrt{3x+4} - 60$$

(D) Using DERIVE to evaluate the integrals, find the particular solutions of the following problems:

(i) $$\frac{dy}{dx} = -x$$
$y(0) = 2$

(ii) $$\frac{dP}{dt} = \frac{1}{t^2 + 1}$$
$P(1) = 0$

(iii) $$\frac{dy}{dx} = \frac{\sin x}{1 + \cos x}$$
$y\left(\frac{\pi}{4}\right) = 0.5$

(iv) $$\frac{ds}{dt} = \frac{5}{t^2 - t - 6}$$
$s(4) = 1$

(v) $$\frac{dy}{dx} = x^2 \cos x$$
$y\left(\frac{\pi}{2}\right) = 0$

(vi) $$\frac{dv}{dt} = \mathrm{e}^{-0.1t} \sin 2t$$
$v(0) = 0$

(E) (i) Consider the differential equation

$$\frac{dy}{dx} = \frac{x+4}{(x-2)(x+1)}$$

Integrating

$$y = \int \frac{x+4}{(x-2)(x+1)} dx + c$$

and the integral can be evaluated using DERIVE in the usual way.

According to DERIVE the general solution of the differential equation is

$$y = 2\ln(x - 2) - \ln(x + 1) + c$$

where we have added the constant of integration c.

(ii) Use **Simplify** VECTOR(2ln(x – 2) – ln(x + 1) + c, c, –2, 2) and in a **New 2D-plot Window** plot the graphs of the functions obtained. The particular solutions are shown in Figure 2.13.

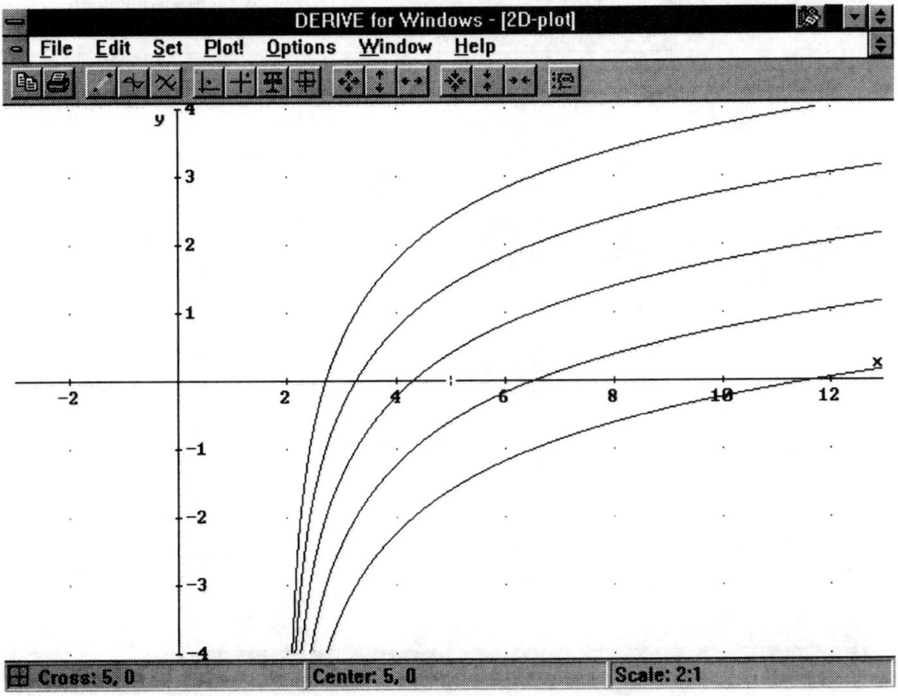

Figure 2.13 Some particular solutions

It looks as if all the solutions occur for $x > 2$. The reason for this is the way DERIVE evaluates the integral. Figure 2.14 shows the direction field for this differential equation, which is obtained using DIRECTION_FIELD($(x + 4)/((x - 2)(x + 1))$, x, –2, 6, 8, y, –4, 4, 8).

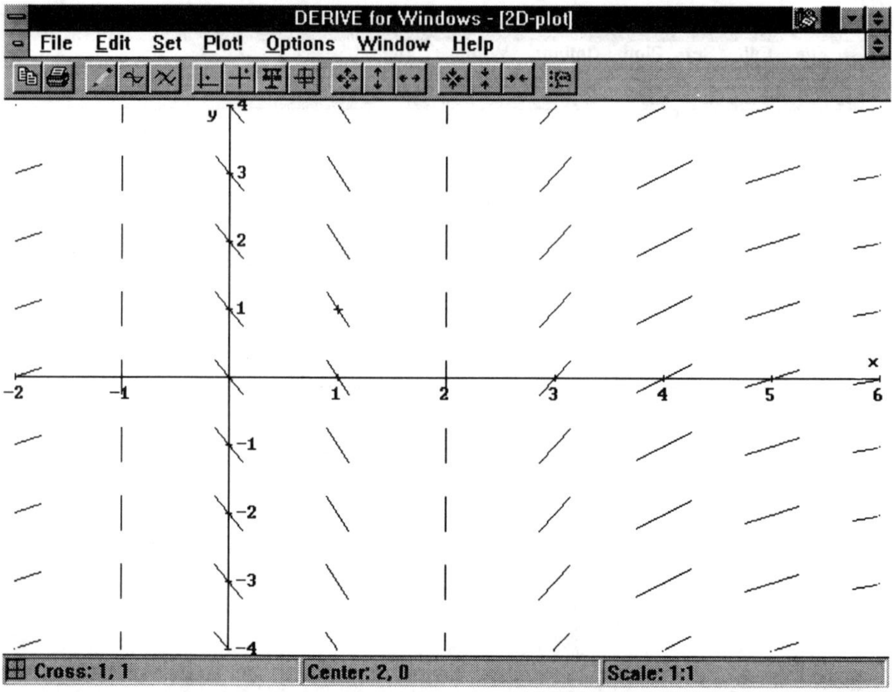

Figure 2.14 Direction field for $\dfrac{dy}{dx} = \dfrac{x+4}{(x-2)(x+1)}$

The direction field suggests that there are solutions for $x < 2$. It looks as if there are three regions to consider, namely $x < -1$, $-1 < x < 2$ and $x > 2$. $\dfrac{dy}{dx}$ has asymptotes $x = -1$ and $x = 2$. The anomaly can be resolved by considering the solution of the integral using partial fractions:

$$\int \frac{x+4}{(x-2)(x+1)}\,dx = \int \left(\frac{2}{x-2} - \frac{1}{x+1} \right) dx$$

$$= 2\ln|x-2| - \ln|x+1| + c$$

We should use modulus signs to take the absolute values of the arguments of the logarithms. The general solution of the differential equation is now

$$y = 2\ln|x-2| - \ln|x+1| + c$$

We need to amend the DERIVE solution by introducing ABS in front of the $(x - 2)$ and $(x + 1)$ expressions.

(iii) In a **2D-plot Window** draw the particular solutions for $c = -2, -1, 0, 1, 2$.

Show that the solution now agrees with the direction field.

(F) Using DERIVE to evaluate the integrals, find the general solution of the following differential equations interpreting your results carefully.

(i) $\dfrac{dy}{dx} = \dfrac{2x-1}{(x+2)(x-3)}$ (ii) $\dfrac{dy}{dx} = \dfrac{1}{x^2-4}$

(iii) $\dfrac{dy}{dx} = \dfrac{x^2+x+2}{x^2-1}$ (iv) $\dfrac{dy}{dx} = \dfrac{x}{x+3}$

In each case plot some of your solutions.

2.4 SEPARATION OF VARIABLES

In many equations the right hand side of the equation

$$\frac{dy}{dx} = f(x,y) \tag{2.6}$$

can be expressed as the product of two functions, one of which, $p(x)$, involves x only and the other of which, $q(y)$, involves y only.

The equation then becomes

$$\frac{dy}{dx} = p(x)q(y) \tag{2.7}$$

For example

Equation	$p(x)$	$q(y)$
$\dfrac{dy}{dx} = x^2 \sin y$	x^2	$\sin y$
$x\dfrac{dy}{dx} = y^4 \quad \left(\text{or } \dfrac{dy}{dx} = \dfrac{y^4}{x}\right)$	$\dfrac{1}{x}$	y^4
$\dfrac{dy}{dx} = 2y^2$	2	y^2

Note that any constant can be included as part of $p(x)$ or $q(y)$. However the equations

$$\frac{dy}{dx} = x^2 + y^3 \quad \text{or} \quad x\frac{dy}{dx} - y = 3x$$

cannot be expressed in the form (2.7) and so cannot be solved by the method of separation of variables.

In order to solve (2.7) we first separate the variables so that, for example, only y terms, and dy, appear on the left hand side and only x terms, and dx, appear on the right hand side. Dividing by $q(y)$ and then multiplying by dx we obtain

$$\frac{1}{q(y)}dy = p(x)dx$$

The variables have now been separated.

Integrating each side gives

$$\int \frac{1}{q(y)}dy = \int p(x)dx + C \tag{2.8}$$

which leads to the general solution.

Note that although there are two integrals, (2.8) only contains one constant of integration. Can you see why this is the case?

If initial values for x and y are known then they can be used to find C and hence to find a particular solution of the equation.

Example 2H

Find the general solution of $\frac{dy}{dx} = \frac{2x}{3y^2}$.

Solution

Separate the variables by multiplying throughout by $3y^2dx$. This gives

$$3y^2dy = 2xdx$$

Integrating

$$\int 3y^2 dy = \int 2x dx + C$$

or

$$y^3 = x^2 + C$$

The general solution is

$$y = (x^2 + C)^{\frac{1}{3}}$$

Note: If a constant had been attached to each integral we would have obtained

$$y^3 + A = x^2 + B$$

Then

$$y^3 = x^2 + (B - A)$$

or, putting $C = B - A$, $y^3 = x^2 + C$ as before. This demonstrates why, for any first order equation, the general solution contains exactly one constant of integration.

□

Example 2I

Find the general solution of $\dfrac{dy}{dx} = 2x(y-1)$, for $y > 1$.

Solution

Dividing by $(y - 1)$ gives

$$\frac{1}{(y-1)}\frac{dy}{dx} = 2x$$

Then multiplying by dx separates the variables

$$\frac{1}{y-1}dy = 2xdx$$

Integrating gives

$$\ln(y-1) = x^2 + C \qquad \text{(providing } y > 1\text{)}$$

To find the general solution explicitly

either (i) take antilogs $y-1 = e^{x^2+C}$

$$y-1 = e^{x^2} \times e^C \qquad \text{(using the rules of indices)}$$

$$y-1 = Ae^{x^2} \qquad \text{(putting } A = e^C\text{)}$$

The general solution is

$$y = 1 + Ae^{x^2}$$

or (ii) let $C = \ln A$ so that

$$\ln(y-1) = x^2 + \ln A$$

$$\ln(y-1) - \ln A = x^2$$

$$\ln\left(\frac{y-1}{A}\right) = x^2 \qquad \text{(using the laws of logarithms)}$$

$$\frac{y-1}{A} = e^{x^2} \qquad \text{(antilogs)}$$

which also rearranges to

$$y = 1 + Ae^{x^2} \qquad \text{(GS)}$$

□

Example 2J

Solve $\tan x \frac{dy}{dx} = y$ given $y = 3$ at $x = \frac{\pi}{6}$.

Solution

Separating the variables

$$\frac{1}{y}dy = \frac{1}{\tan x}dx = \cot x dx$$

Integrating

$$\ln y = \ln \sin x + C \qquad \text{(providing } y > 0, \sin x > 0\text{)}$$

$$\ln y = \ln \sin x + \ln A \qquad \text{(put } C = \ln A\text{)}$$

$$\ln y = \ln(A \sin x) \qquad \text{(laws of logarithms)}$$

The general solution is, taking antilogs,

$$y = A \sin x$$

Using the condition $y = 3$ at $x = \frac{\pi}{6}$ gives

$$3 = A \sin\frac{\pi}{6} = A \times \frac{1}{2}$$

from which $A = 6$. The particular solution is

$y = 6\sin x.$

□

Example 2K

A boat is rowed across a lake. When the oarsmen stop rowing the boat is slowed by the resistance of the water, the resistance being proportional to the speed v of the boat. The equation of motion is $\frac{dv}{dt} = -kv$ (k positive constant). If the speed when the oarsmen stopped rowing was 2ms^{-1} and 5 seconds later it was 1ms^{-1}

(i) what is the speed after 10 seconds?
(ii) how long does it take for the speed to drop to 0.1ms^{-1}?
(iii) how far has the boat travelled in this time?

Solution

The equation of motion is

$$\frac{dv}{dt} = -kv$$

Separating the variables

$$\frac{1}{v}dv = -kdt$$

Integrating

$$\ln v = -kt + C$$

which gives $v = Ae^{-kt}$ (where $C = \ln A$).

Applying the condition $v = 2$ when $t = 0$, $2 = A \times 1$, giving $A = 2$ and $v = 2e^{-kt}$. Applying thc condition $v = 1$ when $t = 5$: $1 = 2e^{-5k}$. This gives $e^{-5k} = \frac{1}{2}$, or $e^{5k} = 2$ or $5k = \ln 2$ and so

$$k = \frac{1}{5}\ln 2 = 0.139 \qquad \text{(correct to 3S)}$$

Hence

$$v = 2e^{-0.139t} \tag{2.9}$$

(i) When $t = 10$ the speed of the boat is $v = 2e^{-0.139 \times 10} = 0.5\text{ms}^{-1}$.

(ii) We need to rearrange $v = 2e^{-0.139t}$ to make t the subject.

$$e^{-0.139t} = \frac{v}{2} \quad \text{or} \quad e^{0.139t} = \frac{2}{v} \quad \text{or} \quad 0.139t = \ln\left(\frac{2}{v}\right)$$

and so

$$t = \frac{1}{0.139}\ln\left(\frac{2}{v}\right)$$

The time taken for the speed to drop to 0.1ms^{-1} is

$$t = \frac{1}{0.139}\ln\left(\frac{2}{0.1}\right) = 21.6 \text{ seconds} \qquad \text{(correct to 3S)}$$

(iii) To find distance x we use (2.9) with v replaced by $\dfrac{dx}{dt}$. Then

$$\frac{dx}{dt} = 2e^{-0.139t}$$

This is a first order equation which can be solved by direct integration. Integrating gives

$$x = \frac{2}{-0.139}e^{-0.139t} + B$$

Using $x = 0$ at $t = 0$: $0 = \dfrac{2}{-0.139} + B$, giving $B = \dfrac{2}{0.139}$. Then

$$x = \frac{2}{0.139}(1 - e^{-0.139t})$$

or

$$x = 14.427(1 - e^{-0.139t})$$

The distance the boat has travelled in the 21.6 seconds since the oarsmen stopped rowing is

$$x = 14.427(1 - e^{-0.139 \times 21.6}) = 13.7 \text{ metres} \qquad \text{(correct to 3S)}$$

□

Exercise 2B

1. Which of the following equations can be solved by separating the variables?

(i) $\dfrac{dy}{dx} = \dfrac{x}{y}$ (ii) $\dfrac{dy}{dx} = x + 3y$ (iii) $\dfrac{dy}{dx} = x + xy$

(iv) $x\dfrac{dy}{dx} = y + y^2$ (v) $\dfrac{dy}{dx} = \dfrac{y}{x} + \dfrac{x}{y}$ (vi) $e^y \dfrac{dy}{dx} = 2\sin x$

2. Find the general solutions of

(i) $\dfrac{dy}{dx} = \dfrac{x}{y}$ (ii) $\dfrac{dy}{dx} = y$ (iii) $\dfrac{dy}{dx} = 4x(1+y)$

(iv) $e^y \dfrac{dy}{dx} = 2\sin x$ (v) $\dfrac{dy}{dx} = \cot 2y$ (vi) $\dfrac{dy}{dx} = 8x^3(1+y^2)$

(vii) $\sqrt{4-x^2}\,\dfrac{dy}{dx} = y$ (viii) $x^3\dfrac{dy}{dx} = y^3$ (ix) $\sqrt{x}\,\dfrac{dy}{dx} = 1 + 3y$

3. Find the particular solutions of

(i) parts (ii), (iii), (vii) and (ix) of question 2 given $y(0) = 2$
(ii) parts (iv), (v) and (vi) of question 2 given $y(0) = 0$.

4. A newly constructed reservoir is stocked with trout. The fish have plenty of space and food and breed quickly according to the equation $\frac{dN}{dt} = kN$, where N is the number of trout at time t years and k is a positive constant. Initially the reservoir was stocked with 1000 trout and one year later the population was estimated at 1350.

(i) How many trout will there be after 5 years?
(ii) How long will it take for the trout population to reach 10000?

(See question 12 for a further investigation into the trout population.)

5. The temperature θ°C at time t minutes of a cup of coffee standing in a room of temperature 20°C satisfies the equation $\frac{d\theta}{dt} = -0.04(\theta - 20)$. If the temperature of the cup of coffee was initially 100°C find

(i) a formula for θ in terms of t and hence the temperature after 20 minutes

(ii) a formula for t in terms of θ and hence the time the coffee takes to cool to 40°C

(iii) sketch the graph of θ against t.

6. The speed v ms^{-1} of a bullet at time t seconds satisfies the equation $\frac{dv}{dt} = -1.2v$. Its initial speed, as it leaves the barrel of the rifle, is 1000ms^{-1}.

(i) Find a formula for v at time t and hence the speed of the bullet after ½ second.

(ii) Use $v = \frac{ds}{dt}$ to find the distance the bullet has covered in the half second.

(iii) The rifle is aimed at a target 600 metres away. Find how long it will take the bullet to reach the target and its speed on impact.

7. The height, h metres, of water in a canal lock as it is being filled satisfies the equation $\frac{dh}{dt} = c(4-h)$, where c is a constant. Initially the lock is empty and when water is let in the level starts to rise at a rate of 50 cm per minute. The lock gates will open when the water has risen to a height of 3.5 metres. How long will this take? Find the height h of water in the lock at time t and comment on the time the solution predicts it will take to fill the lock completely, to a depth of 4 metres.

8. A tea urn is emptied through a tap at the bottom. The depth, h centimetres, of tea in the urn at time t satisfies the equation

$$\frac{dh}{dt} = c\sqrt{h} \qquad (c \text{ constant})$$

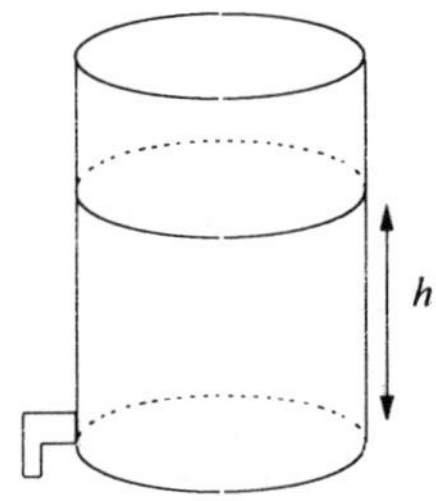

The urn is 80 cm tall and is initially full of tea. 30 seconds after opening the tap the depth of the tea has fallen to 50 cm. Find how long it will take to empty the urn completely.

Show that the time taken to empty the lower half of the urn is about 2.4 times that to empty the upper half.

9. The speed v of a ball thrown vertically upwards from ground level with an initial speed of 20ms^{-1} satisfies the equation $\frac{dv}{dt} = -\frac{(20+v)}{2}$.

(i) Find v at any time t and show that it takes about 1.4 seconds to reach its highest point.

(ii) Find the height h at time t using $v = \frac{dh}{dt}$. Show that the ball reaches a height of about 12.3 metres.

On its way back down to the ground the equation of motion is $\frac{dv}{dt} = \frac{20-v}{2}$. Find the speed with which the ball hits the ground and the total time the ball is in the air.

10. Radioactive material decays so that the amount m present at time t satisfies the equation $\frac{dm}{dt} = -\lambda m$ (1), where λ is a constant depending on the substance involved.

(i) Solve the equation given $m = M$ when $t = 0$.

(ii) The half life T is the time for m to reduce from M to $\frac{1}{2}M$. Show that $T = \frac{1}{\lambda}\ln 2$ and hence find the half life of plutonium, for which $\lambda = 0.00003$.

(iii) Carbon 14 dating is a method for estimating the time since a plant, tree or animal died. Living material contains a small, constant, proportion of carbon 14 (C^{14}) but once the material dies the C^{14} decays according to equation (1). Given that C^{14} has a half life of 5568 years, find λ. For living wood $M = 6.68$. In 1976 a sample was taken from King Arthur's round table, which hangs from a wall in the great hall of Winchester castle. Tests showed that $m = 6.08$. Estimate the date at which the table was made. If King Arthur died in the 5'th century what do you conclude?

11. For a rocket to escape the pull of the earth's gravity its speed must have reached a certain minimum value when its fuel is exhausted. The speed v of the rocket when it is at a distance x from the centre of the earth satisfies the equation $v\frac{dv}{dx} = -\frac{gR^2}{x^2}$, where g is the (constant) acceleration due to gravity and R is the radius of the earth. Given that the rocket is launched from the surface of the earth with speed u show that $v^2 = \frac{2gR^2}{x} + u^2 - 2gR$. Deduce that to escape from the earth's gravitational pull the initial speed u must satisfy $u \geq \sqrt{2gR}$. Find u given $g = 9.81\text{ms}^{-2}$ and $R = 6.368 \times 10^6$m.

12. Consider the trout population in question 4. As the population grows the fish will have to compete for space and food and the number N at time t then satisfies the equation $\dfrac{dN}{dt} = 0.00001N\ (25000 - N)$.

    (i) Solve this equation given $N = 10000$ when $t = 0$. What will be the population after 5 years?

    (ii) How long will it take for the population to double to 20000?

    (iii) What happens to the population ultimately?

13. In a chemical reaction two substances, with initial concentrations α and β, combine to form a third, with concentration x at time t. The rate of increase is given by $\dfrac{dx}{dt} = c(\alpha - x)(\beta - x)$, where c is a positive constant and $\alpha \neq \beta$.

    (i) Find x as a function of t given $x(0) = 0$.

    (ii) What is the limiting value of x when (a) $\alpha > \beta$, (b) $\alpha < \beta$?

    (iii) Investigate the solution when $\alpha = \beta$.

    (iv) Find the limiting value of x and the time it takes x to reach half this value when c, α and β take the following values (in appropriate units).

    (a) $c = 1$, $\alpha = 1$, $\beta = 2$, (b) $c = 1$, $\alpha = 5$, $\beta = 2$, (c) $c = 1$, $\alpha = 4$, $\beta = 4$.

14. The concentration c of a drug in the blood satisfies the equation $\frac{dc}{dt} = -\lambda c$, where λ is a constant depending on the drug used.

(i) Solve this equation given $c(0) = c_0$. The drug ceases to have a therapeutic effect once its concentration is reduced to c_1. How long will this take?

(ii) For the drug aminophylline, given to asthma patients, $\lambda = 0.17\ \text{hr}^{-1}$. The maximum safe concentration is 20mgl^{-1} but the drug ceases to have a beneficial effect once its concentration has reduced to 5mgl^{-1}. How often must a dose of the drug be given to keep the concentration between 5 and 20mgl^{-1}? Describe a regime which leads to equal doses at equal time intervals resulting in the concentration always lying between 10 and 20mgl^{-1}.

(iii) An alternative method is to apply the drug continuously through an intravenous drip. The equation then becomes $\frac{dc}{dt} = I - \lambda c$, where I is the rate of infusion.

(a) Solve the equation given $c(0) = c_0$ and describe the long term behaviour of c.

(b) Suggest suitable values of c_0 and I for the administration of aminophylline so that c lies in the middle of the range from 5 to 20mgl^{-1}.

DERIVE ACTIVITY 2C

This activity introduces the DERIVE library function SEPARABLE(p,q,x,y,x_0,y_0) for solving differential equations of the form

$$\frac{dy}{dx} = p(x)q(y)$$

where x_0 and y_0 are the initial values of x and y.

The library function SEPARABLE_GEN(p,q,x,y,c) gives the general solution if no initial values are given. **File Load** the **Utility** file ode1.mth. Set precision to **Exact** for this activity.

(A) (i) To solve the differential equation

$$\frac{dy}{dx} = x^2 y$$

Author Expression SEPARABLE(x^2,y,x,y,$x0$,$y0$) and **Simplify**. Then **soLve Algebraically** for y to obtain the general solution.

(ii) Use **Simplify Substitute** for **Variables** to substitute $x_0 = 0$ and $y_0 = 12$. Use **Simplify** to obtain a particular solution. **Plot** this solution.

(iii) Use DERIVE to find the value of y when $x = 2$.

(iv) For what value of x is $y = 50$?

(v) **Author Expression** SEPARABLE_GEN(x^2,y,x,y,c) and **Simplify**. This shows the general solution in a different form to (i). How would you find c given $y(0) = 12$?

(vi) **Author Expression** SEPARABLE(x^2,y,x,y,0,12) and **Simplify** to obtain the particular solution immediately.

Figure 2.15 shows the various parts of this activity.

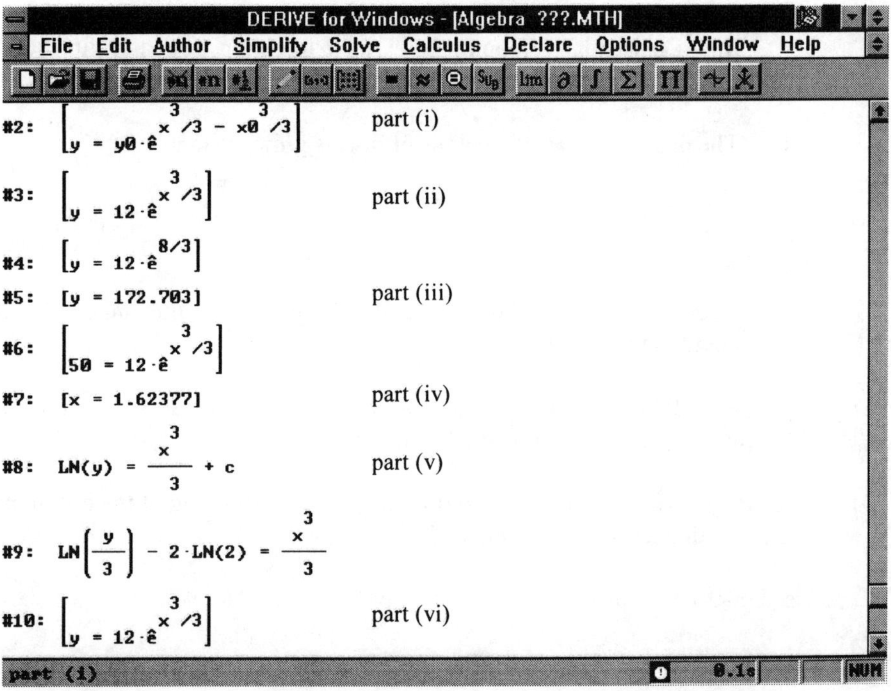

Figure 2.15 Using SEPARABLE to solve o.d.e.'s

(B) Solve each of the problems below using DERIVE.

(i) A diver in the sea falls so that his speed changes according to

$$v\frac{dv}{dx} = -(2+3v)$$

Solve this differential equation given that $v = 15$ when $x = 0$. Plot the solution. How deep does the diver descend?

(ii) The temperature T of an object changes so that

$$\frac{dT}{dt} = -\frac{(T-40)}{2}$$

where t is time measured in minutes. If the object is initially at 80°C, find a particular solution. Solve for t to obtain time as a function of temperature. How long does it take for the temperature to fall to 60°C?

(iii) The population, P, of a colony of insects grows so that

$$\frac{dP}{dt} = \frac{P}{4}$$

where t is time measured in days. Find the time taken for the population to double in size.

(C) Use the SEPARABLE_GEN function to find the general solution of the differential equations in question 2 of Exercise 2B.

(D) Use the SEPARABLE function to find the particular solution of the differential equations in question 3 of Exercise 2B.

(E) Use DERIVE to solve questions 4 to 14 of Exercise 2B.

2.5 FIRST ORDER LINEAR EQUATIONS

The general first order equation $\frac{dy}{dx} = f(x, y)$ is linear if it can be written in the form

$$\frac{dy}{dx} + p(x)y = q(x) \tag{2.10}$$

The differential equation (2.9) is **linear** because $\frac{dy}{dx} = q(x) - p(x)y$ is a **linear function of y**. Note that the term involving y, $p(x)y$, contains y to the power 1. If the equation contains y^2, $\sqrt{y}$, $\frac{1}{y}$, $\sin y$, e^y, $\cos y$ or any other function of y it is **not** linear.

The first derivative in the standard form of the linear differential equation (2.10) has coefficient 1 and the first step in solving such equations is to rewrite it this form. For example, if we are given

$$x\frac{dy}{dx} + x^2y = 3$$

then the standard form is obtained by dividing through by x to give

$$\frac{dy}{dx}+xy=\frac{3}{x}$$

Consider the following differential equations:

(i) $\dfrac{dy}{dx}+x^2y=e^x$ is linear with $p(x) = x^2$, $q(x) = e^x$

(ii) $x\dfrac{dy}{dx}-y=x^3$ can be rearranged to give $\dfrac{dy}{dx}-\dfrac{1}{x}\cdot y=x^2$ and is linear with $p(x)=-\dfrac{1}{x}$ and $q(x) = x^2$

(iii) $\dfrac{dy}{dx}+xy^2=x$ is not linear because of y^2 (not y)

(iv) $y\dfrac{dy}{dx}+xy=x^4$ is not linear as it cannot be put in the form (2.10). If we divide by y we get $\dfrac{dy}{dx}=\dfrac{x^4}{y}-x$ which is not linear in y.

To solve equation (2.10) we multiply it by a function $\alpha(x)$, defined later and called the **integrating factor**. Then

$$\alpha(x)\frac{dy}{dx}+\alpha(x)p(x)y=\alpha(x)q(x)$$

or, for brevity,

$$\alpha\frac{dy}{dx}+\alpha py=\alpha q \tag{2.11}$$

We choose α so that

$$\alpha p=\frac{d\alpha}{dx}$$

Solving this equation by separation, $\frac{1}{\alpha}d\alpha = pdx$, so that $\ln\alpha = \int pdx$, giving $\alpha = e^{\int pdx}$ (ignoring the constant of integration). With this choice of α, (2.11) becomes

$$\alpha\frac{dy}{dx} + \frac{d\alpha}{dx}y = \alpha q$$

or

$$\frac{d}{dx}(\alpha y) = \alpha q$$

Integrating

$$\alpha y = \int \alpha q dx + C$$

and so

$$y = \frac{1}{\alpha}\left\{\int \alpha q dx + C\right\}$$

The general solution of (2.10) is

$$y = e^{-\int p(x)dx}\left\{\int \alpha(x)q(x)dx + C\right\}$$

Comments

1. The method puts the left hand side of (2.10) in the form $u\frac{dv}{dx} + v\frac{du}{dx}$ so that it can be replaced by the single term $\frac{d}{dx}(uv)$ using the product rule for differentiation. For example

$$\frac{d}{dx}(x^3 \sin x) = x^3 \cos x + 3x^2 \sin x$$

so that

$$x^3 \cos x + 3x^2 \sin x = \frac{d}{dx}(x^3 \sin x)$$

Similarly $2x\cos x - x^2 \sin x = \frac{d}{dx}(x^2 \cos x)$ and $x^2 \frac{dy}{dx} + 2xy = \frac{d}{dx}(x^2 y)$.

However $x^2 \frac{dy}{dx} + 4xy$ is not the derivative of a product, as the derivative of x^2 is not $4x$, but we can put it into the required form by multiplying it by x^2. That is,

$$x^4 \frac{dy}{dx} + 4x^3 y = \frac{d}{dx}(x^4 y)$$

Multiplying (2.10) by the integrating factor ensures that the left hand side is in the form $u\frac{dv}{dx} + v\frac{du}{dx}$, which is then replaced by $\frac{d}{dx}(uv)$.

2. The calculation of the integrating factor from $\alpha = e^{\int p dx}$ involves integrating p. If this in turn contains logarithms, α can be simplified. For example

$$e^{\ln 2} = 2,\ e^{\ln x} = x,\ e^{2\ln x} = e^{\ln x^2} = x^2,$$
$$e^{-3\ln x} = x^{-3}, \text{ and, in general, } e^{a\ln[f(x)]} = [f(x)]^a$$

3. When integrating p no constant of integration is needed. To see why put one in and see what happens!

The method of solving linear differential equations using an integrating factor is summarised in the following steps.

1. Put the equation into the form $\frac{dy}{dx} + p(x)y = q(x)$ (2.10).

2. Identify $p(x)$ and find the integrating factor $\alpha(x) = e^{\int p(x)dx}$.

3. Multiply (2.10) by the integrating factor $\alpha(x)$.

4. Rewrite the left hand side as $\frac{d}{dx}[\alpha y]$.

5. Integrate.

6. Rearrange for $y = \ldots$ to find the general solution.

7. Apply the condition to find the constant of integration and the particular solution.

Example 2L

Find the general solution of $\frac{dy}{dx} + y = 1$.

Solution

Step 1: The equation is already in the correct form.

Step 2: $p(x) = 1$ and $\alpha = e^{\int 1 dx} = e^x$

Step 3: $e^x \frac{dy}{dx} + e^x y = e^x$

Step 4: $\frac{d}{dx}(ye^x) = e^x$ Note: LHS is **always** $\frac{d}{dx}(y \times \alpha(x))$

Step 5: $ye^x = e^x + C$

Step 6: $y = 1 + Ce^{-x}$ (GS)

□

Example 2M

Solve $x\frac{dy}{dx} - 3y = x^5$ given $y = 0$ when $x = 1$.

Solution

Step 1: $(\div x)$ $\dfrac{dy}{dx} - \dfrac{3}{x}y = x^4$

Step 2: $p(x) = \dfrac{-3}{x}$ and $\alpha(x) = \mathrm{e}^{\int -\frac{3}{x}dx} = \mathrm{e}^{-3\ln x} = \mathrm{e}^{\ln x^{-3}} = x^{-3}$

Step 3: $x^{-3}\dfrac{dy}{dx} - 3x^{-4}y = x^4 \times x^{-3}$

Step 4: $\dfrac{d}{dx}(yx^{-3}) = x$

Step 5: $yx^{-3} = \dfrac{1}{2}x^2 + C$

Step 6: $y = \dfrac{1}{2}x^5 + Cx^3$ (GS)

Step 7: Using $y(1) = 0$, $0 = \dfrac{1}{2} \times 1 + C \times 1$ from which $C = -\dfrac{1}{2}$.

The particular solution is $y = \dfrac{1}{2}x^5 - \dfrac{1}{2}x^3$

□

Example 2N

Solve $(1+\sin x)\dfrac{dy}{dx} - y\cos x = (1+\sin x)^4$ given $y(0) = 1$.

Solution

Dividing by $(1 + \sin x)$, the equation becomes

$$\frac{dy}{dx} - \frac{\cos x}{1+\sin x}\cdot y = (1+\sin x)^3$$

The integrating factor is

$$\alpha = e^{\int -\frac{\cos x}{1+\sin x}dx} = e^{-\ln(1+\sin x)} = (1+\sin x)^{-1}$$

Then

$$(1+\sin x)^{-1}\frac{dy}{dx} - \cos x(1+\sin x)^{-2}y = (1+\sin x)^2$$

or

$$\frac{d}{dx}[(1+\sin x)^{-1}y] = 1 + 2\sin x + \sin^2 x$$

Integrating

$$\frac{y}{1+\sin x} = x - 2\cos x + \frac{x}{2} - \frac{1}{4}\sin 2x + C$$

Applying $y(0) = 1$, $1 = -2 + C$, so that $C = 3$. Hence

$$y = (1+\sin x)\left(\frac{3x}{2} - 2\cos x - \frac{1}{4}\sin 2x + 3\right)$$

or

$$y = \frac{1}{4}(1+\sin x)(6x - 8\cos x - \sin 2x + 12)$$

□

Example 2O

An electrical circuit consists of a resistance R and an inductance L connected in series to a battery of constant voltage E_0. The current i at time t after closing the circuit satisfies the equation $L\frac{di}{dt} + Ri = E_0$. Given that $i = 0$ at $t = 0$ find i at time t and its ultimate value.

Solution

Dividing by L

$$\frac{di}{dt} + \frac{R}{L}i = \frac{E_0}{L}$$

The integrating factor is $\alpha(t) = e^{\int \frac{R}{L}dt} = e^{\frac{Rt}{L}}$. Then

$$e^{\frac{Rt}{L}}\frac{di}{dt} + \frac{R}{L}e^{\frac{Rt}{L}}i = \frac{E_0}{L}e^{\frac{Rt}{L}}$$

$$\frac{d}{dt}\left(ie^{\frac{Rt}{L}}\right) = \frac{E_0}{L}e^{\frac{Rt}{L}}$$

Integrating

$$ie^{\frac{Rt}{L}} = \frac{E_0}{R}e^{\frac{Rt}{L}} + C$$

$$i = \frac{E_0}{R} + Ce^{-\frac{Rt}{L}}$$

Using $i = 0$ at $t = 0$, $0 = \frac{E_0}{R} + C$, from which $C = -\frac{E_0}{R}$. Hence the current i at time t is given by

$$i = \frac{E_0}{R}\left(1 - e^{-\frac{Rt}{L}}\right)$$

As t increases $e^{-\frac{Rt}{L}}$ decreases towards zero and i approaches $\frac{E_0}{R}$.

□

It is important to note that the differential equation in Example 2O can also be solved using the method of separation of variables. Rearranging the equation $L\frac{di}{dt} = E_0 - Ri$. Separating,

$$\frac{L}{E_0 - Ri}di = dt$$

Integrating

$$\frac{L}{-R}\ln\left|E_0 - Ri\right| = t + A$$

$$\ln\left|E - Ri\right| = -\frac{Rt}{L} + B \qquad \left(\text{where } B = -\frac{RA}{L}\right)$$

Since $i = 0$ when $t = 0$ we have $B = \ln\left|E\right|$ and

$$\ln\left|\frac{E - Ri}{E}\right| = -\frac{Rt}{L}$$

Taking antilogs and solving for i gives

$$i = \frac{E}{R}\left(1 - e^{-\frac{Rt}{L}}\right) \qquad \text{as before.}$$

If you have the choice between the integrating factor and separation of variable methods, then the latter is usually easier to use.

When given a differential equation you have to choose an appropriate method. The following table gives examples of differential equations and possible methods of solution.

Equation	Use separation of variables	Use the integrating factor method
$\frac{dy}{dx} = xy$	Yes	Yes
$\frac{dy}{dx} = \frac{x}{y}$	Yes	No
$\frac{dy}{dx} = \frac{y}{x}$	Yes	Yes
$\frac{dy}{dx} = x + y$	No	Yes
$\frac{dy}{dx} = x^2 + y^2$	No	No
$\frac{dy}{dx} = 2x^2 - 3xy$	No	Yes
$\frac{dy}{dx} = x^4 - 2y^3$	No	No

The above examples show that for some equations separation of variables only can be used while for others the integrating factor method will work. There are examples for which either method will work but there are also those for which neither method is satisfactory.

Exercise 2C

1. Which of the following equations can be solved by the integrating factor method?

(i) $\dfrac{dy}{dx}+y=2e^{-x}$ (ii) $\dfrac{dy}{dx}+\dfrac{x}{y}=x^2$ (iii) $\dfrac{dy}{dx}=x+y^3$

(iv) $x\dfrac{dy}{dx}=y+x^4$ (v) $x^2\dfrac{dy}{dx}+\sin y=1$ (vi) $\cos x\dfrac{dy}{dx}=e^x-3y$

2. Find the general solution of

(i) $\dfrac{dy}{dx}+y=2e^{-x}$ (ii) $\dfrac{dy}{dx}+\dfrac{3y}{x}=x$

(iii) $x\dfrac{dy}{dx}-y=6x^4$ (iv) $(x+1)\dfrac{dy}{dx}+2y=x+1$

(v) $\dfrac{dy}{dx}+y\tan x=\cos x$ (vi) $e^{x^2}\dfrac{dy}{dx}=8x+4xy$

3. Find the particular solution of

(i) parts (i), (iv), (v) and (vi) of question 2 given $y(0)=1$

(ii) parts (ii) and (iii) of question 2 given $y(1)=2$.

4. A ball is projected with speed 40ms^{-1} at an angle of 60° to the horizontal.

(i) The horizontal distance x covered by the ball at time t satisfies the equation $\frac{dx}{dt}+0.1x=20$. Find x at any time t.

(ii) The vertical distance y reached by the ball at time t satisfies the equation $\frac{dy}{dt}+0.1y=20\sqrt{3}-10t$. Find y at any time t.

(iii) Given that the highest point is reached after about 3 seconds find the coordinates of this point.

(iv) Given that the total time of flight is about 6.3 seconds find the horizontal distance covered by the ball.

5. A thick-walled cylinder has internal and external radii a and b and is subjected to internal pressure. The radial stress P at distance r from the axis of the cylinder is given by $r\frac{dP}{dr}+2P=2c$, where c is a constant. Given that $P(a)=P_0$ and $P(b)=0$ show that $P=\frac{P_0a^2}{(b^2-a^2)}\left(\frac{b^2}{r^2}-1\right)$.

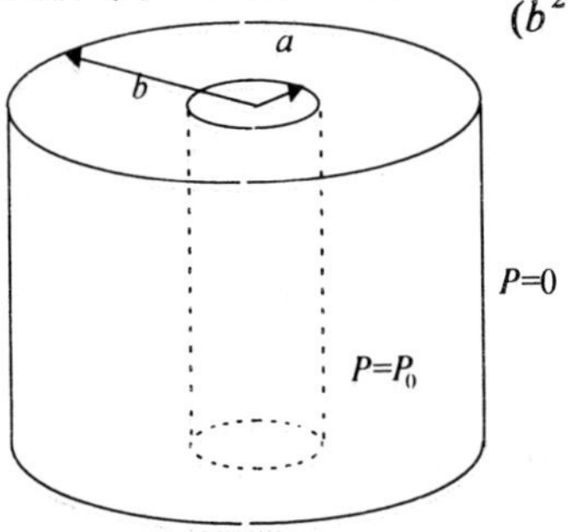

6. A sphere of mass m and radius a is held stationary in a fluid of viscosity η. When released it falls due to the force of gravity with the fluid providing a resistance of $k\eta av$ when the speed of the sphere is v. k is a constant. The equation of motion of the sphere is $m\frac{dv}{dt}=mg-k\eta av$. Find v at any time t.

Show that the speed approaches $\frac{mg}{k\eta a}$ and find the time to attain half this speed.

7. A hot body is placed in an environment whose temperature varies sinusoidally with mean 15° and amplitude 5°. The temperature θ of the body at time t satisfies the equation $\frac{d\theta}{dt}+0.1\theta=1.5+0.5\sin 2t$. Find θ at any time t given $\theta(0) = 100$.

8. In the electrical circuit in Example 2O the constant voltage E_0 is replaced with the alternating voltage $E_0 \sin \omega t$, so that the equation becomes

$$L\frac{di}{dt}+Ri=E_0\sin\omega t$$

Show that, in this case, $i=\frac{E_0}{R^2+\omega^2L^2}\left(R\sin\omega t-\omega L\cos\omega t+\omega Le^{-\frac{Rt}{L}}\right)$.

9. The following equation is a simplified model of the human heart in which $v(t)$ is the volume of blood in the heart, $I(t)$ is the inflow and $F(t)$ is the outflow.

$$\frac{dv}{dt}=I-F$$

The pressure $p(t)$ is related to v and F by

$$p=k(v-v_0) \quad \text{and} \quad F=p/R$$

where k, v_0 and R are constants.

(a) Formulate the differential equation satisfied by $p(t)$.

(b) In the systolic stage, blood is pumped out of the heart for the time interval $t = t_0$ to $t = T$ and $I(t) = 0$. Find $p(t)$ if $p = p_0$ when $t = t_0$.

(c) In the diastolic stage, blood is pumped into the heart for the time interval $t = 0$ to $t = t_0$ during which $I = I_0 =$ constant. Find $p(t)$ if $p = p_0$ at the end of the interval $0 \le t \le t_0$.

(d) The motion of the heart is periodic with period T so that $p(0) = p(T)$. Find an implicit equation for R so that this condition is satisfied.

10. Interest on money invested in a bank or building society is usually added to the capital once or twice a year, but occasionally accounts can be found where interest is added quarterly, monthly or even more frequently. For a given headline interest rate, say 10%, the more often interest is added the better. For example £100 invested at 10% compounded annually will grow to £161.05 after 5 years, whereas if interest is added monthly it will grow to £164.53.

(i) If we assume that interest is added continuously the rate of change $\frac{ds}{dt}$ of the value s of the investment is equal to the rate at which interest accrues which, for an interest rate of r%, is $\frac{rs}{100}$. Hence the differential equation is $\frac{ds}{dt} = \frac{rs}{100}$.

(a) Solve this equation given an initial investment of £P.

(b) £1000 is invested at 8%. Find the value of the investment after 10 years assuming that interest is compounded annually, every six months, monthly, daily or continuously. (If interest is compounded n times a year then $s = P\left(1 + \frac{r}{100n}\right)^{nt}$.)

(ii) If interest is added continuously at r% and, in addition, deposits (or withdrawals) are made at constant rate k then the equation becomes $\frac{ds}{dt} = \frac{rs}{100} + k$.

(a) Solve the equation given $s(0) = P$.

(b) I took out a pension plan at the age of 25 and agreed an initial contribution of £1000 followed by annual payments of £2000 paid in a continuous manner. Assuming an interest rate of 8% what will be the value of my pension fund on retirement at 65? How much of this is interest on the payments? (The size of my pension fund will probably surprise you but remember that it has been assumed that interest has been added continuously at a constant 8% and, crucially, that the ravages of inflation will have a severe effect.)

(c) I have just received an inheritance of £250,000 which I intend to invest in order to produce a continuous income on which to live. The best interest rate available at the moment is 6% (compounded continuously). How long will the money last if I withdraw £25000 per year? How much longer will it last if I decide to withdraw £20000 per year instead?

DERIVE ACTIVITY 2D

This activity introduces the DERIVE library function LINEAR1(p,q,x,y,x_0,y_0) for solving differential equations of the form

$$\frac{dy}{dx}+p(x)y=q(x)$$

where x_0 and y_0 are the initial values of x and y.

The library function LINEAR1_GEN(p,q,x,y,c) simplifies to the general solution in terms of a constant c.

File Load the **Utility** file ODE1.

(A) (i) To solve the differential equation

$$\frac{dy}{dx}-\frac{3}{x}y=x^4$$

given $y = 0$ when $x = 1$ (Example 2M).

Author Expression LINEAR1(–3/x,x^4,x,y,1,0) and **Simplify**.

DERIVE gives the solution $y=\dfrac{x^3(x^2-1)}{2}$.

(ii) To obtain a general solution **Author Expression** LINEAR1_GEN (–3/x,x^4,x,y,c) and **Simplify**.

(B) Solve each of the problems below using DERIVE.

(i) A sky diver falls so that his speed changes according to

$$\frac{dv}{dt}+0.3v=10$$

Solve this differential equation given that $v = 0$ when $t = 0$. **Plot!** the solution and state the terminal speed of the sky diver.

(ii) Find the equation of the graph whose tangent to the x-axis makes an angle θ given by

$$\tan\theta = \frac{y}{2x} - 1$$

and the curve passes through the point (2,3).

(iii) Try to use DERIVE to solve the differential equation

$$\frac{dy}{dx} = \frac{y}{2x^2} - 1$$

Explain why DERIVE cannot proceed to an analytical solution.

(C) Use the LINEAR1_GEN function to find the general solution of the differential equations in question 2 of Exercise 2C.

(D) Use the LINEAR1 function to find the particular solution of the differential equations in question 3 of Exercise 2C.

(E) Use DERIVE to solve questions 4-10 of Exercise 2C.

2.6 BERNOULLI EQUATIONS

A Bernoulli equation is one that can be put in the form

$$\frac{dy}{dx} + p(x)y = y^n q(x) \tag{2.12}$$

If $n = 0$ or 1 then the equation is linear and can be solved using the techniques of Section 2.5. For other values of n we proceed as follows.

Dividing (2.12) by y^n we obtain

$$y^{-n}\frac{dy}{dx} + p(x)y^{-n+1} = q(x) \tag{2.13}$$

We make the substitution $v = y^{-n+1}$, which reduces the equation to a first order linear equation which can be solved by the integrating factor method.

If $v = y^{-n+1}$ then $\dfrac{dv}{dx} = (-n+1)y^{-n}\dfrac{dy}{dx}$ so that

$$y^{-n}\frac{dy}{dx}=\frac{1}{(-n+1)}\frac{dv}{dx}$$

Substitution into (2.13) gives

$$\frac{1}{(-n+1)}\frac{dv}{dx}+p(x)v=q(x)$$

or

$$\frac{dv}{dx}+(-n+1)p(x)v=(-n+1)q(x)$$

which is in the standard form of linear equations in Section 2.5.

Example 2P

Find the general solution of $\frac{dy}{dx}-4y=8xy^{\frac{3}{2}}$.

Solution

Dividing by $y^{\frac{3}{2}}$

$$y^{-\frac{3}{2}}\frac{dy}{dx}-4y^{-\frac{1}{2}}=8x \qquad (2.14)$$

Let $v=y^{-\frac{1}{2}}$. Then $\frac{dv}{dx}=-\frac{1}{2}y^{-\frac{3}{2}}\frac{dy}{dx}$, giving $y^{-\frac{3}{2}}\frac{dy}{dx}=-2\frac{dv}{dx}$.
Substituting in (2.14) we get

$$-2\frac{dv}{dx}-4v=8x$$

Rearranging

$$\frac{dv}{dx}+2v=-4x \qquad (2.15)$$

The integrating factor is $e^{\int 2dx} = e^{2x}$. Multiplying (2.15) by e^{2x} gives

$$e^{2x}\frac{dv}{dx} + 2e^{2x}v = -4xe^{2x}$$

Then

$$\frac{d}{dx}(ve^{2x}) = -4xe^{2x}$$

Integrating

$$ve^{2x} = -4\left(\frac{1}{2}xe^{2x} - \int 1\times\frac{1}{2}e^{2x}dx\right) \qquad \text{(integration by parts)}$$

$$ve^{2x} = -2xe^{2x} + e^{2x} + C$$

$$v = -2x + 1 + Ce^{-2x}$$

If $v = y^{-\frac{1}{2}}$ then $y = v^{-2}$ and so the required solution is

$$y = (1 - 2x + Ce^{-2x})^{-2}$$

□

Example 2Q

The population P at time t of a bird colony is modelled by the logistic equation $\frac{dP}{dt} = \frac{P}{2} - \frac{P^2}{50000}$. Given that the initial population is 10000 find

(i) the population two years later
(ii) how long it will take for the population to double
(iii) the ultimate size of the population.
(iv) Sketch the graph of P against t.

Solution

Rearranging the equation and dividing by P^2

$$P^{-2}\frac{dP}{dt}-\frac{1}{2}P^{-1}=-\frac{1}{50000}$$

Let $v=P^{-1}$. Then $\frac{dv}{dt}=-P^{-2}\frac{dP}{dt}$ and $P^{-2}\frac{dP}{dt}=-\frac{dv}{dt}$. Substitution into the equation gives

$$-\frac{dv}{dt}-\frac{1}{2}v=-\frac{1}{50000}$$

or

$$\frac{dv}{dt}+\frac{1}{2}v=\frac{1}{50000}$$

The integrating factor is $e^{\int \frac{1}{2}dt}=e^{\frac{1}{2}t}$ and, multiplying by $e^{\frac{1}{2}t}$,

$$e^{\frac{1}{2}t}\frac{dv}{dt}+\frac{1}{2}e^{\frac{1}{2}t}v=\frac{1}{50000}e^{\frac{1}{2}t}$$

or

$$\frac{d}{dt}\left(ve^{\frac{1}{2}t}\right)=\frac{1}{50000}e^{\frac{1}{2}t}$$

Integrating

$$ve^{\frac{1}{2}t}=\frac{1}{25000}e^{\frac{1}{2}t}+A$$

or

$$v=\frac{1}{25000}+Ae^{-\frac{1}{2}t}$$

Since $v = P^{-1}$,

$$\frac{1}{P} = \frac{1}{25000} + Ae^{-\frac{1}{2}t}$$

or

$$P = \frac{1}{\frac{1}{25000} + Ae^{-\frac{1}{2}t}}$$

or

$$P = \frac{25000}{1 + Be^{-\frac{1}{2}t}} \qquad \text{(where } B = 25000A\text{)}$$

Using $P = 10000$ at $t = 0$, $10000 = \dfrac{25000}{1+B}$ from which $B = \dfrac{25000}{10000} - 1 = 1.5$.
Hence the population P at time t satisfies the equation

$$P = \frac{25000}{1 + 1.5e^{-\frac{1}{2}t}}$$

(i) After two years the population is

$$P(2) = \frac{25000}{1 + 1.5e^{-1}} = 16110$$

(ii) When the initial population doubles to 20000 we get

$$20000 = \frac{25000}{1 + 1.5e^{-\frac{1}{2}t}} \quad \text{or} \quad 1 + 1.5e^{-\frac{1}{2}t} = \frac{25000}{20000}$$

Then $1.5e^{-\frac{1}{2}t} = \dfrac{1}{4}$ so that $e^{-\frac{1}{2}t} = \dfrac{1}{6}$ and $e^{\frac{1}{2}t} = 6$. Then $\dfrac{1}{2}t = \ln 6$ and $t = 2\ln 6 \approx 3.58$. The population will double in 3.58 years.

(iii) As t increases, $e^{-\frac{1}{2}t}$ decreases towards zero and so P approaches 25000. This is the equilibrium population level.

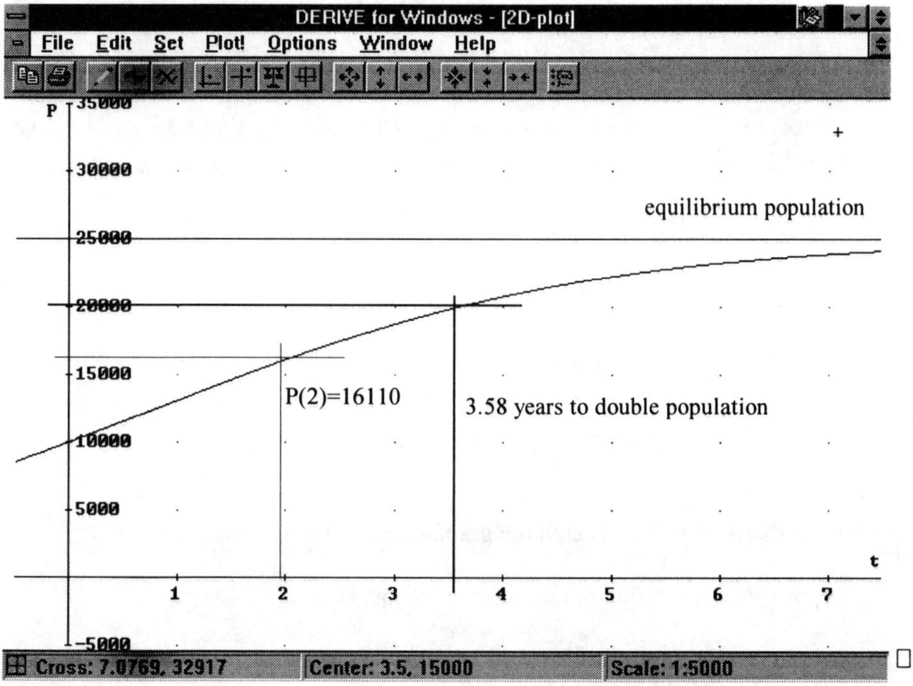

DERIVE ACTIVITY 2E

This activity introduces the DERIVE library function BERNOULLI(p,q,k,x,y,x_0,y_0) for solving differential equations of the form

$$\frac{dy}{dx} + p(x)y = q(x)y^k$$

where k is a constant and x_0 and y_0 are the initial values of x and y.

The library function BERNOULLI_GEN(p,q,k,x,y,c) simplifies to the general solution in terms of a constant c. **File Load** the **Utility** file ODE1.

(A) To solve the differential equation

$$\frac{dy}{dx} - 4y = 8xy^{\frac{3}{2}} \qquad \text{(Example 2P)}$$

Author Expression BERNOULLI_GEN(–4,8x,1.5,x,y,c) and **Simplify**. Check that the solution agrees with Example 2P. **Plot** solution curves for various values of c. **Author Expression** BERNOULLI(–4,8x,1.5,x,y,0,$\frac{1}{9}$) and **Simplify** to find the particular solution of the differential equation with initial condition $y(0) = \frac{1}{9}$.

(B) Use DERIVE to solve the problem in Example 2Q.

(C) Solve the differential equation

$$\frac{dy}{dx} = 3y^3 - \frac{y}{x}$$

given that $y(1) = 0.5$. Sketch the graph of the particular solution.

(D) Investigate the solution of the differential equation

$$\frac{dy}{dx} = 2y\left(1 - \left(\frac{y}{3}\right)^{\alpha}\right)$$

for the cases (a) $0 < \alpha < 1$ and (b) $\alpha > 1$.

2.7 HOMOGENEOUS EQUATIONS

A function $M(x,y)$ is homogeneous of degree n if the sum of the powers of x and y in **each** term is n. Then

$x^2 + xy - 3y^2$ is homogeneous of degree 2

$x^3 - 4xy^2$ is homogeneous of degree 3, but

$2x - 5y^2$ is not homogeneous

A homogeneous differential equation is one which can be written in the form $\dfrac{dy}{dx} = \dfrac{M(x,y)}{N(x,y)}$, where M and N are homogeneous functions of the **same** degree. For example, $\dfrac{dy}{dx} = \dfrac{x+2y}{x-y}$, $\dfrac{dy}{dx} = \dfrac{x^2 - xy}{y^2 + xy}$ and $x^3 \dfrac{dy}{dx} = y^3 - xy^2$ are homogeneous but $\dfrac{dy}{dx} = \dfrac{x^2}{x+y}$ and $(x-1)\dfrac{dy}{dx} = y$ are not.

Homogeneous equations are solved by making the substitution $y = vx$ $\left(\text{or } v = \dfrac{y}{x}\right)$, where v is a function of x. Then $\dfrac{dy}{dx} = v + x\dfrac{dv}{dx}$. The right hand side can be expressed in terms of $\dfrac{y}{x} = v$ by dividing throughout by x^n, where n is the degree of M and N.

Example 2R

Solve $xy^2 \dfrac{dy}{dx} = x^3 + y^3$ given $y = 3$ when $x = 1$.

Solution

Rearranging and then dividing throughout by x^3 as numerator and denominator are homogeneous of degree 3

$$\frac{dy}{dx} = \frac{x^3 + y^3}{xy^2} = \frac{1 + \left(\frac{y}{x}\right)^3}{\left(\frac{y}{x}\right)^2} \tag{2.16}$$

Let $y = vx$. Then $\dfrac{dy}{dx} = v + x\dfrac{dv}{dx}$. Substituting for $\dfrac{dy}{dx}$ and $\dfrac{y}{x}$ in (2.16) gives $v + x\dfrac{dv}{dx} = \dfrac{1+v^3}{v^2} = \dfrac{1}{v^2} + v$, which simplifies to

$$x\frac{dv}{dx} = \frac{1}{v^2}$$

Separating the variables

$$v^2 dv = \frac{1}{x} dx$$

Integrating

$$\frac{1}{3} v^3 = \ln x + C$$

If $y = 3$ when $x = 1$ then $v = 3$ then $x = 1$. Using this, $9 = \ln 1 + C$, from which $C = 9$. Hence

$$\frac{1}{3} v^3 = \ln x + 9$$

or

$$v^3 = 3 \ln x + 27$$

Reverting to y

$$\left(\frac{y}{x}\right)^3 = 3 \ln x + 27$$

Then

$$y^3 = x^3 (3 \ln x + 27)$$

The required solution is

$$y = x(3 \ln x + 27)^{\frac{1}{3}}$$

□

Example 2S

A searchlight has a light source S from which rays of light diverge, strike the reflecting mirror and then form a beam of parallel rays of light.

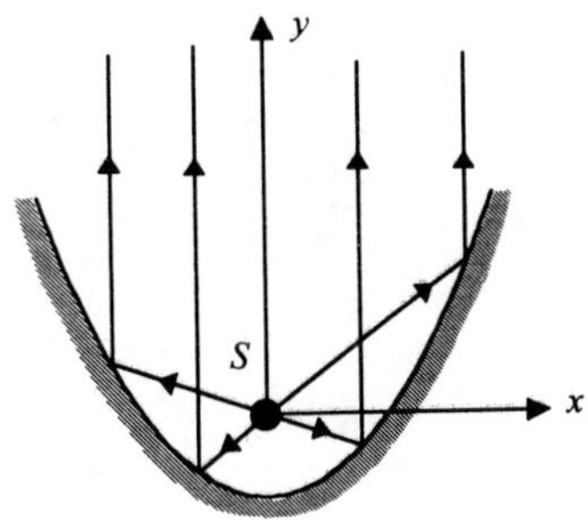

(i) With the origin at S and x and y axes as shown show that the shape of the mirror satisfies the equation $\dfrac{dy}{dx} = \dfrac{y + \sqrt{x^2 + y^2}}{x}$.

(ii) Solve the equation and deduce the shape of the mirror.

Solution

(i) Consider a ray of light striking the reflector at $P(x,y)$. It strikes, and leaves, the mirror at an angle α and, from the diagram, $P\hat{A}S$ is also α. The triangle ASP is isosceles and $AS = SP = \sqrt{x^2 + y^2}$. Also $AS = BC$.

Slope of mirror at P = tan β, where

$$\tan\beta = \frac{PB}{AB} = \frac{y + BC}{x} = \frac{y + \sqrt{x^2 + y^2}}{x}.$$

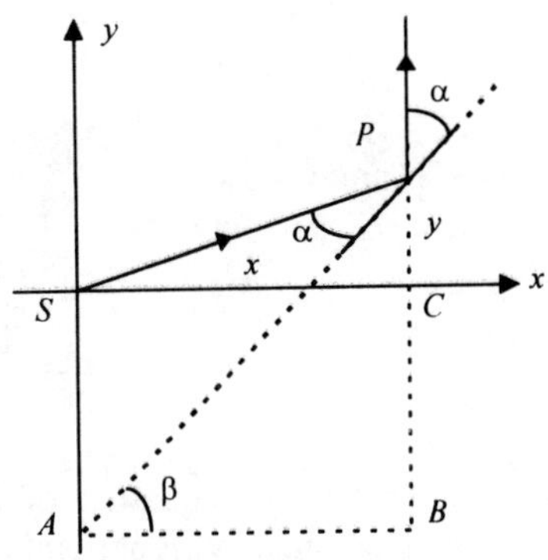

The mirror is tangential to the solution curve and so the slope is $\dfrac{dy}{dx}$.

Hence the shape of the mirror satisfies the equation $\dfrac{dy}{dx} = \dfrac{y + \sqrt{x^2 + y^2}}{x}$.

(ii) Rearranging

$$\frac{dy}{dx} = \frac{y}{x} + \sqrt{1 + \left(\frac{y}{x}\right)^2}$$

Putting $y = vx$ the equation becomes

$$v+x\frac{dv}{dx}=v+\sqrt{1+v^2}$$

Separate

$$\frac{1}{\sqrt{1+v^2}}dv=\frac{1}{x}dx$$

Integrate

$$\ln\left(\sqrt{1+v^2}+v\right)=\ln x+\ln C$$

Simplifying

$$\sqrt{1+v^2}+v=Cx$$

$$\sqrt{1+v^2}=Cx-v$$

$$1+v^2=C^2x^2-2Cxv+v^2$$

Using $y = vx$

$$1=C^2x^2-2Cy$$

$$y=\frac{1}{2}Cx^2-\frac{1}{2C}$$

This is the equation of the reflector; it is a parabola and so the mirror is parabolic. With particular values of C:

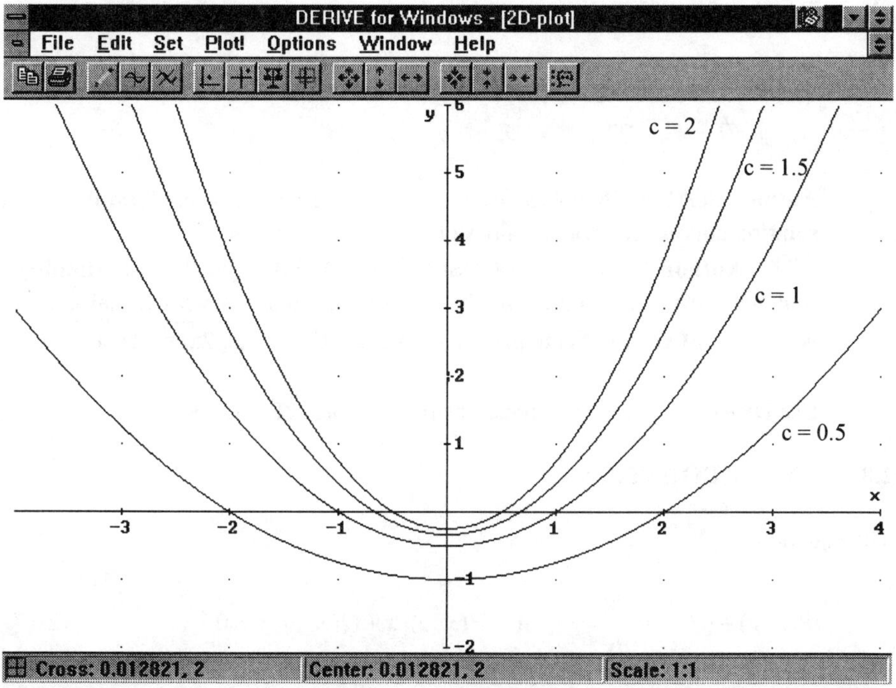

If $C = 2$ the equation becomes $\quad y = x^2 - \frac{1}{4} \quad$ (1)

If $C = 1$ the equation becomes $\quad y = \frac{1}{2}x^2 - \frac{1}{2} \quad$ (2)

If $C = ½$ the equation becomes $\quad y = \frac{1}{4}x^2 - 1 \quad$ (3)

□

DERIVE ACTIVITY 2F

This activity introduces the DERIVE library function HOMOGENEOUS(r,x,y,x_0,y_0) for solving differential equations of the form

$$\frac{dy}{dx} = r(x, y)$$

where r is the ratio of two homogeneous functions of the same degree and x_0 and y_0 are the initial values of x and y.

The library function HOMOGENEOUS_GEN(r,x,y,c) simplifies to the general solution in terms of a constant c. **File Load** the **Utility** file ODE1.

(A) To solve the differential equation

$$\frac{dy}{dx} = \frac{2x^2 + y^2}{2xy}$$

Author HOMOGENEOUS_GEN((2x^2 + y^2)/(2xy),x,y,c) and **Simplify**. **Plot** solution curves for various values of c.

Author HOMOGENEOUS((2x^2 + y^2)/(2xy),x,y,1,2) and **Simplify** to find the particular solution of the differential equation with initial condition $y(1) = 2$. **soLve Algebraically** for y. (Gives IF. $y = \pm\sqrt{2x(x+1)}$.)

(B) Use DERIVE to solve the questions in Examples 2R and 2S.

2.8 EXACT EQUATIONS

The equation

$$P(x,y) + Q(x,y)\frac{dy}{dx} = 0 \quad \text{or} \quad P(x,y)dx + Q(x,y)dy = 0 \tag{2.17}$$

is **exact** if it can be put in the form

$$\frac{d}{dx}(f(x,y)) = 0$$

for some f. The solution, found by direct integration, is $f(x,y) = C$.

(i) $2xy + x^2\dfrac{dy}{dx} = 0$ is exact because $2xy + x^2\dfrac{dy}{dx} = \dfrac{d}{dx}(x^2 y)$.

(ii) $y^2 \cos x dx + 2y \sin x dy$ is exact because $y^2 \cos x dx + 2y \sin x dy = d(y^2 \sin x)$.

(iii) $x^3 y + 6x^2 \dfrac{dy}{dx}$ is not exact.

Example 2T

Solve $2xy + x^2 \dfrac{dy}{dx} = 0$.

Solution

It was shown above that $2xy + x^2 \dfrac{dy}{dx} = \dfrac{d}{dx}(x^2 y)$. Hence the equation becomes $\dfrac{d}{dx}(x^2 y) = 0$. Integrating

$$x^2 y = C$$

The general solution is

$$y = \frac{C}{x^2}$$

□

When is an equation exact?

The total differential of $f(x,y) = 0$ is

$$\frac{\partial f}{\partial x} dx + \frac{\partial f}{\partial y} dy = 0 \tag{2.18}$$

For (2.17) and (2.18) to be the same $\dfrac{\partial f}{\partial x} = P$ and $\dfrac{\partial f}{\partial y} = Q$.

Then $\dfrac{\partial^2 f}{\partial x \partial y} = \dfrac{\partial P}{\partial y}$ and $\dfrac{\partial^2 f}{\partial y \partial x} = \dfrac{\partial Q}{\partial x}$.

Hence (2.17) is exact if

$$\frac{\partial P}{\partial y} = \frac{\partial Q}{\partial x} \tag{2.19}$$

The following examples illustrate the steps that are needed to check if a differential equation is exact.

(i) For $2xydx + x^2dy = 0$. Comparing with (2.17)

$$\left.\begin{aligned} &P = 2xy, \text{ so that } \frac{\partial P}{\partial y} = 2x \\ &\text{and } Q = x^2, \text{ so that } \frac{\partial Q}{\partial x} = 2x \end{aligned}\right\}$$

The condition (2.19) is satisfied and so the equation is exact

(ii) For $y^2\cos x dx + 2y\sin x dy = 0$

$$\left.\begin{aligned} &\text{Here } P = y^2 \cos x \text{ giving } \frac{\partial P}{\partial y} = 2y\cos x \\ &\text{and } Q = 2y\sin x \text{ giving } \frac{\partial Q}{\partial x} = 2y\cos x \end{aligned}\right\}$$

(2.19) is satisfied and the equation is exact

(iii) For $x^3ydx + 6x^2dy = 0$

$$\left.\begin{aligned} &\text{In this case } P = x^2y \text{ and so } \frac{\partial P}{\partial y} = x^3 \\ &\text{and } Q = 6x^2 \text{ from which } \frac{\partial Q}{\partial x} = 12x \end{aligned}\right\}$$

(2.19) is not satisfied and so the equation is not exact

Solving exact equations

The solution is $f(x,y) = C$, once we have found f.

$$\left.\begin{aligned} &\text{Integrate } \frac{\partial f}{\partial x} = P \text{ partially with respect to } x \\ &\text{Integrate } \frac{\partial f}{\partial y} = Q \text{ partially with respect to } y \end{aligned}\right\}$$

choose f so that both are satisfied

Example 2U

Solve the exact equation $2xydx + x^2dy = 0$.

Solution

Integrating $\frac{\partial f}{\partial x} = P = 2xy$ partially w.r.t. x: $f = x^2y + g(y)$

Integrating $\frac{\partial f}{\partial y} = Q = x^2$ partially w.r.t. y: $f = x^2y + h(x)$

} We take $g(y) = h(x) = 0$ and then $f(x,y) = x^2y$

It follows that the solution is $x^2y = C$ or $y = \frac{C}{x^2}$.

□

Example 2V

Solve $\frac{dy}{dx} = -\frac{(3x^2 + 4xy)}{(2x^2 + 2y)}$.

Solution

Rearranging

$$(3x^2 + 4xy)dx + (2x^2 + 2y)dy = 0$$

Comparing with (2.17)

$P = 3x^2 + 4xy$ and so $\frac{\partial P}{\partial y} = 4x$

and $Q = 2x^2 + 2y$ from which $\frac{\partial Q}{\partial x} = 4x$

} (2.19) is satisfied and so the equation is exact

Then integrating

$\frac{\partial f}{\partial x} = P = 3x^2 + 4xy$ gives $f = x^3 + 2x^2y + g(y)$

and integrating

$\frac{\partial f}{\partial y} = Q = 2x^2 + 2y$ gives $f = 2x^2y + y^2 + h(x)$

} To satisfy both choose $g(y) = y^2$, $h(x) = x^3$ and then $f = x^3 + 2x^2y + y^2$

Hence the solution is $x^3 + 2x^2y + y^2 = C$.

□

Example 2W

Show that $(2y^3 + 4x^2)dx + 3xy^2dy = 0$ is not exact, but multiplying it by x makes it exact. Hence solve the equation.

Solution

Comparing with (2.17), $P = 2y^3 + 4x^2$ giving $\frac{\partial P}{\partial y} = 6y^2$ and $Q = 3xy^2$ so that $\frac{\partial Q}{\partial x} = 3y^2$. Hence (2.19) is not satisfied and so the equation is not exact.

Multiplying throughout by x, the equation becomes

$$(2xy^3 + 4x^3)dx + 3x^2y^2dy = 0$$

Now $P = 2xy^3 + 4x^3$ and so $\frac{\partial P}{\partial y} = 6xy^2$ and $Q = 3x^2y^2$ giving $\frac{\partial Q}{\partial x} = 6xy^2$. Hence (2.19) is now satisfied and the equation is exact.

Then $\frac{\partial f}{\partial x} = P = 2xy^3 + 4x^3$ and so $f = x^2y^3 + x^4 + g(y)$ and $\frac{\partial f}{\partial y} = Q = 3x^2y^2$ and so $f = x^2y^3 + h(x)$.

Putting $g(y) = 0$ and $h(x) = x^4$ the solution is

$$x^2y^3 + x^4 = C$$

□

DERIVE ACTIVITY 2G

This activity introduces the DERIVE library function EXACT(p,q,x,y,x_0,y_0) for solving differential equations of thc form

$$p(x,y) + q(x,y)\frac{dy}{dx} = 0$$

if the differential is exact and x_0 and y_0 are the initial values of x and y.

The library function EXACT_GEN(p,q,x,y,c) simplifies to the general solution in terms of a constant c. If the differential equation is not exact, DERIVE returns "inapplicable". **File Load** the **Utility** file ODE1.

(A) Use the library function EXACT to attempt to solve the following differential equations. Comment on the DERIVE responses.

(i) $$x^3 y + 6x^2 \frac{dy}{dx} = 0$$

(ii) $$y^2 \cos x + 2y \sin x \frac{dy}{dx} = 0$$

(iii) $$2xy + x^2 \frac{dy}{dx} = 0$$

(iv) $$\mathrm{e}^{-x} \sin y - \mathrm{e}^{-x} \cos y \frac{dy}{dx} = 0$$

(B) Use DERIVE to solve the problems in Examples 2V and 2W.

(C) Use DERIVE to construct an implicit solution $F(t,x) = c$ to

$$(6tx + 4)dt + (3t^2 + 10x)dx = 0$$

and sketch several of its solution curves.

2.9 MISCELLANEOUS SUBSTITUTIONS

A substitution can sometimes transform a differential equation into one of the standard types. You will have seen this approach in integration where there are a number of standard methods available and a substitution can transform a difficult integral into one of the standard types. The same applies to differential equations but, as in integration, not all equations can be solved using this approach and numerical methods must be used. These methods are discussed in Chapter 3.

Example 2X

Solve $\dfrac{dy}{dx} = (y-x)^2$ using the substitution $v = y - x$. (Assume that $y > x + 1$).

Solution

If $v = y - x$ then $\frac{dv}{dx} = \frac{dy}{dx} - 1$ and so $\frac{dy}{dx} = \frac{dv}{dx} + 1$. The equation then becomes

$$\frac{dv}{dx} + 1 = v^2 \quad \text{or} \quad \frac{dv}{dx} = v^2 - 1$$

Separating

$$\frac{1}{v^2 - 1}\, dv = dx$$

Integrating

$$\frac{1}{2}\ln\frac{v-1}{v+1} = x + C \quad \text{or} \quad \ln\frac{v-1}{v+1} = 2x + A$$

Taking antilogs

$$\frac{v-1}{v+1} = e^{2x+A} = e^{2x} \times e^{A} = Be^{2x}$$

Then $v - 1 = (v + 1)Be^{2x}$ or $v(1 - Be^{2x}) = (1 + Be^{2x})$. Hence

$$v = \frac{1 + Be^{2x}}{1 - Be^{2x}}$$

Using $v = y - x$, the general solution is

$$y = \frac{1 + Be^{2x}}{1 - Be^{2x}} + x$$

□

Example 2Y

Solve the Riccati equation $\frac{dy}{dx} = \frac{y}{x} + \frac{2y^2}{x^2} - 2$ using the substitution $y = x + \frac{1}{v}$.

Solution

If $y = x + \frac{1}{v}$ then $\frac{dy}{dx} = 1 - \frac{1}{v^2}\frac{dv}{dx}$ and the equation becomes

$$1 - \frac{1}{v^2}\frac{dv}{dx} = 1 + \frac{1}{xv} + \frac{2}{x^2}\left(x^2 + \frac{2x}{v} + \frac{1}{v^2}\right) - 2$$

which simplifies to the linear equation

$$\frac{dv}{dx} + \frac{5}{x}v = -\frac{2}{x^2}$$

Multiplying by the integrating factor x^5 and rearranging gives

$$\frac{d}{dx}(x^5 v) = -2x^3$$

Integrating

$$x^5 v = -\frac{1}{2}x^4 + A$$

so that

$$v = \frac{A - \frac{1}{2}x^4}{x^5} = \frac{B - x^4}{2x^5}$$

and

$$\frac{1}{v} = \frac{2x^5}{B - x^4}$$

The solution of the equation is then, using $y = x + \frac{1}{v}$,

$$y = x + \frac{2x^5}{B - x^4}$$

□

Exercise 2D

Solve the following problems "by hand" and check your answers using DERIVE.

1. Which of the following are Bernoulli equations?

(i) $\frac{dy}{dx} - y = y^2$ (ii) $x\frac{dy}{dx} + y = x^2 y^4$

(iii) $\frac{dy}{dx} + 2xy^2 = y^5$ (iv) $x\frac{dy}{dx} + \frac{2x}{y} = 3y^3$

2. Find the general solution of

(i) $\frac{dy}{dx} - y = y^2$ (ii) $x\frac{dy}{dx} + 4y = 10\sqrt{xy}$

3. Solve

(i) $x\frac{dy}{dx} + y = x^2 y^4$ given $y(1) = \frac{1}{3}$

(ii) $2\sin x\frac{dy}{dx} - y\cos x = 4y^3 \cos x$ given $y\left(\frac{\pi}{2}\right) = 1$

4. The number N of trout in a reservoir is initially 10000 and grows according to the equation $\frac{dN}{dt} = 0.25N - 0.00001N^2$. Solve this equation and determine how long it will take the population to double to 20000.

When the population reaches 20000 fishing is allowed and this has the effect of changing the differential equation to $\frac{dN}{dt} = 0.2N - 0.0000125N^2$. What happens to the trout population subsequently?

5. Which of the following equations are homogeneous?

(i) $\dfrac{dy}{dx} = \dfrac{y}{x} + 2\tan\dfrac{y}{x}$ (ii) $\dfrac{dy}{dx} = \dfrac{x^2 + y^2}{x + y}$

(iii) $2xy\dfrac{dy}{dx} = x^2 + y^2$ (iv) $\dfrac{dy}{dx} = \dfrac{y^2}{x^2 + 1}$

6. Find the general solution of

(i) $\dfrac{dy}{dx} = \dfrac{y}{x} + 2\tan\dfrac{y}{x}$ (ii) $x\dfrac{dy}{dx} = 4\sqrt{x^2 + y^2} + y$

7. Solve

(i) $2xy\dfrac{dy}{dx} = x^2 + y^2$ given $y(1) = 2$

(ii) $x^2\dfrac{dy}{dx} = y^2 + xy + 4x^2$ given $y(1) = \dfrac{1}{2}$

8. Which of the following equations are exact?

(i) $2x^4y\dfrac{dy}{dx} + 4x^3y^2 = 0$ (ii) $(2x - 6xy^2)dx + (3y^2 - 3x^2)dy = 0$

(iii) $(2x + \sin y)dx + x^3\cos y\,dy = 0$

(iv) $\dfrac{dy}{dx} = \dfrac{2xy^3 - y^2e^x}{2ye^x - 3y^2x^2}$

9. Find the general solution of

(i) $2x^4y\dfrac{dy}{dx} + 4x^3y^2 = 0$

(ii) $(xe^x + e^x - 2xy)dx + (e^y - x^2)dy = 0$

10. Solve

(i) $$\frac{dy}{dx} = \frac{2xy^3 - 2y^2e^x}{2ye^x - 3y^2x^2} \text{ given } y(0) = -1$$

(ii) $\cos y^2 dx + (1 - 2xy \sin y^2)dy = 0$ given $y(2) = 0$

11. Show that the following equations are not exact, but become exact after multiplying throughout by the $\alpha(x,y)$ given.

(i) $$\frac{dy}{dx} = \frac{3x\cos y + 2}{x^2 \sin y}, \ \alpha(x,y) = x$$

(ii) $$2e^{2x} y \ln y \, dx + e^{2x} \, dy = 0, \ \alpha(x,y) = \frac{1}{y}$$

Hence solve (i) given $y(1) = 1$, and (ii) given $y(0) = e$.

12. Fill in each box in the following table with YES or NO. Do not solve any of the equations.

	Method of solution/type of equation				
Equation	Separation of variables	Integrating Factor	Bernoulli	Homogeneous	Exact
$xy' = y + 1$					
$\sin x . y' = e^x - y^2$					
$y' = \frac{2x - 3y}{3x - 4y}$					
$y' = (x - 3y)^2 - x^2$					
$2xyy' = e^x - y^2$					

13. Solve each of the following equations using the given substitutions.

	Equation	Substitution
(i)	$\dfrac{dy}{dx} = (x+y)^2$	$v = x + y$
(ii)	$e^y \dfrac{dy}{dx} + e^y = 4x$	$v = e^y$
(iii)	$x^2 \dfrac{dy}{dx} - 2xy = 2xy^2$	$y = vx^2$
(iv)	$\dfrac{dy}{dx} = \dfrac{x-y+2}{x-y+4}$	$v = x - y + 2$

14. Solve the equation $x^2 \dfrac{dy}{dx} + xy = y^2$

(i) as a Bernoulli equation
(ii) as a homogeneous equation
(iii) using the substitution $y = \dfrac{v}{x}$

15. A Riccati equation is of the form $\dfrac{dy}{dx} = p(x)y^2 + q(x)y + r(x)$.

(i) What does the equation become in the cases (a) $p(x) = 0$, (b) $r(x) = 0$?

(ii) Given that $y = y_1(x)$ is a solution of the equation show that the substitution $y = y_1 + \dfrac{1}{v}$ reduces it to a first order linear equation in v.

(iii) Show that $y = x$ is a solution of $\dfrac{dy}{dx} = \dfrac{y^2}{x^2} - \dfrac{3y}{x} + 3$ and hence use the substitution $y = x + \dfrac{1}{v}$ to find the general solution.

(iv) Show that $y = 1$ is a solution of $\frac{dy}{dx} = y^2 + \left(\frac{1}{x} - 1\right)y - \frac{1}{x}$ and hence find its general solution.

16. A chain is coiled up near the edge of a smooth table, disturbed, and begins to fall over the edge. The speed v at time t of the falling part of the chain satisfies the equation $xv\frac{dv}{dx} + v^2 = gx$. Use the substitution $y = v^2$ to show that $v = \sqrt{\frac{2gx}{3}}$.

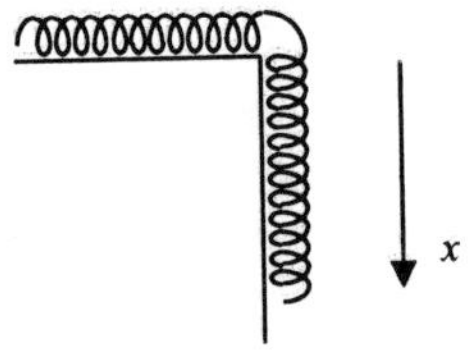

17. A particle of mass m moves along the x axis and is attracted towards the origin O by a force $T = n^2mx$ proportional to its displacement x from O. It is also subject to a resistive force $F = mkv^2$ which varies in magnitude with the square of the speed v. The equation of motion of the particle is $v\frac{dv}{dx} = -(kv^2 + n^2x)$. Use the substitution $y = v^2$ to solve the equation given that $v = u$ when $x = 0$.

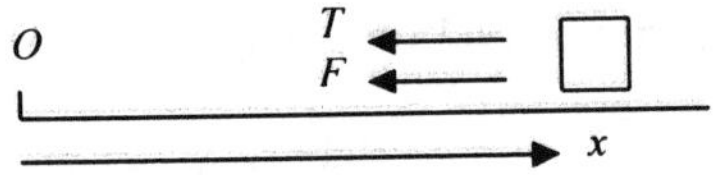

3

Numerical Solution of First Order Differential Equations

3.1 INTRODUCTION

In Chapter 2 we discussed different analytical methods for solving first order differential equations. The general solution is an equation between x and y which involves a constant, while a particular solution can be found if an initial condition is given.

For example the general solution of the equation

$$\frac{dy}{dx} + y = 2e^{-x} \quad \text{or} \quad y' + y = 2e^{-x}$$

(writing $\frac{dy}{dx} = y'$ for brevity) is $y = e^{-x}(2x + C)$ and, given $y(0) = 1$, the particular solution is $y = e^{-x}(2x + 1)$. This is the exact solution. There are, however, many differential equations for which no exact analytical method of solution exists and for which approximate numerical methods must be used. These methods find approximate particular solutions.

In general we wish to solve

$$y' = f(x,y) \quad \text{given} \quad y(x_0) = y_0 \tag{3.1}$$

in order to find y at $x = x^*$.

Numerical methods use the equations and condition to obtain approximations to the solution values at selected values of x. The initial condition tells us that $y = y_0$ at $x = x_0$ and we will then estimate y at $x_1, x_2, x_3, \ldots, x^*$. These x values are equally spaced h apart so that $x_1 = x_0 + h$, $x_2 = x_0 + 2h = x_1 + h$, $x_3 = x_0 + 3h = x_2 + h$, ... The approximate y values at $x_1, x_2, x_3, \ldots$ are denoted by $Y_1, Y_2, Y_3, \ldots$

We can make some progress by integrating (3.1) from x_r to x_{r+1} ($r = 0, 1, 2, \ldots$). Then

$$\int_{x_r}^{x_{r+1}} y'\,dx = \int_{x_r}^{x_{r+1}} f(x,y)dx$$

$$\left[y\right]_{x_r}^{x_{r+1}} = \int_{x_r}^{x_{r+1}} f(x,y)dx$$

$$y(x_{r+1}) - y(x_r) = \int_{x_r}^{x_{r+1}} f(x,y)dx$$

Denoting $y(x_r)$ as y_r and $y(x_{r+1})$ as y_{r+1}, this becomes

$$y_{r+1} = y_r + \int_{x_r}^{x_{r+1}} f(x,y)dx \tag{3.2}$$

Using $y = y_0$ at $x = x_0$, and putting $r = 0$, the integral $\int_{x_0}^{x_1} f(x,y)dx$ is estimated and hence Y_1, our approximation to y_1, is found. This information is used, with $r = 1, 2, 3, \ldots$ to find $Y_2, Y_3, Y_4, \ldots$, stopping when $y(x^*)$ has been estimated.

As the integral can only be evaluated approximately there will be a gradual build up of error. For example, Y_4 will contain errors in finding Y_1, Y_2 and Y_3 as well as in the calculation of Y_4 itself.

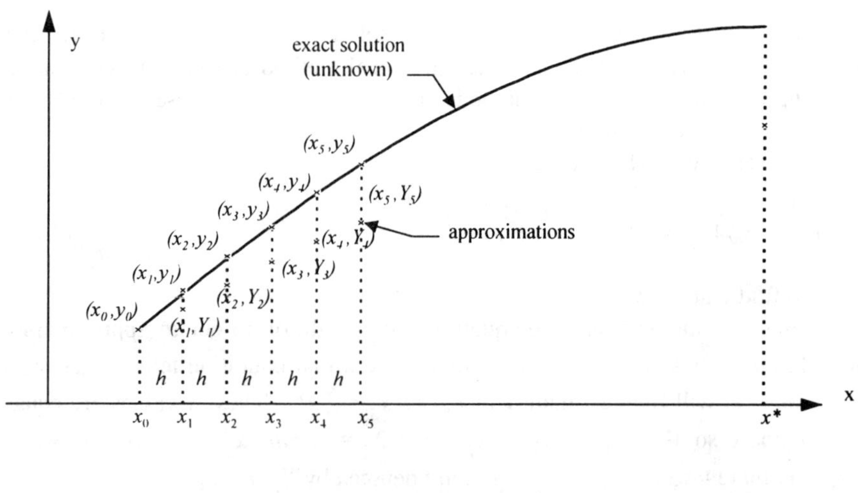

Figure 3.1

The approximate values Y_r will diverge away from the exact values y_r as r increases and it is possible that the error becomes so large as to make the approximations useless.

By using different methods and/or different values of the step length h, different approximations to $y(x^*)$ can be obtained. For the purposes of illustration we will use different numerical methods to solve the equation

$$y' = 2xy \quad \text{given} \quad y(0) = 1$$

The exact solution of this equation is $y = e^{x^2}$ and we can use this to compare different methods and different step lengths for accuracy.

In Section 3.2 a number of different methods are introduced. An analysis of these methods is the subject of Section 3.3.

3.2 NUMERICAL METHODS FOR SOLVING FIRST ORDER DIFFERENTIAL EQUATIONS

3.2.1 Euler's method

Euler's method approximates to the trajectories of the direction field of the differential equation by a series of straight lines, such that each is parallel to the direction field at its starting point. The initial condition $y(x_0) = y_0$ specifies the starting point A. The tangent at A is drawn but, as we move along it, its slope diverges from that of the direction field and an error is introduced. So that the error does not become too large we change direction after x has been increased by a suitable value h to x_1 ($x_1 = x_0 + h$), at the point B. The y ordinate at B, Y_1, is the Euler approximation to $y(x_1)$.

From B the line parallel to the direction field is drawn to C, where $x = x_1 + h$ and the y ordinate of C, Y_2, is the Euler approximation to $y(x_2)$. Continuing in this way we find Y_1, Y_2, Y_3, Y_4, ..., stopping when we find our estimate of $y(x^*)$ (see Figure 3.2).

We shall not, of course, find Y_1, Y_2, Y_3, ... diagramatically and require formulae from which they can be calculated.

Consider one step of the method, from $x = x_r$, where we have already found $Y_r \approx y(x_r)$, to $x = x_{r+1}$ where we wish to find $Y_{r+1} \approx y(x_{r+1})$.

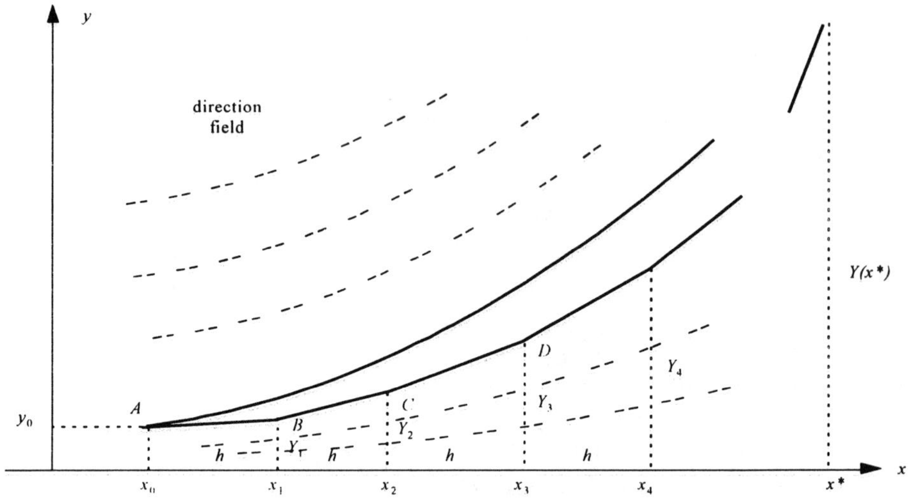

Figure 3.2

From Figure 3.3 we can see that

$$Y_{r+1} = Y_r + a = Y_r + h\tan\theta$$

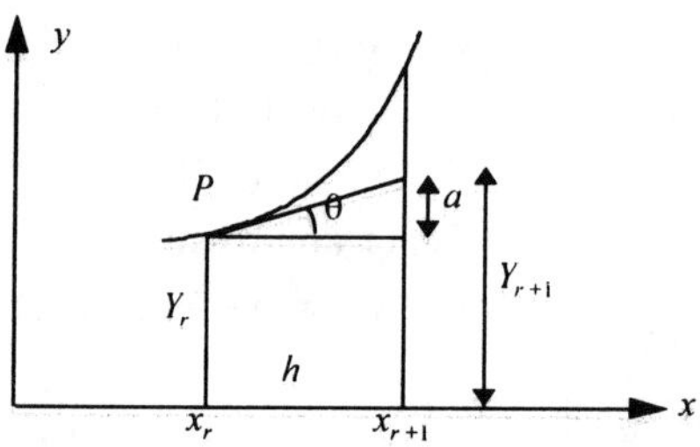

Figure 3.3

As $\tan\theta$ is the gradient of the tangent at P, this also equals the value of $\dfrac{dy}{dx} = y'$ there $(= Y_r')$. Hence we obtain $Y_{r+1} = Y_r + hY_r'$, which is Euler's method.

Alternatively, from the recurrence relation

$$Y_{r+1} = Y_r + \int_{x_r}^{x_{r+1}} f(x, y)dy$$

we need to estimate the integral which, for the curve shown in Figure 3.4, is the area under the $y' = f(x,y)$ curve from x_r to x_{r+1}. Assuming Y_r' is constant, the integral is approximated by the area of the rectangle. That is $\int_{x_r}^{x_{r+1}} f(x,y)dx \approx hf(x_r, Y_r) = hY_r'$

and so, once again, we obtain

$$Y_{r+1} = Y_r + hY_r'$$

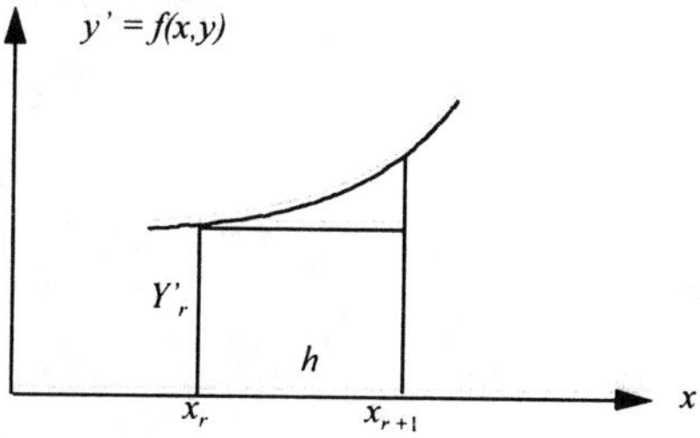

Figure 3.4

Euler's formula is

$$Y_{r+1} = Y_r + hY_r' \qquad r = 0, 1, 2, \ldots \qquad (3.3)$$

This is an **explicit** method as the right hand side contains only quantities that are already known. That is, we use information about the point r only to estimate y at the point $r + 1$.

Example 3A

Solve $y' = 2xy$ given $y(0) = 1$ to estimate $y(0.5)$, $y(1.0)$ and $y(1.5)$ with (i) $h = 0.5$, (ii) $h = 0.1$. Estimate $y(1.5)$ with (iii) $h = 0.01$, (iv) $h = 0.001$ and find the error in each estimate of $y(1.5)$.

Solution

Applying (3.3) to the equation $y' = 2xy$ we obtain

$$Y_{r+1} = Y_r + h2x_rY_r$$

or $\quad Y_{r+1} = (1 + 2hx_r)Y_r \qquad (3.4)$

The initial conditions are $x_0 = 0$, $y_0 = Y_0 = 1$.

(i) With $h = 0.5$, (3.4) becomes $Y_{r+1} = (1 + x_r)Y_r$.

Putting $r = 0$: $Y_1 = (1+x_0)y_0 = 1$ Hence $y(0.5) \approx 1$
$r = 1$: $Y_2 = (1+x_1)Y_1 = 1.5 \times 1 = 1.5$ $y(1) \approx 1.5$
$r = 2$: $Y_3 = (1+x_2)Y_2 = 2 \times 1.5 = 3$ $y(1.5) \approx 3$

We can compare these results with the exact solution $y = e^{x^2}$ (rounded to 3D)

x	0	0.5	1	1.5
Euler ($h = 0.5$)	1	1	1.5	3
e^{x^2} (3D)	1	1.284	2.718	9.488
$\lvert \text{Error} \rvert = \lvert Y(1.5) - y(1.5) \rvert$	0	0.284	1.218	6.488

h is too large to give a solution to any reasonable accuracy.

(ii) With $h = 0.1$, (3.4) becomes $Y_{r+1} = (1 + 0.2x_r)Y_r$.

Putting $r = 0$: $Y_1 = (1+0.2x_0)y_0 = (1+0.2\times0)1 = 1$ Hence $y(0.1) \approx 1$
$r = 1$: $Y_2 = (1+0.2x_1)Y_1 = (1+0.2\times0.1)1 = 1.02$ $y(0.2) \approx 1.02$
$r = 2$: $Y_3 = (1+0.2x_2)Y_2 = (1+0.2\times0.2)1.02 = 1.0608$ $y(0.3) \approx 1.061$
$\vdots$
$r = 4$: $Y_5 = 1.21440384$ $y(0.5) \approx 1.214$
$\vdots$
$r = 9$: $Y_{10} = 2.334633363$ $y(1) = 2.335$
$\vdots$
$r = 14$: $Y_{15} = 6.8353689953$ $y(1.5) \approx 6.835$

Comparing the exact and approximate results (all rounded to 3D)

x	0	0.5	1	1.5
Euler ($h = 0.1$)	1	1.214	2.335	6.835
e^{x^2} (3D)	1	1.284	2.718	9.488
$\lvert \text{Error} \rvert$	0	0.070	0.383	2.653

Although the results are better than in (i) there is still an unacceptably large error in $y(1.5)$.

(iii) With $h = 0.01$, (3.4) becomes $Y_{r+1} = (1 + 0.02x_r)Y_r$ and after 150 steps we find

$$Y_{150} \approx 9.144 \quad \text{(3D)} \qquad \text{so that} \qquad y(1.5) \approx 9.144$$

(iv) With $h = 0.001$, (3.4) becomes $Y_{r+1} = (1 + 0.002x_r)Y_r$ and after 1500 steps we find

$$Y_{1500} \approx 9.452 \quad \text{(3D)} \qquad \text{so that} \qquad y(1.5) \approx 9.452$$

Tabulating the error for in $y(1.5)$ various values of h:

h	Number of steps	Euler approx to $y(1.5)$	$\lvert\text{Error}\rvert$ to 3D
0.5	3	3	6.488
0.1	15	6.835	2.653
0.01	150	9.144	0.344
0.001	1500	9.452	0.036

It is clear that the error decreases as h decreases, but at the cost of a large increase in the amount of computation, but even with a step length of 0.001 we have only found $y(1.5)$ correct to one decimal place.

□

Example 3B

Consider the equation $y' = \sqrt{x^2 + y^2}$, $y(1) = 0$. Estimate $y(2)$ with $h = 0.5, 0.25, 0.2, 0.1$ and 0.05. Plot a graph of the estimates of $y(2)$ against h and hence, assuming that the points for the smaller h values lie on a straight line, estimate the exact value of $y(2)$.

Solution

Applying (3.3) to $y' = \sqrt{x^2 + y^2}$ we obtain the recurrence relation

$$Y_{r+1} = Y_r + h\sqrt{x_r^2 + Y_r^2} \qquad (3.5)$$

The initial conditions are $x_0 = 1$, $y_0 = Y_0 = 0$.

With $h = 0.5$, (3.5) becomes $Y_{r+1} = Y_r + 0.5\sqrt{x_r^2 + Y_r^2}$.

Putting $r = 0$: $Y_1 = Y_0 + 0.5\sqrt{x_0^2 + Y_0^2} = 0 + 0.5\sqrt{1^2 + 0^2} = 0.5$

$r = 1$: $Y_2 = Y_1 + 0.5\sqrt{x_1^2 + Y_1^2} = 0.5 + 0.5\sqrt{1.5^2 + 0.5^2} = 1.2906$ (4D)

Hence $y(2) \approx 1.2906$.

With $h = 0.25$, (3.5) becomes $Y_{r+1} = Y_r + 0.25\sqrt{x_r^2 + Y_r^2}$.

Putting $r = 0$: $Y_1 = Y_0 + 0.25\sqrt{x_0^2 + Y_0^2} = 0 + 0.25\sqrt{1^2 + 0^2} = 0.25$

$r = 1$: $Y_2 = Y_1 + 0.25\sqrt{x_1^2 + Y_1^2} = 0.25 + 0.25\sqrt{1.25^2 + 0.25^2} = 0.5687$ (4D)

$r = 2$: $Y_3 = Y_2 + 0.25\sqrt{x_2^2 + Y_2^2} = 0.5687 + 0.25\sqrt{1.5^2 + 0.5687^2} = 0.9697$ (4D)

$r = 3$: $Y_4 = Y_3 + 0.25\sqrt{x_3^2 + Y_3^2} = 0.9697 + 0.25\sqrt{1.75^2 + 0.9697^2} = 1.4699$ (4D)

Hence $y(2) \approx 1.4699$.

The following results can be obtained in a similar way.

h	0.2	0.1	0.05
Approx to $y(2)$ to 4D	1.5101	1.5962	1.6426

A plot of the values the estimates of $y(2)$ against h is given in Figure 3.5. Apart from the value when $h = 0.5$ the points lie reasonably close to a straight line.

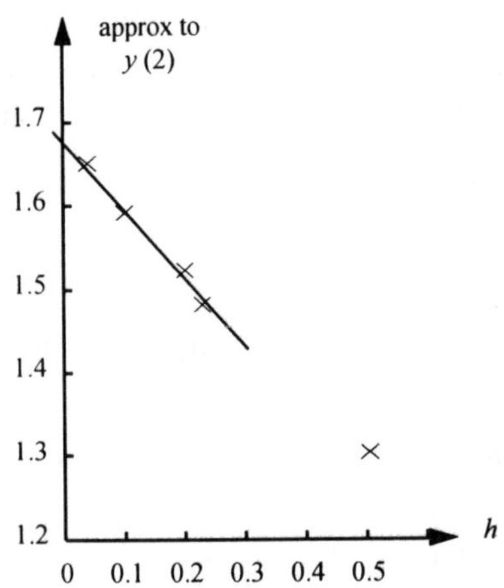

Figure 3.5

If the estimates of $y(2)$ improve as h decreases then a better estimate of the exact value of $y(2)$ is where the line cuts the vertical axis (that is, where $h = 0$). Reading this off from the graph gives $y(2) \approx 1.69$.

3.2.2 Trapezoidal method

In Euler's method $\int_{x_r}^{x_{r+1}} f(x,y)dx$ is approximated by the area of the rectangle A.

A method that is generally more accurate is the trapezium rule, where the area is approximated by the trapezium B. Then

$$\int_{x_r}^{x_{r+1}} f(x,y)dx \approx \frac{1}{2}h(Y'_r + Y'_{r+1})$$

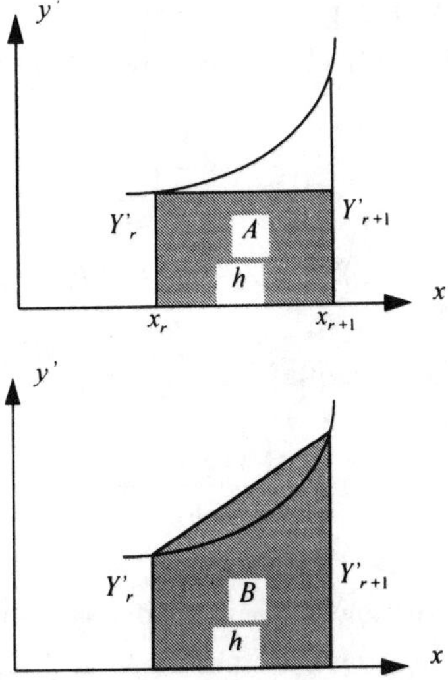

Figure 3.6

The recurrence relation (3.2) in Section 3.1 becomes

$$Y_{r+1} = Y_r + \frac{1}{2}h(Y'_r + Y'_{r+1}) \tag{3.6}$$

This is the **trapezoidal method** for solving $y' = f(x,y)$ with $y(x_0) = y_0$.

There is, however, an inherent difficulty in using (3.6) as in order to find Y_{r+1} we need to know Y'_{r+1}, which cannot be found until Y_{r+1} is known. The trapezoidal method (and any method in which the subscript r+1 appears on both sides of the equation) is called **implicit**.

In order to use the trapezoidal method, (3.6) must be solved for Y_{r+1}. This can be done fairly easily if the equation is linear but not otherwise and for this reason the method is not usually used on its own, but in conjunction with others (see Section 3.2.5).

Example 3C

Solve $y' = 2xy$ given $y(0) = 1$ to estimate $y(1.5)$ using the trapezoidal method with (i) $h = 0.5$, (ii) $h = 0.1$ and (iii) $h = 0.01$.

Solution

At the points x_r and x_{r+1} the equation becomes $Y'_r = 2x_r Y_r$ and $Y'_{r+1} = 2x_{r+1}Y_{r+1}$ respectively. Substitution into the trapezoidal method (3.6) gives

$$Y_{r+1} = Y_r + \frac{1}{2}h(2x_r Y_r + 2x_{r+1}Y_{r+1})$$

$$Y_{r+1} = Y_r + hx_r Y_r + hx_{r+1}Y_{r+1}$$

Rearranging: $(1 - hx_{r+1})Y_{r+1} = (1 + hx_r)Y_r$ giving

$$Y_{r+1} = \frac{(1+hx_r)}{(1-hx_{r+1})}Y_r \tag{3.7}$$

We now have an explicit equation for Y_{r+1} in terms of known quantities. After choosing h we put $r = 0, 1, 2, 3, \ldots$ to generate the approximate solutions Y_1, Y_2, Y_3, Y_4, ...

(i) With $h = 0.5$ (3.7) becomes $Y_{r+1} = \dfrac{(1+0.5x_r)}{(1-0.5x_{r+1})}Y_r$.

Putting $r = 0$: $\quad Y_1 = \dfrac{(1+0.5\times 0)}{(1-0.5\times 0.5)}\times 1 = 1.33\dot{3}$ giving $y(0.5) \approx 1.333$

Putting $r = 1$: $\quad Y_2 = \dfrac{(1+0.5x_1)}{(1-0.5x_2)}Y_1 = \dfrac{(1+0.5\times 0.5)}{(1-0.5\times 1)}\times 1.33\dot{3} = 3.33\dot{3}$

so that $\quad y(1) \approx 3.333$

Putting $r = 2$: $\quad Y_3 = \dfrac{(1+0.5x_2)}{(1-0.5x_3)}Y_2 = \dfrac{(1+0.5\times 1)}{(1-0.5\times 1.5)}\times 3.33\dot{3} = 20$

so that $\quad y(1.5) \approx 20$

$$|\text{Error}| \text{ in } y(1.5) = |20 - 9.488| \approx 10.5$$

With an error of over 100%, h is too large to give a meaningful result.

(ii) With $h = 0.1$ (3.7) becomes $Y_{r+1} = \dfrac{(1+0.1x_r)}{(1-0.1x_{r+1})} Y_r$.

Putting $r = 0$: $Y_1 = \dfrac{(1+0.1\times 0)}{(1-0.1\times 0.1)} \times 1 = 1.0101\ldots$ so that $y(0.1) \approx 1.010$

Putting $r = 1$: $Y_2 = \dfrac{(1+0.1\times 0.1)}{(1-0.1\times 0.2)} \times Y_1 = \dfrac{1.01}{0.98} \times 1.0101\ldots = 1.04102$

so that $y(0.2) \approx 1.041$

$\vdots$

Putting $r = 4$: $Y_5 = 1.2857728953$ so that $y(0.5) \approx 1.286$

$\vdots$

Putting $r = 9$: $Y_{10} = 2.7365975223$ so that $y(1) \approx 2.737$

$\vdots$

Putting $r = 14$: $Y_{15} = 9.6787350243$ so that $y(1.5) \approx 9.679$

$$|\text{Error}| \text{ in } y(1.5) = |9.679 - 9.488| \approx 0.191$$

(iii) With $h = 0.01$ we find $y(1.5) \approx 9.4896041176$ and

$$|\text{Error}| \text{ in } y(1.5) = |9.490 - 9.488| \approx 0.002$$

Note that, as before, the smaller the value of h the better the results, and also that the Trapezoidal method gives better results than Euler's method for the same value of h.

□

Example 3D

Solve $y' = \dfrac{2y}{x} + 5\sin x$, $y(1) = -1$ to estimate $y(2)$ by the trapezoidal method with $h = 0.5$, 0.2, 0.1 and 0.01. Determine the error in each estimate using the exact answer $y(2) = 5.44798$ (5D). Does the error in $y(2)$ change approximately linearly with h, as in Example 3B?

Solution

Here $Y_r' = \dfrac{2Y_r}{x_r} + 5\sin x_r$ and $Y_{r+1}' = \dfrac{2Y_{r+1}}{x_{r+1}} + 5\sin x_{r+1}$.
The trapezoidal rule (3.6) then becomes

$$Y_{r+1} = Y_r + \frac{1}{2}h\left(\frac{2Y_r}{x_r} + 5\sin x_r + \frac{2Y_{r+1}}{x_{r+1}} + 5\sin x_{r+1}\right)$$

$$Y_{r+1}\left(1 - \frac{h}{x_{r+1}}\right) = Y_r\left(1 + \frac{h}{x_r}\right) + \frac{5}{2}h(\sin x_r + \sin x_{r+1})$$

and we obtain the explicit recurrence relation

$$Y_{r+1} = \frac{Y_r\left(1 + \dfrac{h}{x_r}\right) + \dfrac{5}{2}h(\sin x_r + \sin x_{r+1})}{1 - \dfrac{h}{x_{r+1}}}$$

Using the above equation with different values of h results in the following estimates of $y(2)$.

h	number of steps	estimate of $y(2)$ to 5D	$\lvert \text{Error} \rvert$ to 2S
0.5	2	5.30787	0.14
0.2	5	5.42754	0.020
0.1	10	5.44295	0.0050
0.01	100	5.44793	0.000050

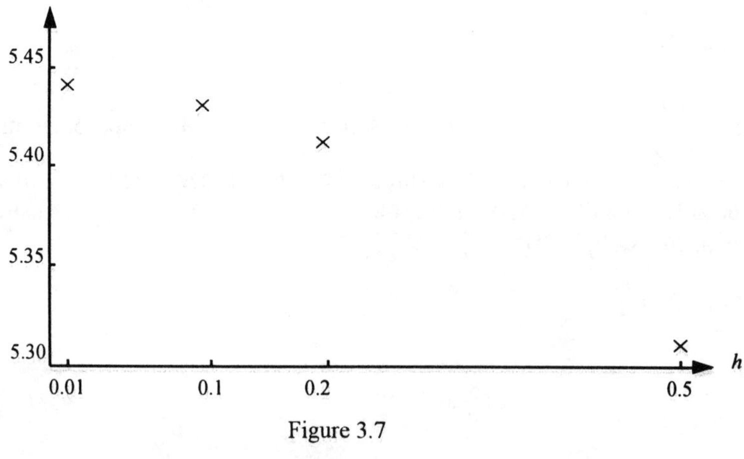

Figure 3.7

Plotting the estimates of $y(2)$ against h shows that the error is **not** linearly dependent on h.

□

3.2.3 Taylor series methods

Taylor's series allows us to evaluate a function f at a point $x + h$ in terms of h and the values of f and its derivatives at the point x. The usual form of the series is

$$f(x+h) = f(x) + hf'(x) + \frac{1}{2!}h^2 f''(x) + \frac{1}{3!}h^3 f'''(x) + \ldots$$

For example, we could use the series to find sin 35° correct to three decimal places using the known values of $\sin 30° = 0.5$, $\cos 30° = \frac{\sqrt{3}}{2}$ and $h = 5° = 5 \times \frac{\pi}{180} = \frac{\pi}{36}$ radians. Then

$$\sin 35° = \sin 30° + \frac{\pi}{36}\cos 30° + \frac{1}{2!}\left(\frac{\pi}{36}\right)^2(-\sin 30°) + \frac{1}{3!}\left(\frac{\pi}{36}\right)^3(-\cos 30°) + \cdots$$

$$= 0.5 + \frac{\pi}{36} \times \frac{\sqrt{3}}{2} - \frac{1}{4}\left(\frac{\pi}{36}\right)^2 - \frac{\sqrt{3}}{12}\left(\frac{\pi}{36}\right)^3 + \ldots$$

$$= 0.5 + 0.07552\ldots - 0.00190\ldots - 0.00010\ldots + \ldots$$

$$= 0.574 \qquad \text{to three decimal places}$$

Using y in place of f, x_r for x and x_{r+1} for $x + h$, Taylor's series becomes $y(x_{r+1}) = y(x_r) + hy'(x_r) + \frac{1}{2!}h^2 y''(x_r) + \frac{1}{3!}h^3 y'''(x_r) + \ldots$ or, in our notation,

$$y_{r+1} = y_r + hy'_r + \frac{1}{2!}h^2 y''_r + \frac{1}{3!}h^3 y'''_3 + \ldots$$

When solving a differential equation the y values will be approximate Y values and we then obtain

$$Y_{r+1} = Y_r + hY'_r + \frac{1}{2!}h^2 Y''_r + \frac{1}{3!}h^3 Y'''_r + \ldots$$

All the terms on the right hand side are known or can be found by differentiating the differential equation as they contain the subscript r only. Hence Taylor series methods are explicit.

Taylor series methods of different orders are found by truncating the series at different points.

$Y_{r+1} = Y_r + hY_r'$ is the 1st order Taylor series method (Euler's method)

$Y_{r+1} = Y_r + hY_r' + \frac{1}{2!}h^2 Y_r''$ is the 2nd order Taylor series method

$Y_{r+1} = Y_r + hY_r' + \frac{1}{2!}h^2 Y_r'' + \frac{1}{3!}h^3 Y_r'''$ is the 3rd order Taylor series method

$Y_{r+1} = Y_r + hY_r' + \frac{1}{2!}h^2 Y_r'' + \frac{1}{3!}h^3 Y_r''' + \frac{1}{4!}h^4 Y_r^{iv}$ is the 4th order Taylor series method

In particular the second and fourth order Taylor series methods TS2, TS4 are, respectively

$$Y_{r+1} = Y_r + hY_r' + \frac{1}{2!}h^2 Y_r'' \tag{3.8}$$

$$Y_{r+1} = Y_r + hY_r' + \frac{1}{2!}h^2 Y_r'' + \frac{1}{3!}h^3 Y_r''' + \frac{1}{4!}h^4 Y_r^{iv} \tag{3.9}$$

When using Taylor series method of order two or higher, second or higher derivatives of y must be found. Example 3E gives some practice at the implicit differentiation required.

Example 3E

(i) Find y'', y''' and y^{iv} for the equations (a) $y' = 2x^2 - 3y$, (b) $y' = 2xy$.

(ii) Find y'' given (a) $y' = \dfrac{x}{2x+3y}$, (b) $y' = y^2\sin(4x - 2y)$.

Solution

Remember that $\dfrac{d}{dx}[f(y)] = f'(y) \times \dfrac{dy}{dx} = y'f'(y)$.

(i) (a) If $y' = 2x^2 - 3y$ then

$$y'' = 4x - 3y', \quad y''' = 4 - 3y'' \text{ and } y^{iv} = -3y'''$$

(b) If $y' = 2xy$ then

$$y'' = 2xy' + 2y \qquad \text{(product rule)}$$

$$y''' = 2y' + 2xy'' + 2y' = 4y' + 2xy''$$

$$y^{iv} = 4y'' + 2xy''' + 2y'' = 6y'' + 2xy'''$$

(ii) (a) If $y' = \dfrac{x}{2x+3y}$ then, by the quotient rule

$$y'' = \frac{(2x+3y)\times 1 - x\times(2+3y')}{(2x+3y)^2}$$

(b) If $y' = y^2 \sin(4x - 2y)$ then

$$y'' = y^2 \times \cos(4x-2y)\times(4-2y') + 2yy'\sin(4x-2y)$$

You can see from these two examples how quickly the determination of higher order derivatives becomes complicated.

□

Example 3F

Solve $y' = 2xy$, $y(0) = 1$ using the Taylor series method of (i) order 2, (ii) order 4, with $h = 0.1$ to estimate $y(1.5)$.

Solution

(i) With $h = 0.1$ equation (3.8) becomes $Y_{r+1} = Y_r + 0.1Y_r' + 0.005Y_r''$, where $Y_r' = 2x_rY_r$ and, differentiating, $Y_r'' = 2Y_r + 2x_rY_r'$.
Putting $r = 0$ we obtain $Y_1 = Y_0 + 0.1Y_0' + 0.005Y_0''$ where

$$Y_0 = 1, \quad Y_0' = 2x_0Y_0 = 2\times 0\times 1 = 0,$$

$$Y_0'' = 2Y_0 + 2x_0Y_0' = 2\times 1 + 2\times 0\times 0 = 2$$

Hence $Y_1 = 1 + 0.1 \times 0 + 0.005 \times 2 = 1.01$ and so $y(0.1) \approx 1.01$

Putting $r = 1$ we obtain $Y_2 = Y_1 + 0.1Y_1' + 0.005Y_1''$ where

$$Y_1 = 1.01, \quad Y_1' = 2x_1Y_1 = 2\times 0.1\times 1.01 = 0.202,$$

$$Y_1'' = 2Y_1 + 2x_1Y_1' = 2\times 1.01 + 2\times 0.1\times 0.202 = 2.0604$$

Hence $Y_2 = 1.01 + 0.1 \times 0.202 + 0.005 \times 2.0604$

so that

$$Y_2 = 1.04052, \quad \text{giving} \quad y(0.2) \approx 1.04052$$

Continuing with $r = 2, 3, \ldots$ we eventually obtain, putting $r = 14$: $Y_{15} = 9.189198$ and then $y(1.5) \approx 9.189198$.

(ii) With $h = 0.1$ equation (3.9) becomes
$Y_{r+1} = Y_r + 0.1Y_r' + 0.005Y_r'' + \dfrac{0.1^3}{6}Y_r''' + \dfrac{0.1^4}{24}Y_r^{iv}$ where Y_r', Y_r'' are as before and

$$Y_r''' = 2Y_r' + 2Y_r' + 2x_rY_r'' = 4Y_r' + 2x_rY_r''$$

$$Y_r^{iv} = 4Y_r'' + 2Y_r'' + 2x_rY_r''' = 6Y_r'' + 2x_rY_r'''$$

Putting $r = 0$ we obtain

$$Y_0 = 1, \quad Y_0' = 0, \quad Y_0'' = 2,$$

$$Y_0''' = 4Y_0' + 2x_0Y_0'' = 4 \times 0 + 2 \times 0 \times 2 = 0$$

$$Y_0^{iv} = 6Y_0'' + 2x_0Y_0''' = 6 \times 0 + 2 \times 0 + 0 = 0$$

Then $Y_1 = 1.01$ (as before).

Continuing, we find

$$Y_{15} = 9.485343 \quad \text{and so} \quad y(1.5) \approx 9.485343.$$

Comparing the results with the exact answer $y = e^{x^2}$:

Method	Approx $Y(1.5)$	Exact $y(1.5)$	$\lvert\text{Error}\rvert$
TS2	9.189	9.488	0.299
TS4	9.485	9.488	0.003

Notice that the higher order method gave better results; this is generally true.

□

Example 3G

Carry out two steps with $h = 0.2$ of the Taylor series (i) second order method, (ii) fourth order method to estimate $y(0.4)$ from $y' = x^2 + y^2$, $y(0) = 1$.

Solution

(i) With $h = 0.2$ the Taylor series second order method (TS2) is, from (3.8),

$$Y_{r+1} = Y_r + 0.2Y_r' + 0.02Y_r''$$

where $Y_r' = x_r^2 + Y_r^2$ and $Y_r'' = 2x_r + 2Y_rY_r'$

Putting $r = 0$: $Y_1 = Y_0 + 0.2Y_0' + 0.02Y_0''$ where

$$Y_0 = y_0 = 1, \quad Y_0' = x_0^2 + Y_0^2 = 0^2 + 1^2 = 1$$

and $\quad Y_0'' = 2x_0 + 2Y_0Y_0' = 2 \times 0 + 2 \times 1 \times 1 = 2$

Hence $Y_1 = 1 + 0.2 \times 1 + 0.02 \times 2 = 1.24$ and so $y(0.2) \approx 1.24$

With $r = 1$: $Y_2 = Y_1 + 0.2Y_1' + 0.02Y_1''$ where

$$Y_1 = 1.24, \quad Y_1' = x_1^2 + Y_1^2 = 0.2^2 + 1.24^2 = 1.5776$$

and $\quad Y_1'' = 2x_1 + 2Y_1Y_1' = 2 \times 0.2 + 2 \times 1.24 \times 1.5776 = 4.312448$

Hence $Y_2 = 1.24 + 0.2 \times 1.5776 + 0.02 \times 4.312448$

$$Y_2 = 1.641769 \quad (6D) \quad \text{and so} \quad y(0.4) \approx 1.641769$$

(ii) With $h = 0.2$ the Taylor series fourth order method (TS4) is, from (3.9),

$$Y_{r+1} = Y_r + 0.2Y_r' + 0.02Y_r'' + \frac{0.004}{3}Y_r''' + \frac{0.0002}{3}Y_r^{iv}$$

where Y_r' and Y_r'' are as before and

$$Y_r''' = 2 + 2Y_rY_r'' + 2Y_r'^2$$

$$Y_r^{iv} = 2Y_rY_r''' + 2Y_r'Y_r'' + 4Y_r'Y_r'' = 2Y_rY_r''' + 6Y_r'Y_r''$$

Putting $r = 0$: $Y_1 = Y_0 + 0.2Y_0' + 0.02Y_0'' + \frac{0.004}{3}Y_0''' + \frac{0.0002}{3}Y_0^{iv}$ where

$$Y_0 = 1, \quad Y_0' = 1, \quad Y_0'' = 2, \quad Y_0''' = 2 \times 1 \times 2 + 2 \times 1^2 = 6$$

and $\quad Y_0^{iv} = 2 \times 1 \times 6 + 6 \times 1 \times 2 = 24$

Hence $Y_1 = 1 + 0.2 \times 1 + 0.02 \times 2 + \frac{0.004}{3} \times 6 + \frac{0.0002}{3} \times 24$

$$Y_1 = 1.2496 \quad \text{giving} \quad y(0.2) \approx 1.2496$$

Putting $r = 1$: $Y_2 = Y_1 + 0.2Y_1' + 0.02Y_1'' + \frac{0.004}{3}Y_1''' + \frac{0.0002}{3}Y_0^{iv}$ where

$$Y_1 = 1.2496, \quad Y_1' = 1.601500, \quad Y_1'' = 4.402469,$$

$$Y_1''' = 7.628806 \quad \text{and} \quad Y_1^{iv} = 61.369242 \qquad \text{(all to 6D)}$$

Substituting these values we find that

$$Y_2 = 1.672212 \quad \text{and so} \quad y(0.4) \approx 1.672212 \qquad \text{(6D)}$$

□

The biggest drawback of Taylor series methods is the need to differentiate the differential equation in order to find y'', y''' and y^{iv}. This difficulty is overcome by using Runge-Kutta methods, which are the subject of Section 3.2.4.

3.2.4 Runge-Kutta methods

The *n*th order Taylor series method

$$Y_{r+1} = Y_r + hY_r' + \frac{1}{2!}h^2 Y_r'' + \ldots + \frac{1}{n!}h^n Y_r^{(n)}$$

has the disadvantage that the derivatives, Y_r'', Y_r''', ..., $Y_r^{(n)}$ need to be found and evaluated. Runge-Kutta methods replace these derivatives, which are all evaluated at x_r, with an equivalent combination of values of $f(x,y)$ at a number of points between x_r and x_{r+1}, this being more convenient for automatic computation. As is the case with Taylor series methods, all Runge-Kutta methods are explicit.

3.2.4.1 Runge-Kutta first order method

The first order Taylor series method is

$$Y_{r+1} = Y_r + hY_r' \quad \text{or} \quad Y_{r+1} = Y_r + hf(x_r, Y_r)$$

This is simply Euler's method.

3.2.4.2 Runge-Kutta second order method (RK2)

The derivation of this method is quite difficult and can be done in more than one way. The following is probably the easiest. The trapezoidal method is

$$Y_{r+1} = Y_r + \frac{1}{2}h(Y_r' + Y_{r+1}')$$

Replacing y' by $f(x,y)$ this becomes

$$Y_{r+1} = Y_r + \frac{1}{2}h\{f(x_r, Y_r) + f(x_{r+1}, Y_{r+1})\}$$

Substitute the Euler expression (3.3) for Y_{r+1} into the right hand side

$$Y_{r+1} = Y_r + \frac{1}{2}h\{f(x_r, Y_r) + f(x_{r+1}, Y_r + hf(x_r, Y_r))\}$$

The method is

$$Y_{r+1} = Y_r + \frac{1}{2}(k_1 + k_2) \tag{3.10}$$

where $k_1 = hf(x_r, Y_r)$ and $k_2 = hf(x_r + h, Y_r + k_1)$

A geometrical interpretation of the method is given in the following diagram.

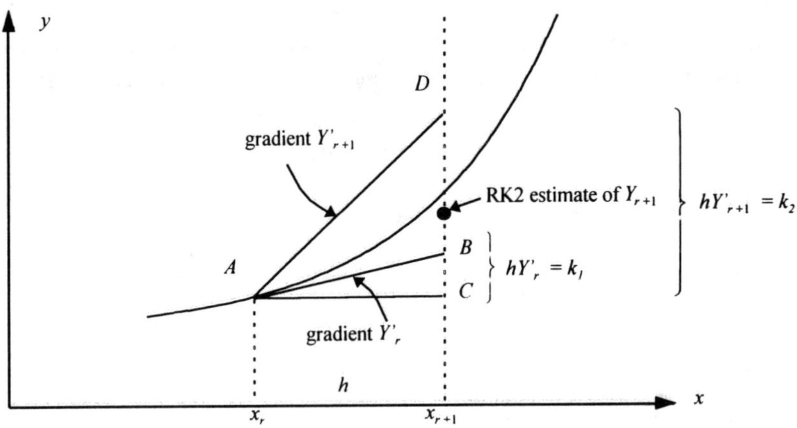

Figure 3.8

The tangent AB at A has slope Y_r' and so $BC = hY_r'$. BE is the Euler estimate of y_{r+1}, and $BC = hY_r' = k_1$. AD is a line drawn parallel to the direction field at B and so has gradient $Y_{r+1}' = f(x_{r+1}, Y_{r+1})$. Then $DC = hY_{r+1}' = k_2$. The mid-point of BD, with ordinate $Y_r + \frac{1}{2}h\{f(x_r, Y_r) + f(x_{r+1}, Y_r + k_1)\}$ is the RK2 estimate of Y_{r+1}.

The method assumes that the gradient of the solution over the interval $[x_r, x_{r+1}]$ is the average of the gradients at x_r and x_{r+1}.

Example 3H

Use the Runge-Kutta second order method to solve $y' = 2xy$, $y(0) = 1$ to find $y(1.5)$ with (i) $h = 0.5$, (ii) $h = 0.1$.

Solution

(i) $h = 0.5$

Putting $r = 0$, (3.10) becomes $Y_1 = Y_0 + \frac{1}{2}(k_1 + k_2)$

where $k_1 = hf(x_0, Y_0) = 0.5f(0,1) = 0.5 \times 2 \times 0 \times 1 = 0$

and $k_2 = hf(x_0 + h, Y_0 + k_1) = 0.5f(0.5,1) = 0.5 \times 2 \times 0.5 \times 1 = 0.5$

Hence $Y_1 = 1 + \frac{1}{2}(0 + 0.5) = 1.25$ and so $y(0.5) \approx 1.25$

Putting $r = 1$, (3.10) becomes $Y_2 = Y_1 + \frac{1}{2}(k_1 + k_2)$

where $k_1 = 0.5f(0.5,1.25) = 0.5 \times 2 \times 0.5 \times 1.25 = 0.625$

and $k_2 = 0.5f(1,1.875) = 0.5 \times 2 \times 1 \times 1.875 = 1.875$

Hence $Y_2 = 1.25 + \frac{1}{2}(0.625 + 1.875) = 2.5$ and so $y(1) \approx 2.5$

Putting $r = 2$, (3.10) becomes $Y_3 = Y_2 + \frac{1}{2}(k_1 + k_2)$

where $k_1 = 0.5f(1,2.5) = 0.5 \times 2 \times 1 \times 2.5 = 2.5$

and $k_2 = 0.5f(1.5,5) = 0.5 \times 2 \times 1.5 \times 5 = 7.5$

Hence $Y_3 = 2.5 + \frac{1}{2}(2.5 + 7.5) = 7.5$ from which $y(1.5) \approx 7.5$

(ii) $h = 0.1$

Putting $r = 0$, (3.10) becomes $Y_1 = Y_0 + \frac{1}{2}(k_1 + k_2)$

where $k_1 = 0.1f(0,1) = 0.1 \times 2 \times 0 \times 1 = 0$

and $k_2 = 0.1f(0.1,1) = 0.1 \times 2 \times 0.1 \times 1 = 0.02$

Hence $Y_1 = 1 + \frac{1}{2}(0 + 0.02) = 1.01$ and so $y(0.1) \approx 1.01$

Putting $r = 1$, (3.10) becomes $Y_2 = Y_1 + \frac{1}{2}(k_1 + k_2)$

where $k_1 = 0.1f(0.1,1.01) = 0.1 \times 2 \times 0.1 \times 1.01 = 0.0202$

and $k_2 = 0.1f(0.2,1.0302) = 0.1 \times 2 \times 0.2 \times 1.0302 = 0.041208$

Hence $Y_2 = 1.01 + \frac{1}{2}(0.0202 + 0.041208) = 1.040704$ giving $y(0.2) \approx 1.04$

Continuing, we find

$Y_5 = 1.283429005$, giving $y(0.5) \approx 1.283$
$Y_{10} = 2.709057014$, giving $y(1) \approx 2.709$
$Y_{15} = 9.3487620768$, giving $y(1.5) \approx 9.349$

Comparing the results with the exact values of $y(1.5)$

h	Approx $Y(1.5)$	Exact $y(1.5)$	$\lvert\text{Error}\rvert$
0.5	7.5	9.488	1.98
0.1	9.349	9.488	0.139

□

Example 3I

Use RK2 with $h = 0.25$ to estimate $y(2)$ from the equation $y' = 1 - 3\sqrt{\frac{y+1}{x}}$ given $y(1) = 5$.

Solution

(3.10) becomes, with $r = 0$, $Y_1 = Y_0 + \frac{1}{2}(k_1 + k_2)$ where

$$k_1 = hf(x_0, Y_0) = 0.25f(1{,}5) = 0.25\left(1 - 3\sqrt{\frac{6}{1}}\right) = -1.5871$$

and $$k_2 = hf(x_0 + h, Y_0 + k_1) = 0.25f(1.25{,}3.4129) = 0.25\left(1 - 3\sqrt{\frac{4.4129}{1.25}}\right) = -1.1592$$

Hence $Y_1 = 5 + \frac{1}{2}(-1.5871 - 1.1592)$

$Y_1 = 3.6268$ and so $y(1.25) \approx 3.6268$

Putting $r = 1$ in (3.10), $Y_2 = Y_1 + \frac{1}{2}(k_1 + k_2)$

where $k_1 = hf(x_1, Y_1) = 0.25f(1.25{,}3.6268) = -1.1929$

and $k_2 = hf(x_1 + h, Y_1 + k_1) = 0.25f(1.5{,}2.4339) = -0.8848$

Hence

$$Y_2 = 3.6268 + \frac{1}{2}(-1.1929 - 0.8848)$$

$Y_2 = 2.5880$ and so $y(1.5) \approx 2.5880$

Putting $r = 2$ and 3 in (3.10) gives, respectively

$Y_3 = 1.7941$ so that $y(1.75) \approx 1.7941$

and $Y_4 = 1.1863$ so that $y(2) \approx 1.1863$

□

3.2.4.3 Runge-Kutta third order method (RK3)

This is obtained by replacing terms up to and including Y_r''' in the Taylor series by an equivalent combination of function values. The method is

$$Y_{r+1} = Y_r + \frac{1}{6}(k_1 + 4k_2 + k_3) \tag{3.11}$$

where $k_1 = hf(x_r, Y_r)$, $\quad k_2 = hf\left(x_r + \frac{1}{2}h, Y_r + \frac{1}{2}k_1\right)$

and $\quad k_3 = hf(x_r + h, Y_r + 2k_2 - k_1)$

3.2.4.4 Runge-Kutta fourth order method (Classical Runge-Kutta method) (RK4)

Terms up to and including Y_r^{iv} in the Taylor series are replaced by equivalent function values. The RK4 method is

$$Y_{r+1} = Y_r + \frac{1}{6}(k_1 + 2k_2 + 2k_3 + k_4) \tag{3.12}$$

where $k_1 = hf(x_r, Y_r)$ $\qquad k_2 = hf\left(x_r + \frac{1}{2}h, Y_r + \frac{1}{2}k_1\right)$

$$k_3 = hf\left(x_r + \frac{1}{2}h, Y_r + \frac{1}{2}k_2\right) \qquad k_4 = hf(x_r + h, Y_r + k_3)$$

Example 3J

Use RK4 to estimate $y(1.5)$ from $y' = 2xy$, $y(0) = 1$ with (i) $h = 0.5$, (ii) $h = 0.1$.

Solution

(i) $\quad h = 0.5$

Putting $r = 0$ in (3.12): $Y_1 = Y_0 + \frac{1}{6}(k_1 + 2k_2 + 2k_3 + k_4)$ where

$$k_1 = hf(x_0, Y_0) = 0.5f(0,1) = 0.5 \times 2 \times 0 \times 1 = 0$$

$$k_2 = hf\left(x_0 + \frac{1}{2}h, Y_0 + \frac{1}{2}k_1\right) = 0.5f(0.25,1) = 0.5 \times 2 \times 0.25 \times 1 = 0.25$$

$$k_3 = hf\left(x_0 + \frac{1}{2}h, Y_0 + \frac{1}{2}k_2\right) = 0.5f(0.25,1.125) = 0.5 \times 2 \times 0.25 \times 1.25 = 0.28125$$

$$k_4 = hf(x_0 + h, Y_0 + k_3) = 0.5f(0.5, 1.28125)$$

$$= 0.5 \times 2 \times 0.5 \times 1.28125 = 0.640625$$

Hence $Y_1 = 1 + \frac{1}{6}(0 + 2 \times 0.25 + 2 \times 0.28125 + 0.640625)$

$= 1.283654$, so that $y(0.5) \approx 1.284$

Continuing, we find $y(1.0) \approx 2.713$, $y(1.5) \approx 9.372$.

(ii) With $h = 0.1$ we find $y(0.5) \approx 1.284$, $y(1) \approx 2.718$, $y(1.5) \approx 9.487$. Comparing the exact and approximate values of $y(1.5)$:

h	Approx $Y(1.5)$	Exact $y(1.5)$	$\lvert$Error$\rvert$
0.5	9.372	9.488	0.116
0.1	9.487	9.488	0.001

□

3.2.5 Predictor-Corrector methods

Explicit methods, such as Euler, Taylor series or Runge-Kutta, predict y_{r+1} from past information whereas implicit methods, such as the trapezoidal method, cannot easily be used. Predictor-corrector formulae use two methods:

(i) an explicit formula to predict y_{r+1}, and
(ii) a more accurate implicit formula to improve, or correct, this first prediction. This corrector formula can be used once, or more than once if required, at each step.

The simplest predictor-corrector method uses the explicit Euler method as predictor and the more accurate implicit trapezoidal method as corrector, and is usually known as the Modified Euler method.

3.2.5.1 Modified Euler predictor-corrector method

The predictor (Euler) is used to obtain a first estimate of y_{r+1}; this will be referred to as Y^*_{r+1}. The corrector (trapezoidal) can then be used to improve (correct) it and obtain Y_{r+1} (when the corrector is used once only). Then the method is

$$\begin{aligned} &\text{Predictor:} \quad Y^*_{r+1} = Y_r + hY'_r \\ &\text{Corrector:} \quad Y_{r+1} = Y_r + \frac{1}{2}h(Y'_r + Y^{*\prime}_{r+1}) \end{aligned} \qquad (3.13)$$

The steps in the method are

(i) Find Y'_r from the differential equation $y' = f(x,y)$.

(ii) Predict: $Y^*_{r+1} = Y_r + hY'_r$.

(iii) Use the differential equation to find $Y^{*\prime}_{r+1}$.

(iv) Correct: $Y_{r+1} = Y_r + \frac{1}{2}h(Y'_r + Y^{*\prime}_{r+1})$.

If required steps (iii) and (iv) can be repeated to further improve Y_{r+1}.

Example 3K

Use the Modified Euler method to estimate $y(1.5)$ with $h = 0.1$ from $y' = 2xy$, $y(0) = 1$. Use the corrector formula once at each step.

Solution

The initial values are $x_0 = 0$, $Y_0 = y_0 = 1$, $h = 0.1$.

Putting $r = 0$: DE: $Y'_0 = 2x_0Y_0 = 2 \times 0 \times 1 = 0$

P: $Y^*_1 = Y_0 + hY'_0 = 1 + 0.1 \times 0 = 1$

DE: $Y^{*\prime}_1 = 2x_1Y^*_1 = 2 \times 0.1 \times 1 = 0.2$

C: $Y_1 = Y_0 + \frac{1}{2}h(Y'_0 + Y^{*\prime}_1) = 1 + \frac{1}{2} \times 0.1(0 + 0.2) = 1.01$,

giving $y(0.1) \approx 1.01$.

Putting $r = 1$: DE: $Y_1' = 2x_1Y_1 = 2 \times 0.1 \times 1.01 = 0.202$

P: $Y_2^* = Y_1 + hY_1' = 1.01 + 0.1 \times 0.202 = 1.0302$

DE: $Y_2^{*\prime} = 2x_2Y_2^* = 2 \times 0.2 \times 1.0302 = 0.41208$

C: $Y_2 = Y_1 + \frac{1}{2}h(Y_1' + Y_2^{*\prime}) = 1.01 + \frac{1}{2} \times 0.1(0.202 + 0.41208)$

$= 1.040704,$

so that $y(0.2) \approx 1.040704$.

Continuing, we find

$Y_{15} = 9.348763$ so that $y(1.5) \approx 9.348763$

□

Example 3L

Solve $y' = (y - x)^2$, $y(0) = 0$, to estimate $y(1)$ using the Modified Euler method with $h = 0.2$. At each step use the corrector repeatedly, until successive applications differ by less than 0.001. The exact solution is 0.238406 (6D).

Solution

The initial values are $x_0 = 0$, $Y_0 = 0$, $h = 0.2$.

Putting $r = 0$: DE: $Y_0' = (Y_0 - x_0)^2 = 0$

P: $Y_1^* = Y_0 + 0.2Y_0' = 0$

DE: $Y_1^{*\prime} = (Y_1^* - x_1)^2 = 0.04$

C: $Y_1^{(1)} = Y_0 + 0.1(Y_0' + Y_1^{*\prime}) = 0.004$

DE: $Y_1^{*\prime} = (Y_1^{(1)} - x_1)^2 = 0.038416$

C: $Y_1^{(2)} = Y_0 + 0.1(Y_0' + Y_1^{*\prime}) = 0.003842$

As the two corrected values differ by more than 0.001 we correct again:

$$\text{DE:} \quad Y_1^{*'} = (Y_1^{(2)} - x_1)^2 = 0.038478$$

$$\text{C:} \quad Y_1^{(3)} = Y_0 + 0.1(Y_0' + Y_1^{*'}) = 0.003848$$

As the difference between $Y_1^{(2)}$ and $Y_1^{(3)}$ is less than 0.001 we take $y(0.2) \approx 0.003848$.

Continuing with $r = 1, 2, 3, 4$ the following results are obtained:

x	y predicted	y corrected
0.4	0.011543	0.022785
		0.021924
		0.021989
0.6	0.050568	0.066466
		0.064745
		0.064929
		0.064909
0.8	0.122173	0.139486
		0.137169
		0.137476
		0.137435
1.0	0.225233	0.241361
		0.238888
		0.239263
		0.239206

Hence $y(1) \approx 0.239206$, with error 0.000800 (3S).

□

Examples 3K and 3L illustrate two ways in which predictor-corrector formulae can be used:

Method A: use the predictor and corrector once only at each step (as in Example 3K);

Method B: after using the predictor apply the corrector repeatedly at each step, stopping when two successive corrected values differ by some preset tolerance (as in Example 3L).

The second approach, which involves far more calculation, should, intuitively, give a better final answer, but this is not always the case. For some problems Method A will give a better answer than Method B, for any chosen tolerance. For other examples Method B will be better but, perversely, the error in the final answer can increase as the tolerance is decreased. These phenomena are illustrated in Example 3M.

The reasons for this are concerned with the package or program and computer used, and the accuracy with which data is stored. The results given in Example 3M may well be different if other applications or machines are used.

Example 3M

(a) Solving $y' = 2xy$, $y(0) = 1$ with $h = 0.1$ in order to find $y(1.5)$ gives the following results.

(i) Method A: $y(1.5) \approx 9.348763$ with error -0.139 (3S).

(ii) Method B:

Tolerance	$y(1.5)$ (7S)	error (3S)
0.01	9.663931	0.176
0.001	9.677088	0.189
0.0001	9.678563	0.191

In this example Method A gives the better result.

(b) For $y' = (y - x)^2$, $y(0) = 0$ it can be shown that $y(1) = 0.238406$ (6S). Estimating $y(1)$ with $h = 0.2$ gives the following results.

(i) Method A: $y(1) \approx 0.244290$, with error 0.005884.

(ii) Method B:

Tolerance	$y(1)$	error
0.01	0.238406	0.000149
0.001	0.239271	0.000865
0.0001	0.239206	0.000800

In this example Method B gives a better answer for any tolerance used, but the error increases when the tolerance is reduced from 0.01 to 0.001 or 0.0001.

□

It is suggested that Method A is used in preference to Method B and as an alternative to the extra calculation in Method B it would be better to repeat Method A using a smaller value for the step length h.

3.2.6 Other numerical methods

We have looked in detail at a number of single step methods for estimating Y_{r+1} from Y_r. There are, however, others available and a number of these are given below for completeness.

3.2.6.1 Simpson's Method

Simpson's rule approximates an integral over two strips:

$$\int_{x_{r-1}}^{x_{r+1}} y\,dx \approx \frac{1}{3}h\left\{y(x_{r-1}) + 4y(x_r) + y(x_{r+1})\right\}$$

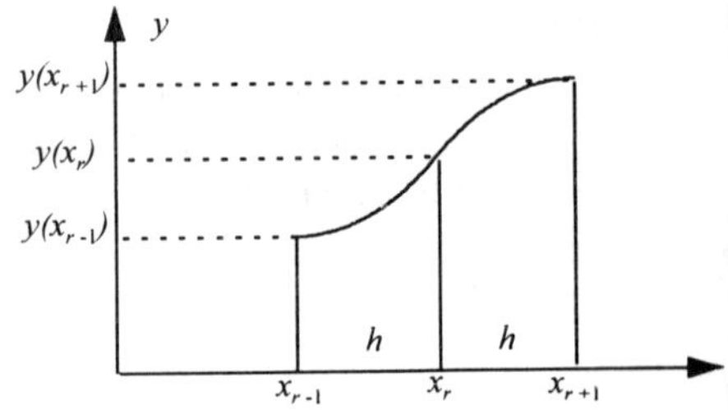

Figure 3.9

To solve $y' = f(x,y)$ we integrate from x_{r-1} to x_{r+1}. Then

$$\int_{x_{r-1}}^{x_{r+1}} y'\,dx = \int_{x_{r-1}}^{x_{r+1}} f(x,y)dx$$

from which

$$y_{r+1} - y_{r-1} = \int_{x_{r-1}}^{x_{r+1}} f(x,y)dx$$

The integral is approximated by Simpson's rule

$$\int_{x_{r-1}}^{x_{r+1}} f(x,y) \approx \frac{1}{3}h(Y'_{r-1} + 4Y'_r + Y'_{r+1})$$

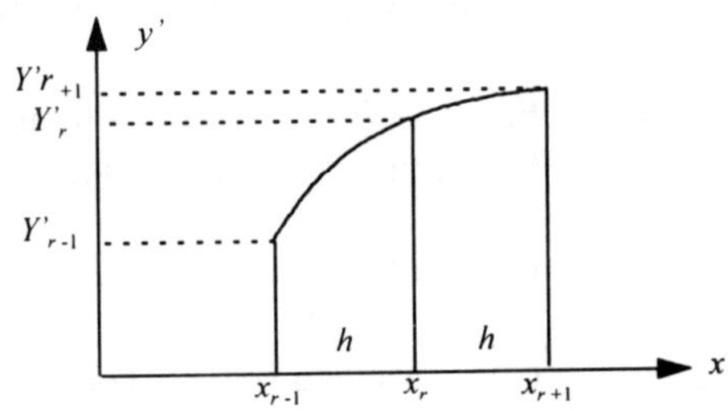

Figure 3.10

It follows that Simpson's method for solving $y' = f(x,y)$ is

$$Y_{r+1} = Y_{r-1} + \frac{1}{3}h(Y'_{r-1} + 4Y'_r + Y'_{r+1}) \qquad (3.14)$$

Applying the method with r = 1, 2, 3, ... finds Y_2, Y_3, Y_4, ..., so long as Y_0 and Y_1 are known. Y_0 will be given by the initial conditions but a one step method such as Runge-Kutta must be used for Y_1. (3.14) is a second order difference equation, and Simpson's method a two step method, as Y_{r+1} is found from Y_{r-1}.

Simpson's method is, like the trapezoidal method, implicit and is not usually used on its own but as the corrector in Milne's predictor-corrector method.

3.2.6.2 Milne's method

This is a predictor-corrector method in which the explicit predictor is obtained by truncating the Newton-Gregory backward difference interpolation formula and the implicit corrector is Simpson's method. Milne's method is

$$\text{Predictor:} \quad Y_{r+1}^{*} = Y_{r-3} + \frac{4}{3}h(2Y'_{r-2} - Y'_{r-1} + 2Y'_r)$$

$$\text{Corrector:} \quad Y_{r+1} = Y_{r-1} + \frac{1}{3}h(Y'_{r-1} + 4Y'_r + Y'^{*}_{r+1}) \tag{3.15}$$

A one step method is used with the initial condition to estimate Y_1, Y_2 and Y_3 and (3.15) can then be used successively with $r = 3, 4, 5$ to find $Y_4, Y_5, Y_6, \ldots$

Example 3N

Use Milne's method with $h = 0.1$ to estimate $y(1.5)$ from $y' = 2xy$, $y(0) = 1$.

Solution

The initial values are $x_0 = 0$, $Y_0 = y_0 = 1$. Applying the Runge-Kutta fourth order method, RK4, we find, to 6D,

$$Y_1 = 1.010050, \quad Y_2 = 1.040811, \quad Y_3 = 1.094174$$

$$Y'_1 = 0.202010, \quad Y'_2 = 0.416324$$

Putting $r = 3$: DE: $Y'_3 = 2x_3Y_3 = 2 \times 0.3 \times 1.094174 = 0.656504$

P: $$Y_4^{*} = Y_0 + \frac{4}{3}h(2Y'_1 - Y'_2 + 2Y'_3)$$

$$= 1 + \frac{4}{3} \times 0.1(2 \times 0.202010 - 0.416324 + 2 \times 0.656504)$$

$$= 1.173427$$

DE: $Y_4^{*'} = 2x_4Y_4^{*} = 2 \times 0.4 \times 1.173427 = 0.938742$

$$\text{C:} \quad Y_4 = Y_2 + \frac{1}{3}h(Y_2' + 4Y_3' + Y_4^{*'})$$

$$= 1.040811 + \frac{1}{3} \times 0.1(0.416324 + 4 \times 0.656504 + 0.938742)$$

$$= 1.173514, \quad \text{giving } y(0.4) \approx 1.173514$$

Putting $r = 4$: DE: $Y_4' = 2x_4Y_4 = 0.938811$

$$\text{P:} \quad Y_5^* = Y_1 + \frac{4}{3}h(2Y_2' - Y_3' + 2Y_4') = 1.283886$$

$$\text{DE:} \quad Y_5^{*'} = 2x_5Y_5^* = 1.283886$$

$$\text{C:} \quad Y_5 = Y_3 + \frac{1}{3}h(Y_3' + 4Y_4' + Y_5^{*'}) = 1.284028$$

giving $y(0.5) \approx 1.284028$.

Continuing, we find

$$Y_{15} = 9.486864 \quad \text{so that} \quad y(1.5) \approx 9.486864$$

with an error of $9.486864 - e^{1.5^2} \approx -0.00087$.

3.2.6.3 Adams-Bashforth formulae

This is a family of explicit formula obtained from Taylor's series. They may require Y_1, Y_2, ... from one step methods to start them off. The first few Adams-Bashforth formulae are

$$\left.\begin{aligned}
Y_{r+1} &= Y_r + hY_r' \qquad \text{(Euler's method)} \\
Y_{r+1} &= Y_r + \frac{1}{2}h(3Y_r' - Y_{r-1}') \\
Y_{r+1} &= Y_r + \frac{1}{12}h(23Y_r' - 16Y_{r-1}' + 5Y_{r-2}') \\
Y_{r+1} &= Y_r + \frac{1}{24}h(55Y_r' - 59Y_{r-1}' + 37Y_{r-2}' - 9Y_{r-3}')
\end{aligned}\right\} \quad (3.16)$$

Example 3O

(i) Obtain the Adams-Bashforth method $Y_{r+1} = Y_r + \frac{1}{2}h(3Y_r' - Y_{r-1}')$.

(ii) Solve $y' = 2xy$, $y(0) = 1$ to estimate $y(1.5)$ using the above formula with $r = 2, 3, 4, \ldots, 14$. Estimate $y(0.1)$ using the Taylor series second order method (TS2).

Solution

(i) The Taylor series expression for y_{r+1} is

$$y_{r+1} = y_r + hy_r' + \frac{1}{2!}h^2 y_r'' + \frac{1}{3!}h^3 y_r''' + \ldots$$

or, truncating after the term in h^2,

$$y_{r+1} \approx y_r + hy_r' + \frac{1}{2}h^2 y_r'' \tag{3.17}$$

We can use this to estimate y_{r-1} by replacing h with $-h$. Then

$$y_{r-1} \approx y_r - hy_r' + \frac{1}{2}h^2 y_r''$$

Differentiating this series:

$$y_{r-1}' \approx y_r' - hy_r'' + \frac{1}{2}h^2 y_r'''$$

Ignoring terms of order h^2 this gives

$$y_{r-1}' \approx y_r' - hy_r'' \qquad \text{from which} \qquad hy_r'' \approx y_r' - y_{r-1}'$$

Substituting for hy_r'' in (3.17):

$$y_{r+1} \approx y_r + hy_r' + \frac{1}{2}h(y_r' - y_{r-1}')$$

giving $y_{r+1} \approx y_r + \frac{1}{2}h(3y'_r - y'_{r-1})$

Using Y_r as our approximation to y_r, the method is

$$Y_{r+1} = Y_r + \frac{1}{2}h(3Y'_r - Y'_{r-1}) \qquad (3.18)$$

(ii) Using $y' = 2xy$ in (3.18) gives

$$Y_{r+1} = Y_r + \frac{1}{2}h(6x_rY_r - 2x_{r-1}Y_{r-1})$$

and, with $h = 0.1$, this becomes

$$Y_{r+1} = Y_r + 0.1(3x_rY_r - x_{r-1}Y_{r-1}) \qquad (3.19)$$

Initially $x_0 = 0$ and $Y_0 = y_0 = 1$. We saw in Example 3F that TS2 gives $Y_2 = 1.01$.

Putting $r = 1$ in (3.19):

$$Y_2 = Y_1 + 0.1(3x_1Y_1 - x_0Y_0)$$

$$= 1.01 + 0.1(3 \times 0.1 \times 1.01 - 0 \times 1)$$

$$= 1.0403 \quad \text{so that} \quad y(0.2) \approx 1.0403$$

Putting $r = 2$ in (3.19):

$$Y_3 = Y_2 + 0.1(3x_2Y_2 - x_1Y_1)$$

$$= 1.0403 + 0.1(3 \times 0.2 \times 1.0403 - 0.1 \times 1.01)$$

$$= 1.092618 \quad \text{giving} \quad y(0.3) \approx 1.092618$$

Continuing with $r = 3, 4, 5, \ldots, 14$ we find

$$Y_{15} = 8.812767 \quad \text{and so} \quad y(1.5) \approx 8.812767$$

with an error of about 0.67 in magnitude.

□

3.2.6.4 Adams-Moulton formulae

This is a family of implicit formulae, the first few of which are

$$Y_{r+1} = Y_r + \frac{1}{2}h(Y'_r + Y'_{r+1}) \qquad \text{(trapezoidal method)}$$

$$Y_{r+1} = Y_r + \frac{1}{12}h(5Y'_{r+1} + 8Y'_r - Y'_{r-1})$$

$$Y_{r+1} = Y_r + \frac{1}{24}h(9Y'_{r+1} + 19Y'_r - 5Y'_{r-1} + Y'_{r-2})$$

The formulae in 3.2.6.3 and 3.2.6.4 are sometimes used in pairs to give a predictor-corrector formula. An Adams-Bashforth formula will be the predictor and a more accurate Adams-Moulton formula will be the corrector.

3.2.7 Summary of results

A number of methods have been used to estimate $y(1.5)$ with $h = 0.1$ from the equation $y' = 2xy$, $y(0) = 1$. From the exact solution $y = e^{x^2}$, the exact value is $e^{1.5^2} = 9.488$ to three decimal places. The approximations found, and the magnitude of the error in each case, are tabulated below.

Method	Approx value of $y(1.5)$	$\lvert$Error$\rvert$
Euler	6.835	2.653
Trapezoidal	9.679	0.191
Taylor Series 2	9.189	0.299
TS4	9.485	0.003
Runge-Kutta 2	9.349	0.139
RK4	9.487	0.001
Modified Euler	9.349	0.139
Milne	9.4765	0.00087
Adams-Bashforth	8.813	0.675

It is clear that, for this equation, Euler's method and, to a lesser extent the Adams-Bashforth method, have given poor results. The other results are better and the magnitudes of their errors fall into two groups. The errors in the Trapezoidal, TS2, RK2 and the Modified Euler methods are all in the first decimal place, while those in TS4, RK4 and the Milne method are, after rounding, in the third decimal place, and range from 0.001 to 0.003.

We need to know whether the above pattern is peculiar to the equation $y' = 2xy$ or is typical of what happens with other equations, and this is the subject of the next section.

Exercise 3A

1. Solve each of the following equations using (i) Euler's method, (ii) Taylor series second order method, (iii) Runge-Kutta second order method, (iv) Modified Euler method, using the corrector once only at each step.

   (a) $y' = 6x + 2y$, $y(0) = 1$. Estimate y at $x = 0.3$ with $h = 0.1$. Compare each estimate of $y(0.3)$ with the value found from the exact solution $y = \frac{5}{2}e^{2x} - 3x - \frac{3}{2}$ and find the error in each case.

   (b) $y' = \sin(y)$, $y(0) = 2$. Estimate $y(1)$ correct to 4D with $h = 0.25$.

   (c) $y' = \dfrac{1}{x+y}$, $y(1) = 0$. Estimate $y(2)$ correct to 5D with $h = 0.2$.

2. Solve the equation $y' = 6x + 2y$, $y(0) = 1$ to estimate y correct to 6D at $x = 0.3$ with $h = 0.1$ using (i) the Trapezoidal method, (ii) Taylor series fourth order method, (iii) Runge-Kutta fourth order method.

3. Apply Methods A and B of the Modified Euler method to each of the following equations with the step size h given. Determine which method gives the better answer and find it as accurately as possible.

	Equation	Initial condition	h	Find	Exact answer (6D)
(i)	$y' = \dfrac{y}{2\sqrt{x-1}}$	$y(2) = 3$	0.25	$y(3)$	4.539541
(ii)	$y' = \dfrac{y^3}{x^3}$	$y(1) = 0.5$	0.4	$y(5)$	0.573539

(iii)	$y' = \dfrac{y^2}{x^2} - \dfrac{y}{x}$	$y(1) = 1$	0.1	$y(2)$	0.8
(iv)	$y' = \dfrac{x+1}{y^2}$	$y(0) = 2$	0.25	$y(1)$	2.320794

4. Runge-Kutta methods can be obtained by rearranging terms in the Taylor series as a linear combination of values of $y' = f(x,y)$. Obtain the Runge-Kutta second order method as follows:

(i) Assume $Y_{r+1} = Y_r + \alpha k_1 + \beta k_2$ (3.20)

where $k_1 = hf(x_r, Y_r) = hY_r'$ and $k_2 = hf(x_r + \gamma h, Y_r + \delta k_1)$.

(ii) Expand $f(x_r + \gamma h, Y_r + \delta k_1)$ about the point (x_r, Y_r) using Taylor's series for two variables, substitute in (3.20) and obtain

$$Y_{r+1} = Y_r + (\alpha + \beta) hY_r' + \beta\gamma h^2 \frac{\partial f_r}{\partial x} + \beta\delta h^2 Y_r' \frac{\partial f_r}{\partial y} \tag{3.21}$$

(iii) Show that the Taylor series second order method can be put in the form

$$Y_{r+1} = Y_r + hY_r' + \frac{1}{2} h^2 \frac{\partial f_r}{\partial x} + \frac{1}{2} h^2 Y_r' \frac{\partial f_r}{\partial y} \tag{3.22}$$

(iv) Show that for (3.21) and (3.22) to be identical we require

$$\alpha + \beta = 1, \quad \beta\gamma = \frac{1}{2} \quad \text{and} \quad \beta\delta = \frac{1}{2}$$

(a set of three equations in four unknowns which have an infinite number of solutions).

(v) Choose $\alpha = 1$ and hence obtain the Runge-Kutta second order method.

(vi) Choose $\alpha = 0$ and obtain the Euler-Cauchy method

$$Y_{r+1} = Y_r + hf\left(x_r + \frac{1}{2}h, Y_r + \frac{1}{2}k_1\right)$$

(vii) Choose $\alpha = \frac{1}{4}$ and obtain

$$Y_{r+1} = Y_r + \frac{1}{4}(k_1 + 3k_2)$$

where $k_1 = hf(x_r, Y_r)$ and $k_2 = hf\left(x_r + \frac{2}{3}h, Y_r + \frac{2}{3}k_1\right)$.

DERIVE ACTIVITY 3A

The aim of this activity is to introduce the DERIVE commands for solving differential equations numerically. For each method we use the differential equation $y' = 2xy$ given $y(0) = 1$.

DERIVE has three of the numerical methods available in the utility file ode_appr.mth. Load this file. The methods that we will use are Euler's method, a Taylor series method and the fourth-order Runge-Kutta method.

(A) EULER(r,x,y,x_0,y_0,h,n) approximates to the solution of $y' = r(x,y)$ with $y = y_0$ at $x = x_0$ with step length h; n is the number of steps required.

(i) To solve the problem

$$y' = 2xy \quad \text{given} \quad y(0) = 1$$

Author Expression EULER(2*xy*,*x*,*y*,0,1,0.2,10) and **Simplify Approximate**. **Plot!** the numerical solution.

Figure 3.11 shows the graphs of the Euler approximations and the analytical solution $y = e^{x^2}$.

As we would expect Euler's method with such a large step length of 0.2 is not very accurate. For example, for $x = 1$ the relative error is about 25%.

(ii) Use Euler's method with smaller steps to find an approximation to $y(2)$. What value of h gives the value to within 5% of the analytical solution e^4?

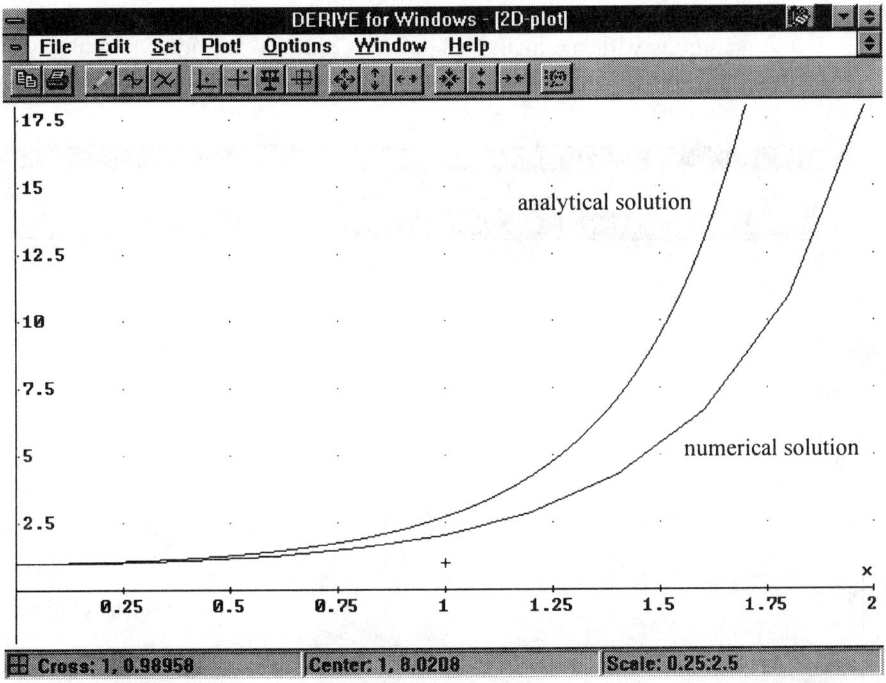

Figure 3.11 Euler's method with $h = 0.2$

(B) TAY_ODE1(r,x,y,x_0,y_0,n) approximates to the Taylor series solution of order n of $y' = r(x,y)$ with $y = y_0$ at $x = x_0$.

Note that this approach is different to that described in Section 3.2.3. There we used a number of steps of a Taylor series of a particular order (e.g. 10 steps of TS2 with $h = 0.1$) whereas DERIVE obtains its answers in one step, increasing the order of the Taylor series to improve accuracy.

(i) To solve the problem

$$y' = 2xy \quad \text{given} \quad y(0) = 1$$

Author Expression TAY_ODE1($2xy,x,y,0,1,4$) and **Simplify Approximate**. This gives the fourth order Taylor series solution $0.5x^4 + x^2 + 1$.

Figure 3.12 shows the screen dump for the analytical and Taylor series solutions. The two solutions differ by 8% when $x = 1$.

(ii) Use the Taylor series method of order 5, 6, 7 and 8. Compare your solutions with the analytical solution $y(2)$. What order is required for the error in the numerical solution to be within 5% of the analytical solution?

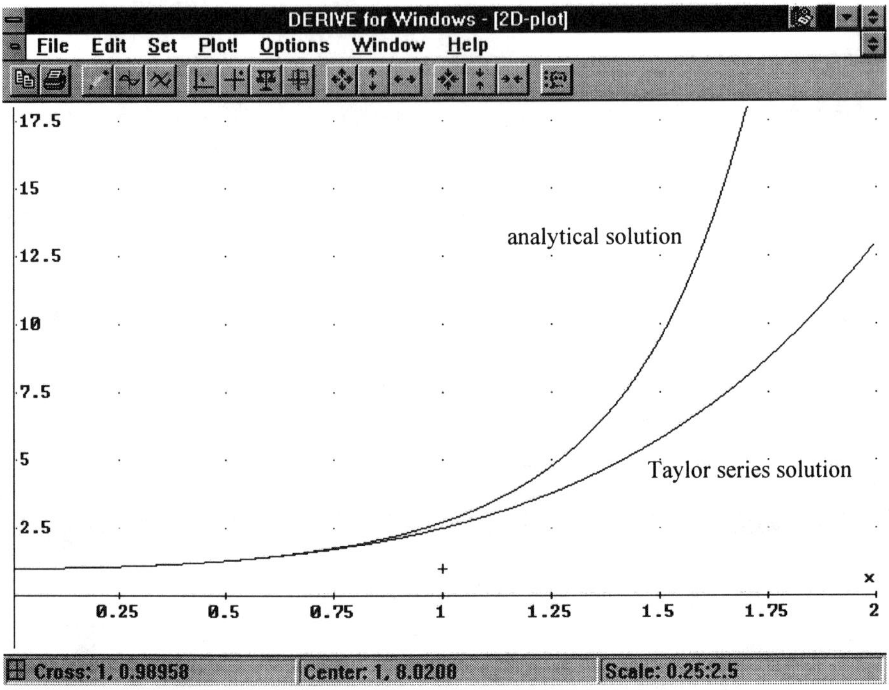

Figure 3.12 Taylor series solution of order 4

(C) RK(r,v,v_0,h,n) approximates to the solution of a system of first order differential equations

$$y_i' = r_i(x, y_i)$$

using the fourth-order Runge-Kutta method. v is a vector $[x, y_1, y_2, \ldots]$, v_0 is corresponding vector of initial values, h is the step length and n is the number of steps.

To solve the problem

$$y' = 2xy \quad \text{given} \quad y(0) = 1$$

Author Expression RK([2*xy*], [*x*,*y*], [0,1], 0.2, 10) and **Simplify Approximate**. Figure 3.13 shows the analytical and numerical solutions.

The two solutions are very close. For $x = 2$ the relative error is 0.26%. If we choose $h = 0.5$ the error is still within 5%.

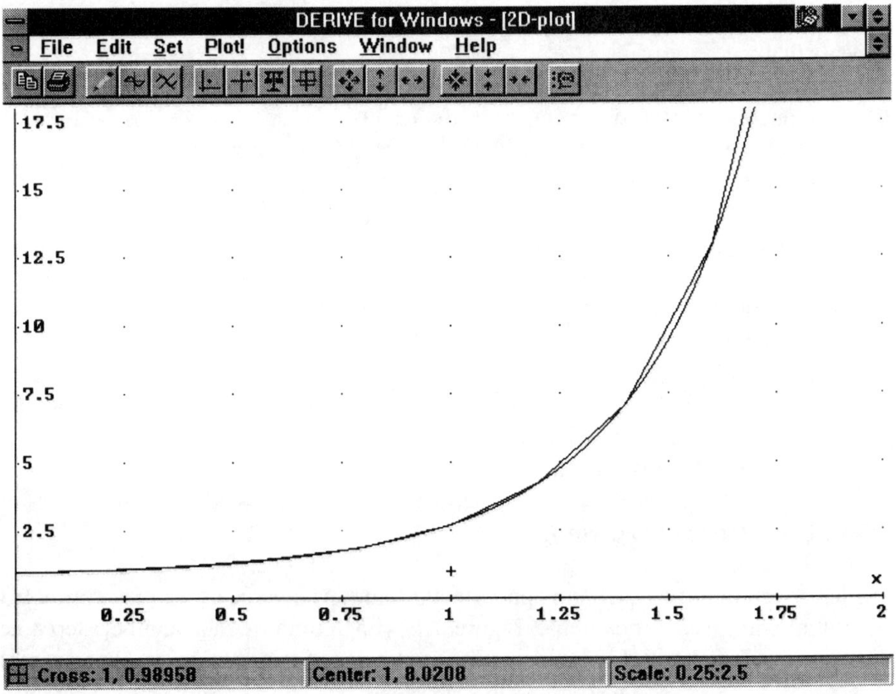

Figure 3.13 Runge-Kutta method with $h = 0.2$

(D) Solve each of the problems below using DERIVE and Euler's methods. Find the step size so that you can deduce the answer to the accuracy given.

(i) Find $y(1)$ to 3S from $\dfrac{dy}{dx} = x + \ln(y)$ given $y(0) = 1$

(ii) Find $y(2)$ to 4S from $\dfrac{dy}{dx} = \dfrac{1}{x^2 + y^2}$ given $y(1) = 2$

(E) Repeat activity (D) using the Runge-Kutta method. What value of h is required to find $y(1)$ in (i) and $y(2)$ in (ii) correct to 8S?

(F) Repeat activity (D) using the Taylor series method. Find the order of the method that gives the accuracy asked for in (i) and (ii).

3.3 THE ANALYSIS OF NUMERICAL METHODS

We saw in Section 3.2 that there are a number of different methods that can be applied to a differential equation and that each involves solving a difference equation. We obviously hope that the solution of the difference equation accurately matches that of the differential equation. On the basis of the examples solved in Section 3.2 it appears that

(i) reducing the step size h improves the results
(ii) some methods give better results than others for the same h
(iii) if h is too large the results may be useless.

In the next two sections we shall look at the methods theoretically so that we can see how to generalise the above observations and also predict what step size is required to obtain the desired accuracy in the final answer.

Some new concepts will be considered. They are

(a) convergence, local truncation error, global truncation error and round-off error, which are related to (i) and (ii) above, and
(b) stability, which is related to (iii) above.

3.3.1 Local truncation error (*L*)

We have seen that the exact and approximate solutions diverge as we move away from the initial point (see, for example, Figure 3.1). Each time we take another step a new error is introduced and this increases the total error at the next point. For example, there will be an error in estimating Y_6 from Y_5, but Y_5 will also contain an accumulation of previous errors. Hence the overall error in Y_6 will be greater than that in Y_5.

The local truncation error L is concerned with the extra error introduced in **one** step. If we use a method which introduces a comparatively small error at each step then it is likely that the overall error in the final answer will be comparatively smaller.

We will assume that there is no error at $x = x_r$, so that $Y_r = y_r$. The local truncation error L measures the error in calculating Y_{r+1}, and is the difference between the approximate solution Y_{r+1} and the exact solution y_{r+1}.

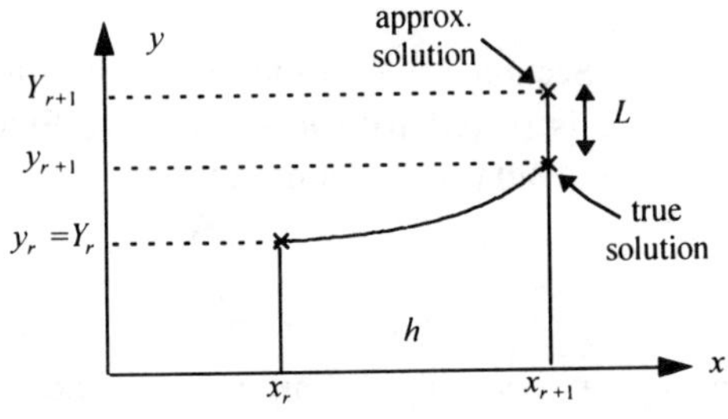

Figure 3.14

We define

Local truncation error = computed value at x_{r+1} – exact value at x_{r+1}

(assuming no error at $x = x_r$). Then

$$L = Y_{r+1} - y_{r+1}$$

The Taylor series expansion of y_{r+1}, that is

$$y_{r+1} = y_r + hy'_r + \frac{1}{2!}h^2 y''_r + \frac{1}{3!}h^3 y'''_r + \dots$$

will be useful when we look at methods in detail. Hence

$$L = Y_{r+1} - \left(y_r + hy'_r + \frac{1}{2!}h^2 y''_r + \frac{1}{3!}h^3 y'''_r + \dots \right) \tag{3.23}$$

We will find that L is a series which involves increasing powers of h. For small h the first term will be the largest and we take this as a measure of L. This first term is called the principal term in the local truncation error. When comparing methods it will be the principal terms that will be compared, and in particular the power of h in that term. The method with the higher power of h in its principal term will give the smaller value of L; this power is the **order** of L.

Example 3P

Referring to the examples in Section 3.2 find the local truncation error in the **first** step when $y' = 2xy$, $y(0) = 1$ is solved with $h = 0.1$ by (i) Euler's method, (ii) the trapezoidal method, (iii) Runge-Kutta second order method.

Solution

The analytical solution of $y' = 2xy$, $y(0) = 1$ is $y = e^{x^2}$ and so the exact value is $y(0.1) = e^{0.1^2} \approx 1.010050$ (6D). The local truncation error in the first step is $L = Y_1 - 1.010050$ and we can obtain the following results:

Method	Approx value of $y(0.1)$	Local truncation error L (2*S*)
Euler	1	–0.010
Trapezoidal	1.010101...	–0.000051
RK2	1.01	–0.000050

□

Example 3Q

Find the principal term in the local truncation error for Euler's method.

Solution

Euler's method is $Y_{r+1} = Y_r + hY'_r$, but if we assume that the solution is correct at x_r, this becomes

$$Y_{r+1} = y_r + hy'_r$$

From (3.23), the local truncation error is

$$L = (y_r + hy'_r) - \left(y_r + hy'_r + \frac{1}{2!}h^2 y''_r + \frac{1}{3!}h^3 y'''_r + \ldots \right)$$

$$L = -\frac{1}{2}h^2 y''_r - \frac{1}{6}h^3 y'''_r - \ldots$$

The principal term in L is $-\frac{1}{2}h^2y_r''$ and the local truncation error is of order h^2. That is, $L = O(h^2)$.

□

Example 3R

Find the principal term in the local truncation error for the Taylor series methods of order (i) 2, and (ii) 4.

Solution

(i) For TS2, (3.23) becomes

$$L = \left(y_r + hy_r' + \frac{1}{2}h^2y''\right) - \left(y_r + hy_r' + \frac{1}{2!}h^2y_r'' + \frac{1}{3!}h^3y_r''' + \ldots\right)$$

$$L = -\frac{1}{6}h^3y_r''' - \ldots$$

The principal term in L is $-\frac{1}{6}h^3y_r'''$ and $L = O(h^3)$.

(ii) For TS4, (3.23) becomes

$$L = \left(y_r + hy_r' + \frac{1}{2}h^2y_r'' + \frac{1}{3!}h^3y_r''' + \frac{1}{4!}h^4y_r^{iv}\right)$$

$$-\left(y_r + hy_r' + \frac{1}{2!}h^2y_r'' + \frac{1}{3!}h^3y_r''' + \frac{1}{4!}h^4y_r^{iv} + \frac{1}{5!}h^5y_r^{v} \ldots\right)$$

$$L = -\frac{1}{5!}h^5y_r^{v} - \ldots$$

The principal term in L is $-\frac{1}{5!}h^5y_r^{v}$ and $L = O(h^5)$.

□

Example 3S

Find the principal term in the local truncation error for the Adams-Bashforth method $Y_{r+1} = Y_r + \frac{1}{2}h(3Y'_r - Y'_{r-1})$.

Solution

From (3.23),

$$L = \left(y_r + \frac{1}{2}h(3y'_r - y'_{r-1}) \right) - \left(y_r + hy'_r + \frac{1}{2!}h^2 y''_r + \frac{1}{3!}h^3 y'''_r + \ldots \right)$$

In order to simplify this expression we need the same subscript in each term, so the term y'_{r-1} has to be transformed.

The Taylor series for y_{r+1} is

$$y_{r+1} = y_r + hy'_r + \frac{1}{2!}h^2 y''_r + \frac{1}{3!}h^3 y'''_r + \ldots$$

and finds y_{r+1} at x_{r+1} from y_r at x_r. If we replace h by $(-h)$ we will step back from x_r to x_{r-1} and then find y_{r-1}. Hence

$$y_{r-1} = y_r + (-h)y'_r + \frac{1}{2!}(-h)^2 y''_r + \frac{1}{3!}(-h)^3 y'''_r + \ldots$$

$$y_{r-1} = y_r - hy'_r + \frac{1}{2!}h^2 y''_r - \frac{1}{3!}h^3 y'''_r \ldots$$

Differentiating this series:

$$y'_{r-1} = y'_r - hy''_r + \frac{1}{2!}h^2 y'''_r - \frac{1}{3!}h^3 y^{iv}_r \ldots$$

Substituting for y'_{r-1} in the expression for L gives

$$L = \left(y_r + \frac{3}{2}hy'_r - \frac{1}{2}h\left(y'_r - hy''_r + \frac{1}{2!}h^2 y'''_r - \frac{1}{3!}h^3 y^{iv}_r \ldots \right) \right)$$

$$- \left(y_r + hy'_r + \frac{1}{2!}h^2 y''_r + \frac{1}{3!}h^3 y'''_r \ldots \right)$$

$$L = -\frac{5}{12}h^3 y_r''' + \ldots$$

The principal term in L is $-\frac{5}{12}h^3 y_r'''$ and $L = O(h^3)$.

□

Example 3T

For each of the methods considered in Examples 3Q, 3R and 3S, by what factor would you expect the local truncation error to decrease if the step size was halved?

Solution

(i) The principal term in the local truncation error for Euler's method is $L = -\frac{1}{2}h^2 y_r''$. If the step size is halved to $h_1 = \frac{1}{2}h$, then

$$L_1 = -\frac{1}{2}h_1^2 y_r'' = -\frac{1}{2}\left(\frac{1}{2}h\right)^2 y_r'' = -\frac{1}{8}h^2 y_r'' = \frac{1}{4}L$$

and L has been reduced by a factor of four.

(ii) For TS2, $L = -\frac{1}{6}h^3 y_r'''$ and, when h is halved, this becomes

$$L_1 = -\frac{1}{6}\left(\frac{1}{2}h\right)^3 y_r''' = -\frac{1}{48}h^3 y_r''' = \frac{1}{8}L$$

and L is reduced by a factor of $2^3 = 8$.

(iii) For TS4, $L = -\frac{1}{5!}h^5 y^v$ and will be reduced to $L_1 = -\frac{1}{5!}\frac{h^5}{2^5}y^v$, a factor of $2^5 = 32$.

(iv) Similarly, for the Adams-Bashforth method $L = -\frac{5}{12}h^3 y_r'''$ and will be reduced by a factor of $2^3 = 8$ to $L_1 = -\frac{5}{96}h^3 y_r'''$.

Remember, however, that halving h means that there will be twice as many calculations in order to estimate y at a particular value of x, so we should not expect the accumulated error at this x value to be reduced by the same factor. (See Global truncation error.)

□

3.3.2 Global truncation error (G)

The local truncation error L discussed in the last section is concerned with the error in **one** step of a method.

To go from an initial point x_0 to a final point x^* will require a number of steps, each with its own local truncation error. These errors will accumulate to give an error in the final estimate of $y(x^*)$.

If there are n steps from x_0 to x^*, so that $x^* = x_0 + nh$, then the approximate and exact values of y at x^* are Y_n and $y_n = y(x^*)$ respectively.

The global truncation error is defined to be the error in the final value due to the accumulation of local truncation errors. That is,

$$\text{Global truncation error} = \text{approx value at } x = x^* - \text{exact value at } x = x^*$$

or $$G = Y_n - y(x^*)$$

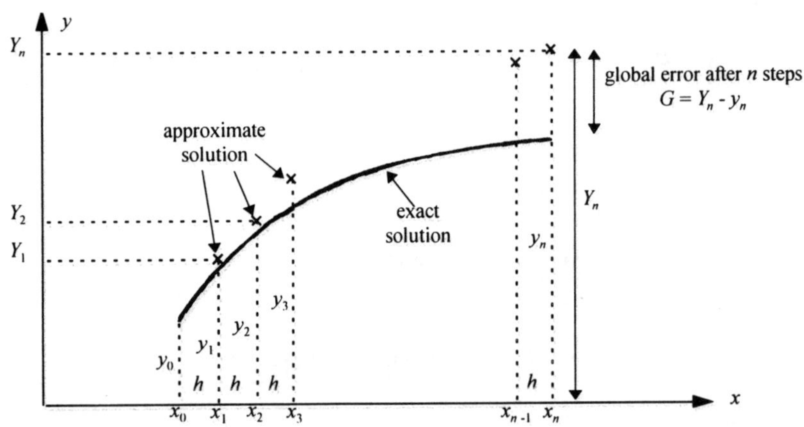

Figure 3.15

We saw in Section 3.3.1 that the local truncation error L for Euler's method is of order h^2 and so L reduces as h reduces. We intuitively expect the global truncation error G to decrease with h as well.

Suppose we wish to solve an equation by Euler's method in order to find $y(2)$, starting from $y(0) = 1$. A step size of h will require $\frac{2}{h}$ steps and, at each step, a local truncation error of order h^2 is introduced. The global truncation error in $y(2)$ will be the sum of the local truncation errors in each of the $\frac{2}{h}$ steps and so the global truncation error is of order $\frac{2}{h} \times h^2 = 2h$; that is, it is of order h. Hence the global truncation error in Euler's method is $G = O(h)$.

A similar argument holds for other methods. If the local truncation error is of order h^p then the number of steps required to find $y(x^*)$ is proportional to $\frac{1}{h}$, resulting in a global truncation error of order h^{p-1}.

Hence if for a method, and for small h, the local truncation error is $L = O(h^p)$, then the global truncation error is $G = O(h^{p-1})$.

Example 3U

What happens to the global truncation error for (i) Euler, (ii) RK2 when h is reduced by a factor or (a) 2, (b) 10?

Solution

(i) For Euler, Global error $\propto h$, and so $G = Ch$

(a) Halving h halves G.

(b) Reducing h to $\frac{1}{10}h$ divides G by 10.

(ii) For RK2, Global error $\propto h^2$, and so $G = Ch^2$

(a) Halving h reduces G to $\left(\frac{1}{2}\right)^2 = \frac{1}{4}$ of what it was.

(b) Reducing h to $\frac{1}{10}h$ reduces G to $\left(\frac{1}{10}\right)^2 = \frac{1}{100}$ of its previous value.

□

Summary of results

Method	Rule	Type	L	G
Euler	$Y_{r+1} = Y_r + hY'_r$	explicit	$O(h^2)$	$O(h)$
Trapezoidal	$Y_{r+1} = Y_r + \frac{1}{2}h(Y'_r + Y'_{r+1})$	implicit	$O(h^3)$	$O(h^2)$
Simpson	$Y_{r+1} = Y_r + \frac{1}{3}h(Y'_{r-1} + 4Y'_r + Y'_{r+1})$	implicit	$O(h^5)$	$O(h^4)$
TS2	$Y_{r+1} = Y_r + hY'_r + \frac{1}{2}h^2 Y''_r$	explicit	$O(h^3)$	$O(h^2)$
TS4	$Y_{r+1} = Y_r + hY'_r + \frac{1}{2}h^2 Y''_r + \frac{1}{3!}h^3 Y''' + \frac{1}{4!}h^4 Y_r^{iv}$	explicit	$O(h^5)$	$O(h^4)$
RK2	$Y_{r+1} = Y_r + \frac{1}{2}(k_1 + k_2)$	explicit	$O(h^3)$	$O(h^2)$
RK3	$Y_{r+1} = Y_r + \frac{1}{6}(k_1 + 4k_2 + k_3)$	explicit	$O(h^4)$	$O(h^3)$
RK4	$Y_{r+1} = Y_r + \frac{1}{6}(k_1 + 2k_2 + 2k_3 + k_4)$	explicit	$O(h^5)$	$O(h^4)$
Modified Euler (P–C)	$\begin{cases} Y^*_{r+1} = Y_r + hY'_r \\ Y_{r+1} = Y_r + \frac{1}{2}h\left(Y'_r + Y^{*\prime}_{r+1}\right) \end{cases}$	explicit	$O(h^3)$	$O(h^2)$

Example 3V

For $y' = 2xy$, $y(0) = 1$, Euler's method gives the following estimates of $y(1)$:

h	0.05	0.025	0.02	0.0125	0.01
$Y(1)$ (3D)	2.511	2.610	2.631	2.663	2.674

Draw a suitable graph and hence estimate the exact value of $y(1)$.

Solution

For Euler's method the global error satisfies, approximately

$$Y(1) - y(1) = Ch$$

or $$Y(1) = Ch + y(1)$$

The graph of $Y(1)$ against h should be approximately linear, with gradient C and intercept $y(1)$.

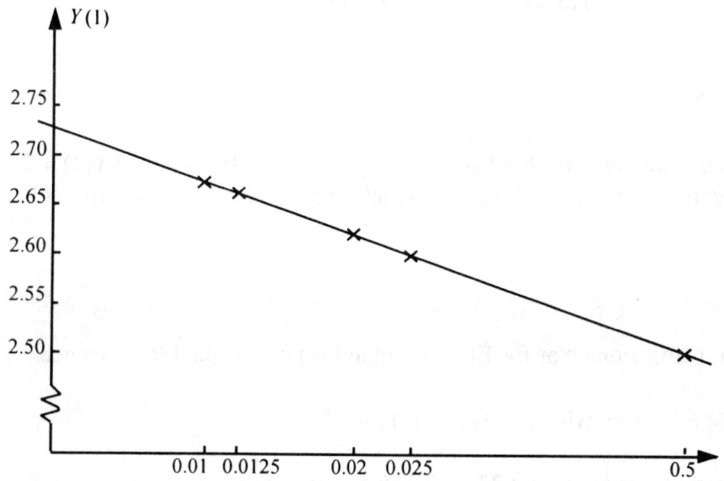

Figure 3.16

The value of $y(1)$ is where $h = 0$ so that $y(1) \approx 2.72$.

Note: This example indicates why the similar graph drawn in Example 3B was linear.

□

Example 3W

Estimates $Y(a)$, for y at $x = a$ are found with different values of h. How would you obtain a straight line graph if the method used is (i) TS2, (ii) RK4?

Solution

(i) For TS2, $G \propto h^2$, so that $Y(a) - y(a) = Ch^2$ and then

$$Y(a) = Ch^2 + y(a)$$

Comparing with $y = mx + c$, a graph of $Y(a)$ against h^2 should be linear, with intercept $y(a)$.

(ii) For RK4, $G \propto h^4$, so that $Y(a) - y(a) = Ch^4$ and then $Y(a) = Ch^4 + y(a)$
A graph of $Y(a)$ against h^4 should be linear, with intercept $y(a)$.

□

Example 3X

Using Euler's method with $h = 0.1$ to solve $y' = 2xy$, $y(0) = 1$, gives $y(1) \approx 2.335$. The exact value is 2.718 (3D). What step length would be required to find $y(1)$ correct to 2D?

Solution

The global truncation error for Euler's method is proportional to h. Hence

$$G \propto h \qquad \text{from which} \qquad Y(1) - y(1) = Ch$$

Using the result for $h = 0.1$: $2.335 - 2.718 = C(0.1)$ and, solving for C we find

$$C = -\frac{0.383}{0.1} = -3.83$$

Hence the global truncation error is $G = -3.83h$.

For accuracy to 2D to need $|G| < 0.005$. It follows that we require $h < \dfrac{0.005}{3.83} \approx 0.0013$. The actual value chosen for h must be such that there is an integer number of steps; that is, $\dfrac{1}{h}$ is an integer. Hence suitable values of h would be

$$h = 0.00125 \text{ (800 steps)} \qquad \text{or} \qquad h = 0.001 \text{ (1000 steps)}$$

□

Example 3Y

The Trapezoidal method was used to estimate $y(1.5)$ from $y' = 2xy$, $y(0) = 1$. With $h = 0.1$ we found $y(1.5) \approx 9.679$. With $h = 0.05$ we found $y(1.5) \approx 9.535$.

(i) Estimate how small h must be to find $y(1.5)$ correct to 4D.
(ii) Obtain a better estimate of $y(1.5)$ from the given results.

Solution

For the Trapezoidal method the global truncation error is

$$G = Y(1.5) - y(1.5) = Ch^2$$

With $h = 0.1$: $9.679 - y(1.5) = C(0.1)^2$

With $h = 0.05$: $9.535 - y(1.5) = C(0.05)^2$ (3.24)

(i) Subtracting: $0.144 = C(0.1^2 - 0.05^2)$. Hence

$$C = \frac{0.144}{0.1^2 - 0.05^2} \approx 19.2$$

and so $G = 19.2h^2$.

For accuracy to 4D we need $|G| < 0.00005$, which means that we need

$$19.2h^2 < 0.00005 \quad \text{or} \quad h < \sqrt{\frac{0.00005}{19.2}} \approx 0.0016$$

For example, we could use $h = 0.0015$ (1000 steps).

(ii) Using (3.24) which, having the smaller h will be more accurate,

$$y(1.5) = 9.535 - C(0.05)^2 = 9.535 - 19.2(0.05)^2$$

$$= 9.487 \qquad \text{(3D)}$$

which is a better estimate of $y(1.5)$ from the given results.

□

Example 3Z

(i) A differential equation is solved using Euler's method with a step size h and again with a step size $\frac{1}{2}h$ to obtain approximations $Y(h)$ and $Y\left(\frac{1}{2}h\right)$ respectively. Show that a better approximation is $2Y\left(\frac{1}{2}h\right)-Y(h)$. (This technique is known as Richardson extrapolation.)

(ii) Solving $y' = 2xy$, $y(0) = 1$ using Euler's method gave the following results (to 3D).
With $h = 0.1$, $y(1.5) \approx 6.835$.
With $h = 0.05$, $y(1.5) \approx 7.967$.
Find a better approximation to $y(1.5)$.

Solution

(i) The global truncation error for Euler's method is proportional to h. Hence

$$G = Y - y = Ch$$

With a step size of h: $Y(h) - y = Ch$ (3.25)

With a step size of $\frac{1}{2}h$: $Y\left(\frac{1}{2}h\right)-y=C\left(\frac{1}{2}h\right)$ (3.26)

Eliminate C by forming (3.25) – 2 × (3.26). Then

$$Y(h)-2Y\left(\frac{1}{2}h\right)+y=0$$

from which $y=2Y\left(\frac{1}{2}h\right)-Y(h)$, as required.

(ii) Using the above result with $h = 0.1$:

$$y(1.5) = 2Y(0.05) - Y(0.1)$$
$$y(1.5) = 2 \times 7.967 - 6.835$$
$$y(1.5) = 9.099 \qquad \text{(3D)}$$

Hence a better approximation to $y(1.5)$ is 9.099. (This is clearly an improvement as the exact answer, to 3D, is 9.488.)

□

Example 3AA

Use Richardson extrapolation to obtain a better approximation to $y(1.5)$ from the following results, obtained by applying RK2 to $y' = 2xy$, $y(0) = 1$.

With $h = 0.1$, $y(1.5) \approx 9.3488$. With $h = 0.05$, $y(1.5) \approx 9.4504$.

Solution

The global truncation error for RK2 is proportional to h^2. Then

$$G = Ch^2 \quad \text{or} \quad Y(h) - y = Ch^2 \tag{3.27}$$

For a step size $\frac{1}{2}h$ this becomes

$$Y\left(\frac{1}{2}h\right) - y = C\left(\frac{1}{2}h\right)^2 \tag{3.28}$$

Eliminate C by forming (3.27) – 4 × (3.28):

$$Y(h) - 4Y\left(\frac{1}{2}h\right) + 3y = 0$$

giving $y = \frac{1}{3}\left(4Y\left(\frac{1}{2}h\right) - Y(h)\right)$

Using the given results a better approximation to $y(1.5)$ is

$$y(1.5) = \frac{1}{3}(4 \times 9.4504 - 9.3488) = 9.4843 \qquad \text{(4D)}$$

□

3.3.3 Global truncation errors and global round off errors

We saw in Section 3.2 that the order of the global truncation error G was h^k, where k varies from method to method, and $k \geq 1$. For example, the global truncation error G in Euler's method, RK2 and TS4 is $O(h)$, $O(h^2)$ and $O(h^4)$ respectively.

With all methods G decreases with h and the graph of $|G|$ against h will qualitatively resemble that in Figure 3.17 To achieve a high accuracy a method with a relatively larger value of k is used, but reference to the examples in Section 3.2 confirms that for larger values of k more calculations have to be carried out at each step.

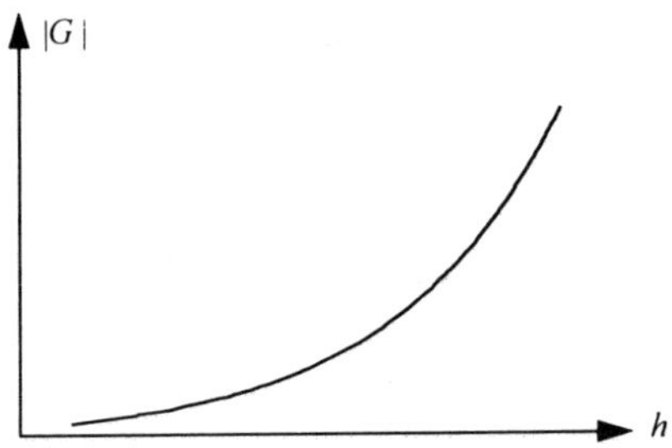

Figure 3.17

There is, however, a second source of error, as round-off errors occur in each calculation due to the limitations of the computer used. If, for example, $Y_1 = 2.1497528391$ but the computer can only work to eight significant figures then it will store Y_1 as 2.1497528, the last three digits being rounded off. In addition, if $f(x,y)$ contains trigonometric, exponential or other functions there will be round-off errors in their calculation.

As h decreases and/or k increases the number of calculations increases and the global round-off error R increases. The graph of $|R|$ against h will resemble that in Figure 3.18

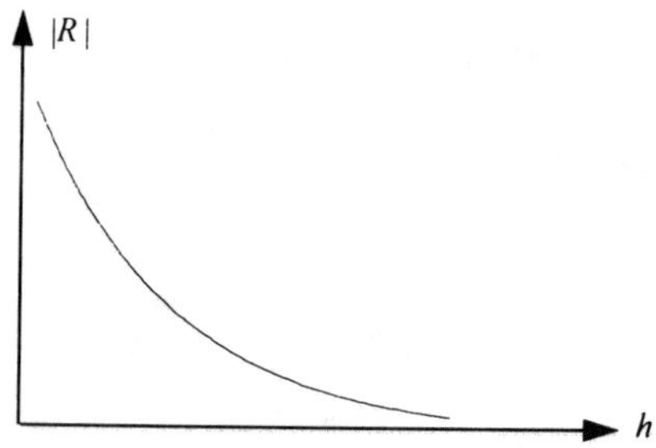

Figure 3.18

The total error T combines the global truncation error G with the global round-off error R.

The graph of $T = |G| + |R|$ against h will resemble that shown in Figure 3.19. The total error will have a minimum value at $h = h_0$. Hence, in practice, reducing h indefinitely will not give the exact answer to any required accuracy. There is a smallest value of h, h_0, depending on the method used and the accuracy of the computer beyond which the total error will increase rather than decrease.

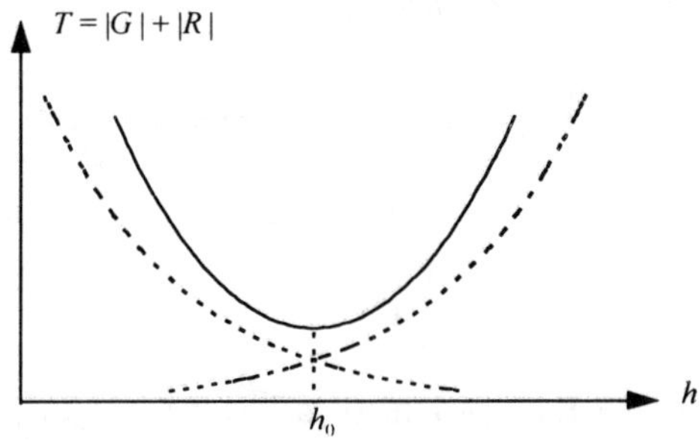

Figure 3.19

Example 3BB

We can simulate the above phenomenon by solving $y' = 2xy$, $y(0) = 1$ by Euler's method with different values of h, rounding **all** calculations to four significant figures. We find

h	0.1	0.05	0.025	0.01	0.005	0.0025	0.0015	0.001	0.0005
$Y(1.5)$	6.835	7.967	8.667	9.141	9.308	9.377	9.367	9.310	8.904
Total Error = $\lvert Y(1.5) - 9.488 \rvert$	2.653	1.521	0.821	0.347	0.180	0.111	0.121	0.178	0.584

The total error decreases with h until $h = 0.0025$, but then starts to increase if h is decreased further. With $h = 0.0025$ we find $y(1.5) \approx 9.377$ with total error 0.111. Hence, using Euler's method and a machine which rounds all calculations to four significant figures, the best result that can be achieved is $y(1.5) \approx 9.377$; this answer is not even accurate to one decimal place.

□

With other methods, equations and/or accuracies the same problems can arise. In general it must be remembered that reducing h indefinitely need not result in continual improvement in the answer. There will be a smallest value of h beyond which the results get worse rather than better due to the build up of round-off error in the calculations.

Exercises 3B

1. Show that the principal term in the local truncation error L for

   (i) the Trapezoidal method $Y_{r+1} = Y_r + \frac{h}{2}(Y_r' + Y_{r+1}')$ is $\frac{1}{12}h^3 Y_r'''$,

   (ii) the corrector formula $Y_{r+1} = Y_r + hY_{r+1}'$ is $\frac{1}{2}h^2 Y_r''$,

   (iii) the Modified Euler method $Y_{r+1} = Y_r + \frac{1}{2}h(Y_r' + Y_{r+1}^{*\prime})$, where $Y_{r+1}^* = Y_r + hY_r'$ is $-\frac{1}{6}h^3 Y_r'''$,

   (iv) the Adams Bashforth method $Y_{r+1} = Y_{r-1} + 2hY_r'$ is $\frac{1}{3}h^3 Y_r'''$,

   (v) the Adams Moulton method $Y_{r+1} = Y_r + \frac{1}{12}h(5Y_{r+1}' + 8Y_r' - Y_{r-1}')$ is $\frac{1}{24}h^4 Y_r^{iv}$,

   (vi) Simpson's method $Y_{r+1} = Y_{r-1} + \frac{1}{3}h(Y_{r-1}' + 4Y_r' + Y_{r+1}')$ is $-\frac{1}{90}h^5 Y_r^{v}$.

2. For each method in question 1, what happens to the local truncation error L when h is reduced by a factor of (i) 4, (ii) 10?

3. Determine the global truncation G for each method in question 1.

4. For each method in question 1, what happens to the global truncation error G when h is reduced by a factor of (i) 4, (ii) 10?

5. Show that using Euler's method with h = 0.05 to solve $\frac{dy}{dx} = 1 + y^2$ with $y(0)$ = 0 gives $y(0.5) \approx 0.5384$ (4D). The exact value is 0.5463 (4D). What step length would be required to find $y(0.5)$ correct to 3D?

6. The following results were obtained when solving $\frac{dy}{dx} = x + y$, $y(0) = 0$, with $h = 0.1$ in order to find $y(1)$:

Method	Approximate value of $y(1)$, correct to 6D	Accuracy required
Euler	0.593742	2D
Trapezoidal	0.720551	3D
Taylor series 2nd order	0.714081	4D
Taylor series 4th order	0.718280	6D
Runge-Kutta 2nd order	0.714081	4D
Runge-Kutta 4th order	0.718280	8D
Modified Euler	0.714081	7D

The exact value is 0.718282 (6D). For each method, find an upper bound on h in order to be able to find $y(1)$ correct to the accuracy required. Specify a suitable value of h in each case and indicate the number of steps needed to find $y(1)$.

7. For the equation $\frac{dy}{dx} = 1 - x^2 y$, $y(0) = 1$, use DERIVE and Euler's method to find estimates for $y(1)$ when $h = 0.01$, 0.005, 0.0025 and 0.001.

The exact value and approximate value are related by $Y(1) - y(1) \approx Ch$. By plotting a graph of $Y(1)$ against h estimate the exact value of $y(1)$.

8. For the equation $\frac{dy}{dx} = 9.81 - 0.02y^2$, $y(0) = 0$, the Modified Euler method gives the following estimates for $y(0.5)$:

h	0.125	0.1	0.0625	0.05
$Y(0.5)$	4.823805	4.824724	4.825716	4.825944

The global truncation error for this method is proportional to h^2. By plotting a suitable graph estimate the exact value of $y(0.5)$.

9. The equation $\frac{dy}{dx} = \cos y$, $y(0) = 1$ was solved using the Modified Euler method. It was found that with $h = 0.1$, $y(0.4) \approx 1.1821556$ and with $h = 0.05$, $y(0.4) \approx 1.1823405$. Using the result that the global truncation error for this method is proportional to h^2

(i) use the above results to obtain a better estimate of $y(0.4)$,

(ii) estimate how small the step length must be to ensure that the approximation to the value of $y(0.4)$ obtained by this method is accurate to 6D. Specify a suitable value for h and the number of steps required.

10. Using DERIVE and the Runge-Kutta method estimate $y(2)$ from the problem

$$\frac{dy}{dx} = \sqrt{x} + \sqrt{y} \qquad \text{with } y(1) = 0$$

and step sizes $h = 0.1$ and $h = 0.04$. Use these results to obtain a better estimate of $y(2)$ and to determine a suitable value of h so that $y(2)$ will be found correct to 8D.

11. (i) The equation $y' = f(x,y)$ is solved with a step length h and with a step length $\frac{1}{4}h$ to obtain approximations $Y(h)$ and $Y\left(\frac{1}{4}h\right)$ respectively.

(a) If the method has global truncation error proportional to h show that a better approximation is $\frac{1}{3}\left(4Y\left(\frac{1}{4}h\right) - Y(h)\right)$.

(b) If the method has global truncation error proportional to h^2 show that a better approximation is $\frac{1}{15}\left(16Y\left(\frac{1}{4}h\right) - Y(h)\right)$.

(ii) Solving $y' = \frac{x}{\sqrt{y}} - \frac{y}{\sqrt{x}}$, $y(1) = 2$, gave the following results:

(a) Using Euler's method:
With $h = 0.1$, $y(3) \approx 2.234416$.
With $h = 0.025$, $y(3) \approx 2.252359$.

(b) Using the Runge-Kutta second order method:
With $h = 0.2$, $y(3) \approx 2.262145$.
With $h = 0.05$, $y(3) \approx 2.258551$.

Find a better approximation to $y(3)$ in each case.

12. Use DERIVE and the Runge-Kutta fourth order method to find an approximation for $y(3)$ for $y' = \dfrac{1}{\sqrt{y^2 - x^2}}$, $y(0) = 1$ with $h = 0.3$ and $h = 0.15$.

    Use Richardson extrapolation to find a better approximation to $y(3)$.

13. Solve $y' = \sqrt{x^2 + y^2}$, $y(1) = 0$ to find $y(2)$ using Euler's method and rounding **all** calculations to 3D. Show that the magnitude of the total error decreases with h until $h \approx 0.01$; but then starts to increase. The exact solution is $y(2) = 1.691$ (3D).

14. The error in the Runge-Kutta methods depends on the form of the equation as well as on the solution itself.

    (a) Verify that $y = (x - 1)^2$ is an analytical solution of both the initial value problems

    $$y' = 2(x-1) \qquad\qquad y(0) = 1$$

    and
    $$y' = \frac{2y}{(x-1)} \qquad\qquad y(0) = 1$$

    (b) Use DERIVE to show that the fourth-order Runge-Kutta method gives an exact solution to the first problem but not for the second problem.

3.4 STABILITY

3.4.1 Difference equations (recurrence relations)

We will need to solve certain difference equations in the next sections and the following is a summary of the required results.

(i) The **order** of a recurrence relation is the difference between the highest and lowest subscripts present. Thus

$u_{r+1} = 4u_r - 2$ is a first order recurrence relation

$u_{r+1} = 3u_r + 2u_{r-1}$ is a second order recurrence relation.

(ii) The general solution of the first order equation $u_{r+1} = au_r + b$ is

$$u_n = C \times a^n + \frac{b}{1-a} \qquad \text{if } a \neq 1$$

or $$u_n = C + nb \qquad \text{if } a = 1$$

If the first term u_0 is known then the particular solution can be found.

Example 3CC

Solve

(i) $u_{r+1} = 2u_r - 1$ given $u_0 = 2$, (ii) $u_{r+1} = u_r + 2$ given $u_0 = 1$.

Solution

(i) Since $a = 2$ ($\neq 1$) and $b = -1$ the general solution is

$$u_n = C \times 2^n + \frac{-1}{1-2} = C \times 2^n + 1$$

Using $u_0 = 2$: $2 = C \times 1 + 1$, so that $C = 1$ and the particular solution is

$$u_n = 2^n + 1$$

(The sequence is 2, 3, 5, 9, 17, ...)

(ii) Since $a = 1$ and $b = 2$ the general solution is

$$u_n = C + 2n$$

Using $u_0 = 1$: $1 = C + 2 \times 0$, so that $C = 1$ and the particular solution is

$$u_n = 1 + 2n$$

(The sequence is 1, 3, 5, 7, 9, ...)

□

Example 3DD

Find the general solution of $u_{r+1} = 3u_r$.

Solution

Here $a = 3$ and $b = 0$ and so the general solution is

$$u_n = C \times 3^n$$

□

(iii) A second order homogeneous recurrence relation with constant coefficients is of the form

$$u_{r+1} = au_r + bu_{r-1}$$

The solution depends on the type of roots of the auxiliary equation

$$m^2 = am + b$$

If the roots are real and different (say λ and μ) then the general solution is

$$u_n = A \times \lambda^n + B \times \mu^n$$

The particular solution can be found if u_0 and u_1 are known.

Example 3EE

Solve $u_{r+1} = 5u_r - 6u_{r-1}$ given $u_0 = 0$ and $u_1 = 1$.

Solution

The auxiliary equation is $m^2 = 5m - 6$ or $m^2 - 5m + 6 = 0$. Factorising:

$$(m - 2)(m - 3) = 0$$

from which the roots are 2 and 3. Hence the general solution is

$$u_n = A \times 2^n + B \times 3^n$$

Using $u_0 = 0$: $\quad 0 = A + B$
Using $u_1 = 1$: $\quad 1 = 2A + 3B$

Solving these equations gives $A = -1$ and $B = 1$ and so the particular solution is

$$u_n = 3^n - 2^n$$

(The sequence is 0, 1, 5, 19, 65, ...)

□

3.4.2 Introduction to stability

We have seen that as the step size h decreases the numerical solution approaches the true solution (assuming that all calculations can be carried out exactly), but we do not know whether a particular step size will give good results.

All numerical methods replace the first order differential equation $y' = f(x,y)$ with a difference equation (recurrence relation), and we hope that the solution of the latter closely resembles that of the former.

For example, Euler's method models

$$y' = f(x,y) \tag{3.29}$$

with the first order difference equation

$$Y_{r+1} = Y_r + hY_r' \tag{3.30}$$

Does the solution to (3.30) always agree with that of (3.29)?

For example, Simpson's method replaces (3.29) with

$$Y_{r+1} = Y_{r-1} + \frac{1}{3}h(Y_{r-1}' + 4Y_r' + Y_{r+1}')$$

which is a second order difference equation with two solutions. We hope that one solution corresponds to that of $y' = f(x,y)$, but what effects does the other solution have?

Instabilities can arise from the nature of the differential equation itself (that is, be inherent) or be introduced by the numerical method used (that is, be induced).

The following example illustrates what can happen.

Example 3FF

Solve $y' = -15y$, $y(0) = 1$ using Euler's method with (i) $h = 0.2$, (ii) $h = 0.05$. Compare with the exact solution $y = e^{-15x}$.

Solution

Euler's method is $Y_{r+1} = Y_r + hY'_r$

With $y' = -15y$ this becomes $Y_{r+1} = Y_r - 15hY_r$ or $Y_{r+1} = (1 - 15h)Y_r$ (3.31)

(i) With $h = 0.2$ we obtain $Y_{r+1} = -2Y_r$

This is a first order difference equation with general solution $Y_n = A(-2)^n$

Using $Y_0 = 1$ we find $A = 1$ and so the particular solution is $Y_n = (-2)^n$

Then $Y_1 = -2$, $Y_2 = 4$, $Y_3 = -8$, $Y_4 = 16$, ... Y_n increases in magnitude and oscillates in sign as n increases. However the exact solution $y = e^{-15x}$ decreases quickly towards zero and so Euler's method with $h = 0.2$ has failed.

(ii) With $h = 0.05$ the recurrence relation (3.31) becomes $Y_{r+1} = 0.25Y_r$

with general solution $Y_n = A(0.25)^n$

Applying $Y_0 = 1$ again gives $A = 1$ and so the particular solution is

$$Y_n = (0.25)^n$$

This time $Y_n \to 0$ as n increases, which is in agreement with the exact solution.
With $h = 0.05$ the approximate solution follows the same trend as the exact solution, whereas with $h = 0.2$ it does not. This suggests that there is a maximum value of h for the true and approximate solutions to behave similarly (but whether this is small enough for the result to have the desired accuracy is another matter).

□

3.4.3 Ill-conditioned equations (inherent instability)

A differential equation is ill-conditioned when small changes in the data result in much larger changes in the solution (otherwise it is well-conditioned). These changes can be measured absolutely or relatively.

Example 3GG

Determine the conditioning of $y' = y - x$ given $y(0) = 1$. The general solution is $y = Ae^x + x + 1$.

Solution

Using $y(0) = 1$: $1 = A \times 0 + 0 + 1$, from which $A = 0$ and the particular solution is $y = x + 1$. However, any numerical process will introduce an error and the solution will follow (A) or (B), rapidly diverging away from the correct solution.

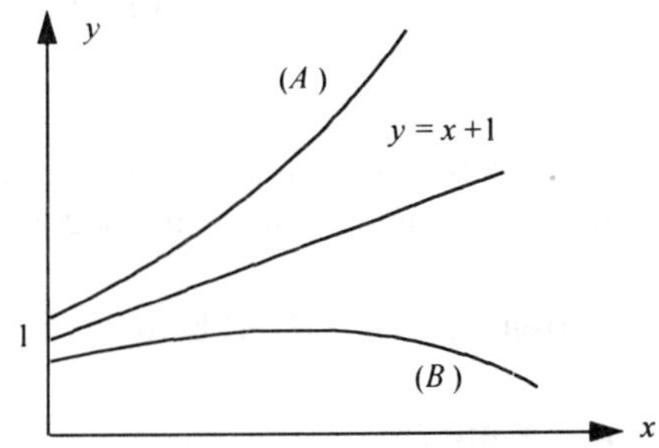

Figure 3.20

We can model this by changing the initial condition to $y(0) = 1 + \varepsilon$. With $y(0) = 1 + \varepsilon$: $1 + \varepsilon = A + 0 + 1$, giving $A = \varepsilon$ and the approximate solution is $Y = \varepsilon e^x + x + 1$

Absolute error $= Y - y = \varepsilon e^x$ which increases with x and the problem is absolutely ill-conditioned.

Relative error $= \dfrac{Y - y}{y} = \dfrac{\varepsilon e^x}{x+1}$ which increases with x and the problem is relatively ill-conditioned. (However small ε is, the term εe^x will eventually dominate the exact solution $x + 1$, so that Y will not closely model y.)

□

Example 3HH

Solve $y' + 2xy = 1 + 2x^2$ with (i) $y(0) = 0$, (ii) $y(0) = \varepsilon$. Is the problem ill-conditioned or well-conditioned?

Solution

The general solution is $y = Ae^{-x^2} + x$.
Using $y(0) = 0$ gives

$$A = 0 \quad \text{and} \quad y = x \qquad \text{(exact particular solution)}$$

Using $y(0) = \varepsilon$ gives

$$A = \varepsilon \quad \text{and} \quad Y = \varepsilon e^{-x^2} + x \qquad \text{(approximate particular solution)}$$

Absolute error $= Y - y = \varepsilon e^{-x^2}$ which decreases as x increases and the problem is absolutely well-conditioned.

Relative error $= \dfrac{Y-y}{y} = \dfrac{\varepsilon e^{-x^2}}{x}$ which decreases as x increases and the problem is relatively well-conditioned.

□

If a differential equation is ill-conditioned then we cannot hope to avoid the build-up of errors, whatever numerical method is used, although it is possible that while the absolute error increases the relative error does not.

When we apply a numerical method to a well-conditioned equation we must investigate the associated recurrence relation; is this equation well-conditioned as well?

3.4.4 Induced instability

A numerical method applied to a differential equation is said to be **absolutely stable** if the difference equation is well-conditioned. It is absolutely unstable if the difference equation is ill-conditioned.

Example 3II

Investigate the stability of Euler's method applied to the equation $y' = \lambda y$, with general solution $y = Ae^{\lambda x}$.

Solution

Consider first the exact solution $y = Ae^{\lambda x}$. If $\lambda > 0$ then $y \to \infty$ as $x \to \infty$. If $\lambda = 0$ then $y = A$ for all x. If $\lambda < 0$ then $y \to 0$ as $x \to \infty$.

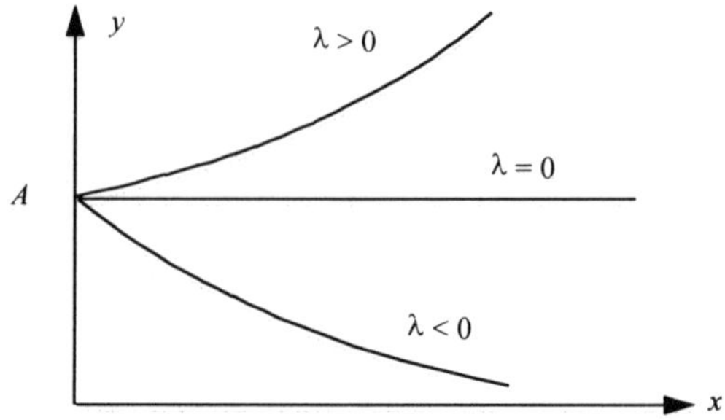

Euler's method is $Y_{r+1} = Y_r + hY'_r$

Using $y' = \lambda y$, this becomes $Y_{r+1} = Y_r + h\lambda Y_r$ or $Y_{r+1} = (1 + \lambda h)Y_r$

This is a first order recurrence relation and has general solution

$$Y_n = C(1 + \lambda h)^n \tag{3.32}$$

(i) If $\lambda > 0$ then $(1 + \lambda h) > 1$ and so Y_n increases with n or x (as $x = x_0 + nh$), as does the exact solution $y = Ae^{\lambda x}$. Both the differential equation and recurrence relation are absolutely ill-conditioned and so the method is absolutely unstable. It may be, however, that despite this the growth of the relative errors will be tolerable. See, for instance, the relative error in Y_{10} in Example 3NN.

(ii) If $\lambda < 0$ then the exact solution $y = Ae^{\lambda x}$ tends to zero as x increases. If the approximate solution is to follow suit then Y_n must decrease as n increases. Then

$$|Y_{n+1}| < |Y_n| \quad \text{or} \quad \left|\frac{Y_{n+1}}{Y_n}\right| < 1$$

Using (3.32): $\left|\dfrac{C(1+\lambda h)^{n+1}}{C(1+\lambda h)^n}\right| < 1$ or $|1 + \lambda h| < 1$

This is satisfied provided

$$-1 < 1 + \lambda h < 1 \quad \text{or} \quad -2 < \lambda h < 0 \quad \text{or} \quad -\frac{2}{\lambda} > h > 0$$

Hence if $\lambda < 0$ the positive step size h must be such that $h < -\frac{2}{\lambda}$. Combining (i) and (ii), Euler's method applied to $y' = \lambda y$ is only stable if h satisfies the inequality $0 < h < -\frac{2}{\lambda}$.

□

Example 3JJ

For what values of h is Euler's method stable, when applied to (i) $y' = 15y$, (ii) $y' = -15y$?

Solution

From Example 3HH (and the equation $y' = \lambda y$), h must satisfy $0 < h < -\frac{2}{\lambda}$ for stability.

(i) When $\lambda = 15$ the inequality becomes $0 < h < -\frac{2}{15}$, which is not satisfied by any positive h. Hence, Euler's method applied to $y' = 15y$ is unstable for all h.

(ii) With $\lambda = -15$, h must satisfy $0 < h < \frac{2}{15} \approx 0.13$ and so Euler's method is stable for $h = 0.1$, 0.05 or smaller.

The conclusions in (ii) explain the results of Example 3FF.

□

Example 3KK

Investigate the stability of the Taylor series second order method (TS2) applied to $y' = \lambda y$.

Solution

TS2 is $Y_{r+1} = Y_r + hY_r' + \frac{1}{2}h^2 Y_r''$

Differentiating $y' = \lambda y$ we obtain $y'' = \lambda y' = \lambda^2 y$ and then TS2 becomes

$$Y_{r+1} = Y_r + h\lambda Y_r + \frac{1}{2}h^2\lambda^2 Y_r \quad \text{or} \quad Y_{r+1} = \left(1 + \lambda h + \frac{1}{2}\lambda^2 h^2\right)Y_r$$

This is a first order recurrence relation with general solution

$$Y_n = C\left(1 + \lambda h + \frac{1}{2}\lambda^2 h^2\right)^n \tag{3.33}$$

(i) If $\lambda > 0$ then, as for Euler's method, both the differential equation the recurrence relation are ill conditioned and so TS2 is absolutely unstable.

(ii) If $\lambda < 0$ then for the approximate solution to decrease, as does the exact solution we require

$$|Y_{n+1}| < |Y_n| \quad \text{or} \quad \left|\frac{Y_{n+1}}{Y_n}\right| < 1$$

From (3.33) this becomes

$$\left|\frac{C\left(1 + \lambda h + \frac{1}{2}\lambda^2 h^2\right)^{n+1}}{C\left(1 + \lambda h + \frac{1}{2}\lambda^2 h^2\right)^n}\right| < 1 \quad \text{or} \quad \left|1 + \lambda h + \frac{1}{2}\lambda^2 h^2\right| < 1$$

This is satisfied if

$$-1 < 1 + \lambda h + \frac{1}{2}\lambda^2 h^2 < 1 \quad \text{or} \quad -2 < \lambda h + \frac{1}{2}\lambda^2 h^2 < 0$$

To determine when this inequality is satisfied consider $y = x + \frac{1}{2}x^2$, or $y = \frac{1}{2}(x+1)^2 - \frac{1}{2}$. The graph of y is a parabola with its minimum value at $x = -1$, and $y(-1) = -\frac{1}{2}$. Also $y = 0$ when $x = 0$ and when $x = -2$.

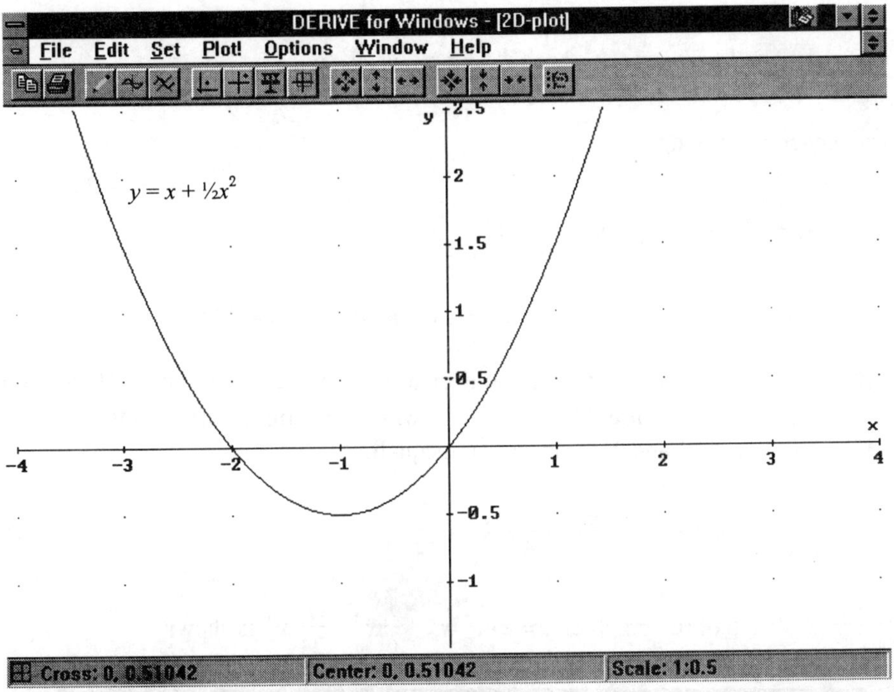

Hence $-2 < x + \frac{1}{2}x^2 < 0$ is satisfied by $-2 < x < 0$ and it follows that $-2 < \lambda h + \frac{1}{2}\lambda^2 h^2 < 0$ is satisfied by $-2 < \lambda h < 0$ or, for positive h, $h < -\frac{2}{\lambda}$.

Combining (i) and (ii) it follows that, when solving $y' = \lambda y$, TS2 is stable for all h's satisfying the inequality $0 < h < -\frac{2}{\lambda}$.

Example 3LL

Investigate the stability of the Taylor Series fourth order method (TS4) applied to $y' = \lambda y$.

Solution

TS4 is $Y_{r+1} = Y_r + hY_r' + \frac{1}{2}h^2 Y_r'' + \frac{1}{3!}h^3 Y_r''' + \frac{1}{4!}h^4 Y_r^{iv}$. From $y' = \lambda y$ it follows that $y'' = \lambda y' = \lambda^2 y$, $y''' = \lambda^2 y' = \lambda^3 y$ and $y^{iv} = \lambda^3 y' = \lambda^4 y$. Substituting into TS4 and simplifying:

$$Y_{r+1} = \left(1+\lambda h+\frac{1}{2}\lambda^2 h^2+\frac{1}{6}\lambda^3 h^3+\frac{1}{24}\lambda^4 h^4\right)Y_r$$

with general solution

$$Y_n = C\left(1+\lambda h+\frac{1}{2}\lambda^2 h^2+\frac{1}{6}\lambda^3 h^3+\frac{1}{24}\lambda^4 h^4\right)^n$$

(i) If $\lambda > 0$ then, as for Euler and TS2, the method is unstable.

(ii) If $\lambda < 0$ then the exact solution decreases. For the approximate solution to do likewise we require $|Y_{n+1}| < |Y_n|$ which, by similar arguments to those in Examples 3II and 3JJ leads to the inequality

$$-2 < \lambda h+\frac{1}{2}\lambda^2 h^2+\frac{1}{6}\lambda^3 h^3+\frac{1}{24}\lambda^4 h^4 < 0 \qquad (3.34)$$

A sketch of the graph of $y = x+\frac{1}{2}x^2+\frac{1}{6}x^3+\frac{1}{24}x^4$ is shown.

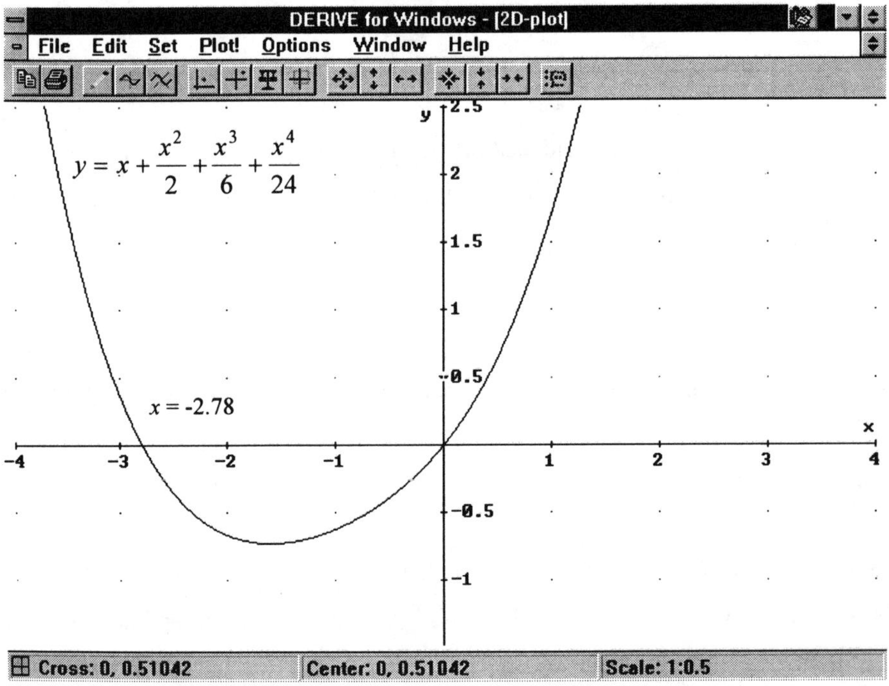

$y = 0$ when $x = 0$ and when $x = -2.78$ (approx.). Hence y lies between -2 and 0 for $-2.78 < x < 0$ and so (3.34) is satisfied, for positive h, when $\lambda h > -2.78$ or $h < -\dfrac{2.78}{\lambda}$. Hence when TS2 is applied to $y' = \lambda y$ we require $0 < h < -\dfrac{2.78}{\lambda}$ for stability.

□

Example 3MM

Obtain and solve the difference equation when Simpson's method with $h = 0.1$ is applied to the equation $y' = 15y$. Compare with the exact solution $y = Ae^{15x}$.

Solution

Simpson's method is $Y_{r+1} = Y_{r-1} + \frac{1}{3}h(Y'_{r-1} + 4Y'_r + Y'_{r+1})$. Using $y' = 15y$ and $h = 0.1$ this becomes

$$Y_{r+1} = Y_{r-1} + \frac{1}{3} \times 0.1(15Y_{r-1} + 60Y_r + 15Y_{r+1})$$

$$Y_{r+1} = Y_{r-1} + 0.5Y_{r-1} + 2Y_r + 0.5Y_{r+1}$$

$$0.5Y_{r+1} = 1.5Y_{r-1} + 2Y_r$$

$$Y_{r+1} = 3Y_{r-1} + 4Y_r$$

This is a second order recurrence relation, similar to that in Example 3DD. The auxiliary equation is $m^2 = 3 + 4m$ or $m^2 - 4m - 3 = 0$ with roots

$$m = \frac{4 \pm \sqrt{28}}{2} = 4.64575, -0.64575 \qquad \text{(5D)}$$

The general solution is

$$Y_n = A(4.64575)^n + B(-0.64575)^n \qquad (3.35)$$

The exact solution of $y' = 15y$ is $y = Ce^{15x}$, which can be expressed as $y_r = Ce^{15x_r}$ and then

$$y_{r+1} = Ce^{15x_{r+1}} = Ce^{15(x_r+h)} = Ce^{15x_r+1.5} = Ce^{1.5}e^{15x_r}$$

or $\quad y_{r+1} = e^{1.5} y_r = 4.48169 y_r \qquad \text{(5D)}$

The general solution of this recurrence relation is

$$y_n = D(4.48169)^n \tag{3.36}$$

Comparing (3.35) and (3.36) there is a similarity between the first term in Y_n and y_n, but the second term in (3.35), $B(-0.64575)^n$, is an extra solution introduced by Simpson's method and which does not correspond to anything in the original differential equation. This term is called **spurious** or **parasitic**.

□

Example 3NN

Consider again the approximate solution of $y' = 15y$ using Simpson's method. Compare the exact and approximate solutions using $y(0) = 1$ and, for Simpson's method, $y(0.1) = 4.4817$ (found by some other method). Determine the effect of the spurious solution. Compare the exact and approximate solutions at $x = 1$.

Solution

The exact solution corresponding to $y = Ae^{15x}$ is $y_n = D(4.48169)^n$. Applying $y(0) = 1$ shows that $D = 1$ and the exact particular solution is $y_n = (4.48169)^n$.

The approximate solution from Simpson's method is equation (3.35). Applying $y(0) = 1$, so that $Y_0 = 1$, gives $A + B = 1$. Applying $y(0.1) = 4.4817$, so that $Y_1 = 4.4817$ gives $4.64575A - 0.64575B = 4.4817$. Solving these equations we find $A = 0.969$ and $B = 0.031$ (exactly) and so the particular solution of (3.35) is

$$Y_n = 0.969(4.64578)^n + 0.031(-0.64575)^n$$

The first term in Y_n is very similar to the exact solution. As n increases the spurious solution $0.031(-0.64575)^n$ decreases towards zero and so its effects reduce. Hence the spurious solution introduced by Simpson's method does not have a significant effect on the result.

When $x = 1$ (so that $n = 10$) the true and approximate solutions are

$$y_{10} = 4.48169^{10} = 3269017 \qquad (= e^{15})$$

and $\quad Y_{10} = 0.969(4.64578)^{10} + 0.031(-0.64575)^{10} = 4538197$

The absolute error is $Y_{10} - y_{10} = 1269180$, which is very large, as is the relative error $\frac{Y_{10} - y_{10}}{y_{10}} \approx 0.388$.

□

Example 3OO

Solve $y' = -15y$ using Simpson's method with $h = 0.1$ and determine the effect of the spurious solution.

Solution

Substituting $y' = -15y$ and $h = 0.1$ into Simpson's method and simplifying, as in Example 3MM, results in the difference equation

$$Y_{r+1} = \frac{1}{3}Y_{r-1} - \frac{4}{3}Y_r$$

The auxiliary equation is $m^2 = \frac{1}{3} - \frac{4}{3}m$ or $3m^2 + 4m - 1 = 0$ with roots 0.21525 and −1.54858 (5D) and so the general solution is

$$Y_n = A(0.21525)^n + B(-1.54858)^n$$

The true solution is $y = Ce^{-15x}$ which corresponds to $y_n = C(e^{-1.5})^n = C(0.2231)^n$. The first term in Y_n corresponds to the exact solution and the second term, $B(-1.54858)^n$, is spurious.

As n increases the true solution decreases, but the spurious solution increases in magnitude. This will eventually swamp the true solution and the results will be useless. Hence Simpson's method is unstable for $y' = -15y$ and should not be used.

□

3.4.5 Induced instability in general

So far we have looked at what happens when a number of methods are applied to a few specific equations, all of the form $y' = \lambda y$.

We shall now generalise the discussion to any first order equation $y' = f(x,y)$. Euler's method is $Y_{r+1} = Y_r + hY'_r$ or, when applied to $y' = f(x,y)$

$$Y_{r+1} = Y_r + hf(x_r, Y_r)$$

We wish to see whether or not errors are magnified as the solution progresses.

Suppose that there is an error ε in Y_r so that the computed value of Y_r is

$$\overline{Y}_r = Y_r + \varepsilon$$

Then the computed value for Y_{r+1} is

$$\overline{Y}_{r+1} = \overline{Y}_r + hf(x_r, \overline{Y}_r) = Y_r + \varepsilon + hf(x_r, Y_r + \varepsilon) \tag{3.37}$$

We can expand $f(x_r, Y_r + \varepsilon)$ about $f(x_r, Y_r)$ is ascending powers of ε by Taylor's series. Then

$$f(x_r, Y_r + \varepsilon) = f(x_r, Y_r) + \varepsilon\left(\frac{\partial f}{\partial y}\right)_{(x_r, Y_r)} + \frac{1}{2}\varepsilon^2\left(\frac{\partial^2 f}{\partial y^2}\right)_{(x_r, Y_r)} + \ldots$$

Truncating after the first two terms

$$f(x_r, Y_r + \varepsilon) \approx f(x_r, Y_r) + \varepsilon\left(\frac{\partial f}{\partial y}\right)_{(x_r, Y_r)}$$

Using this approximation in (3.37) we have

$$\overline{Y}_{r+1} \approx Y_r + \varepsilon + hf(x_r, Y_r) + h\varepsilon\left(\frac{\partial f}{\partial y}\right)_{(x_r, Y_r)}$$

$$\overline{Y}_{r+1} \approx (Y_r + hf(x_r, Y_r)) + \varepsilon\left(1 + h\left(\frac{\partial f}{\partial y}\right)_{(x_r, Y_r)}\right)$$

$$\overline{Y}_{r+1} \approx Y_{r+1} + \varepsilon\left(1 + h\left(\frac{\partial f}{\partial y}\right)_{(x_r, Y_r)}\right)$$

Hence the error in the approximation Y_{r+1} is approximately $\varepsilon\left(1 + h\left(\frac{\partial f}{\partial y}\right)_{(x_r, Y_r)}\right)$ and has been magnified by a factor $\left(1 + h\left(\frac{\partial f}{\partial y}\right)_{(x_r, Y_r)}\right)$ in computing $\overline{Y}_{r+1}$ from $\overline{Y}_r$.

For the errors to decrease we require $\left|1+h\left(\frac{\partial f}{\partial y}\right)_{(x_r,Y_r)}\right| < 1$ or

$$-1 < 1+h\left(\frac{\partial f}{\partial y}\right)_{(x_r,Y_r)} < 1 \quad \text{or} \quad -2 < h\left(\frac{\partial f}{\partial y}\right)_{(x_r,Y_r)} < 0$$

or, for positive h, $\left(\frac{\partial f}{\partial y}\right)_{(x_r,Y_r)}$ must be negative and then $h < -\frac{2}{\left(\frac{\partial f}{\partial y}\right)_{(x_r,Y_r)}}$.

This inequality must be satisfied for all steps and so Euler's method is stable provided $0 < h < -\frac{2}{\frac{\partial f}{\partial y}}$ throughout the interval for which the solution is required.

Example 3PP

For what values of h is Euler's method applied to $y' = \lambda y$ stable? (See also Examples 3HH and 3II).

Solution

Here $f(x,y) = \lambda y$ and $\frac{\partial f}{\partial y} = \lambda$. For stability we require $0 < h < -\frac{2}{\lambda}$ (as in Example 3II).

If $\lambda > 0$ the inequality cannot be satisfied and so Euler's method is unstable for all h.

If $\lambda < 0$ the inequality can be satisfied. For example if $\lambda = -15$ we need $0 < h < \frac{2}{15}$ (as in Example 3JJ).

□

Example 3QQ

Investigate the stability of Euler's method applied to $y' = 1 - x^2 y$ with $y(1) = 1$ over the interval $1 \le x \le 4$.

Solution

Here $f(x,y) = 1 - x^2y$ and then $\dfrac{\partial f}{\partial y} = -x^2$. Hence, for stability, $0 < h < \dfrac{-2}{-x^2}$ or $0 < h < \dfrac{2}{x^2}$ for all x between 1 and 4. The largest value of x will give the smallest value of h and so we need

$$h < \frac{2}{4^2} = 0.125 \qquad \text{(for example } h = 0.1 \text{ or } 0.05\text{) for stability.}$$

□

Summary of results

We have found that the interval of stability for Euler's method applied to $y' = f(x,y)$ is $0 < h < \dfrac{-2}{\dfrac{\partial f}{\partial y}}$. Other methods can be investigated in a similar way and, although the analysis is not included here, the results are summarised in the following table.

Method	Interval of stability for $\dfrac{dy}{dx} = f(x,y)$
Euler	$0 < h < -2\Big/\dfrac{\partial f}{\partial y}$
Trapezoidal	all $h > 0$
TS2	$0 < h < -2\Big/\dfrac{\partial f}{\partial y}$
TS4	$0 < h < -2.78\Big/\dfrac{\partial f}{\partial y}$
RK2	$0 < h < -2\Big/\dfrac{\partial f}{\partial y}$
RK4	$0 < h < -2.78\Big/\dfrac{\partial f}{\partial y}$
Modified Euler	$0 < h < -2\Big/\dfrac{\partial f}{\partial y}$

Example 3RR

Investigate the stability of $y' = \dfrac{12x}{y}$, $y(0) = 1$ in the interval $0 \le x \le 2$. It is known that all y values in this interval lie between 1 and 7.

Solution

Here $f(x,y) = \dfrac{12x}{y}$ and so $\dfrac{\partial f}{\partial y} = -\dfrac{12x}{y^2}$.

(i) For Euler, TS2, RK2 or the Modified Euler method we require

$$h < -\frac{2}{\dfrac{\partial f}{\partial y}} \quad \text{or} \quad h < \frac{-2}{-\dfrac{12x}{y^2}} \quad \text{or} \quad h < \frac{y^2}{6x}$$

The largest value of h will be when $y = 1$ (smallest value) and $x = 2$ (largest value). Hence we require $h < \dfrac{1^2}{6 \times 2} = \dfrac{1}{12} = 0.0833\ldots$ For example the methods are stable for $h = 0.05$.

(ii) The trapezoidal method is stable for all $h > 0$.

(iii) For TS4 and RK4 we require

$$h < \frac{-2.78}{\dfrac{\partial f}{\partial y}} \quad \text{or} \quad h < \frac{2.78y^2}{12x}$$

The worst case is when $y = 1$ and $x = 2$ and then $h < \dfrac{2.78 \times 1^2}{12 \times 2} = 0.1158$ for stability. For example, the methods are stable for $h = 0.1$.

3.5 SYSTEMS OF LINEAR FIRST ORDER DIFFERENTIAL EQUATIONS

So far all differential equations considered have contained a single variable y dependent on an independent variable x. Here we are interested in a system of two (or more) equations that involve two (or more) variables x and y which are both dependent on the independent variable t.

A system of two equations in the variables x and y which are dependent on the independent variable t is of the form

$$\frac{dx}{dt} = f(t, x, y)$$

$$\frac{dy}{dt} = g(t, x, y)$$

with initial conditions $x(t_0) = x_0$, $y(t_0) = y_0$.

We shall see in Section 5.6 how equations of this type can be solved analytically if the equations are linear, but if they are not linear numerical techniques must be used. All the numerical methods studied earlier in this chapter can be extended to apply to simultaneous equations, but DERIVE uses only the classical fourth order Runge-Kutta method, which is in the utility file ODE_APPR.

The individual methods will not be described and discussion is restricted to how to use DERIVE to solve equations. You will recall that the Runge-Kutta fourth order method converges quickly to the true solution as the step length h is decreased.

The DERIVE command for a pair of simultaneous equations is

$$RK(\underbrace{[f(t,x,y), g(t,x,y)]}_{\text{RHS's of equations}},\ \underbrace{[(t,x,y)]}_{\text{variables}},\ \underbrace{[(t_0, x_0, y_0)]}_{\text{initial conditions}},\ \underset{\substack{\text{step}\\\text{length}}}{h},\ \underset{\substack{\text{number of}\\\text{steps}}}{n})$$

approXimating this gives a set if n $[t,x,y]$ triples at t_0, $t_0 + h$, $t_0 + 2h_1$, ..., $t_0 + nh$.

To plot a graph of x against t or y against t (or even x against y) two columns must first be removed from the $[t,x,y]$ triples. The command

EXTRACT_2_COLUMNS(#*k*,1,2)

extracts columns 1 and 2 (the t and x values) from the triples, where k is the label number of the Runge-Kutta RK line.

Example 3SS

Solve the equations $\frac{dx}{dt} = t^2 - x + 2y^2$, $\frac{dy}{dt} = t - xy$ given $x(0) = 5$, $y(0) = 2$ to find $x(1)$ and $y(1)$ correct to six significant figures (6S).

Solution

After loading the utility file using **File Load Utility** ode_appr.mth the DERIVE command is

$$\text{RK}([t^2 - x + 2y^2, t - xy], [t,x,y], [0,5,2], h, n)$$

where values of h and n have to be substituted.

Use **Simplify Approximate** and choose 8 **Digits** for **Precision**. DERIVE gives the following estimates of $x(1)$ and $y(1)$.

Step size h	Number of steps n	$x(1)$	$y(1)$
0.1	10	2.4953368	0.28367342
0.05	20	2.4948414	0.28368220
0.02	50	2.4948140	0.28368287
0.01	100	2.4948134	0.28368288

h was reduced until two successive estimates of $x(1)$ and $y(1)$ agree to 6S, and we are satisfied there will be no further change. With $h = 0.02$ and $h = 0.01$ the method gives

$$x(1) = 2.49481 \quad \text{and} \quad y(1) = 0.283683 \qquad \text{(6S)}$$

The EXTRACT_2_COLUMNS command was used twice to obtain (t,x) and (t,y) pairs (for $h = 0.01$). The plots of x and y against t are shown in Figure 3.21.

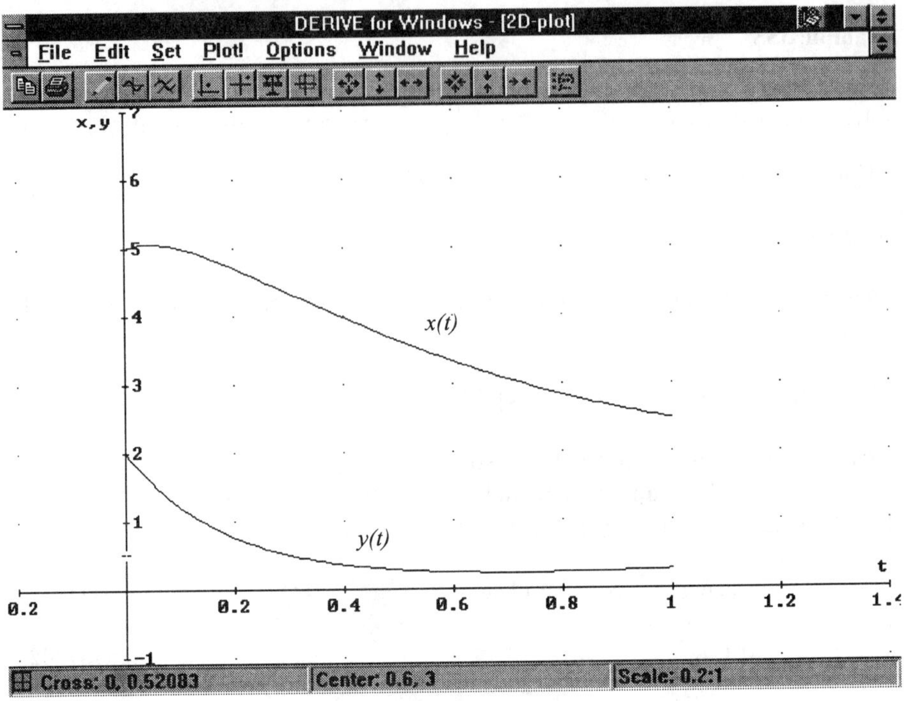

Figure 3.21

Exercises 3C

1. Solve the following difference equations and write down the first five terms of each sequence.

   (i) $u_{r+1} = 5u_r + 20,\ u_0 = 1.$

   (ii) $u_{r+1} = u_r - 5,\ u_0 = 50.$

   (iii) $u_{r+1} = \frac{5}{2}u_r - u_{r-1},\ u_0 = 8,\ u_1 = 7.$

2. Five differential equations, together with their general solutions and initial conditions, are given below. Which of them are (a) absolutely, (b) relatively ill-conditioned (inherently unstable) for numerical solution?

	Equation	General solution	Initial condition
(i)	$\frac{dy}{dx} = y - 1$	$y = Ae^{x} + 1$	$y(0) = 1$.
(ii)	$\frac{dy}{dx} = 2 - y$	$y = 2 - Ae^{-x}$	$y(0) = 2$.
(iii)	$\frac{dy}{dx} + 2xy = 1 - 2x^2$	$y = Ae^{-x^2} + x$	$y(0) = 0$.
(iv)	$\frac{dy}{dx} - 2xy = 1 - 2x^2$	$y = Ae^{x^2} + x$	$y(0) = 0$.
(v)	$\frac{dy}{dx} = 4x + 2y$	$y = Ae^{2x} - 2x - 1$	$y(0) = 1$.

3. Obtain the first order difference equation when each of the following methods is applied to $y' = \lambda y$ and hence find values of h for which each method is stable. (i) Trapezoidal method, (ii) Runge-Kutta second order method, (iii) Modified Euler method.

4. Consider (a) Simpson's method $Y_{r+1} = Y_{r-1} + \frac{1}{3}h(Y'_{r-1} + 4Y'_r + Y'_{r+1})$, (b) the Adams-Bashforth method $Y_{r+1} = Y_r + \frac{1}{2}h(3Y'_r - Y'_{r-1})$ applied with $h = 0.05$ to (i) $y' = 6y$, (ii) $y' = -6y$. In each case obtain and solve the second order difference equation, identify the spurious solution and determine its effect.

5. Consider the method $Y_{r+1} = Y_{r-1} + 2hY'_r$. Show that the application of this method to $y' = \lambda y$ leads to a recurrence relation with solution

$$Y_n = A(\lambda h + \sqrt{1+\lambda^2 h^2})^n + B(\lambda h - \sqrt{1+\lambda^2 h^2})^n$$

Show that the method is unstable, for any h, when $\lambda < 0$.

6. Show that the application of Simpson's method to $y' = \lambda y$ results in the difference equation

$$\left(1-\frac{1}{3}\lambda h\right)Y_{r+1} - \frac{4}{3}h\lambda Y_r - \left(1+\frac{1}{3}\lambda h\right)Y_{r-1} = 0$$

Solve this equation and show that, as $h \to 0$,

$$Y_n \approx A(e^{\lambda h})^n + B(-e^{-\lambda h})^n$$

Deduce that the method is stable, for any h, when $\lambda > 0$.

7. Consider (a) $y' = e^x - 5y + \cos y$, $y(0) = 1$ (b) $y' = x^2 - xy$, $y(1) = 1$ over the interval $1 \le x \le 10$, (c) $y' = \dfrac{6\sqrt{x}}{y} - y$, $y(0) = 1$ over the interval $0 \le x \le 2$, where it is known that $1 \le y \le 2$.

   (i) In each case, and for each numerical method in the Summary of Results, find the values of h for which the method is stable.

   (ii) In (b) estimate the largest value of x for which the method is stable if $h = 0.1$.

8. Use the RK command in DERIVE to solve the following systems of equations.

   (i) $\dfrac{dx}{dt} = e^t - \dfrac{x}{y}$, $\dfrac{dy}{dt} = x^2 - \sqrt{y} - 4t$ given $x(0) = 1$, $y(0) = 4$.
   Find $x(2)$ and $y(2)$ correct to 5D.

   (ii) $\dfrac{dx}{dt} = y$, $\dfrac{dy}{dt} = z$, $\dfrac{dz}{dt} = t + x - yz$ given $x(0) = 1$, $y(0) = 2$, $z(0) = 3$.
   Find $x(1)$, $y(1)$ and $z(1)$ correct to 4D.

4

Applications of First Order Differential Equations

We saw in Chapter 1 how a mathematical model of a real world problem can result in a differential equation. The problems looked at were

1. Emptying the bath
2A. Exponential model of population growth
2B. Logistic model of population growth
3. Throwing a ball in the air
4. Car suspensions.

Applications 1, 2A, 2B and 3 resulted in first order differential equations which we are now in a position to solve. Application 4, car suspensions, resulted in a second order differential equation and its solution is discussed in Chapter 6.

A number of other problems which lead to first order differential equations, or to a set of such equations, will also be looked at. We will use DERIVE to solve the more difficult equations and to display the solutions. The topics include heating and cooling problems and the motion of a particle in one dimension.

4.1 EMPTYING THE BATH

In Application 1 of Chapter 1 we set up a model for calculating the time it takes for a bath of water to empty. The mathematical problem was shown to be:

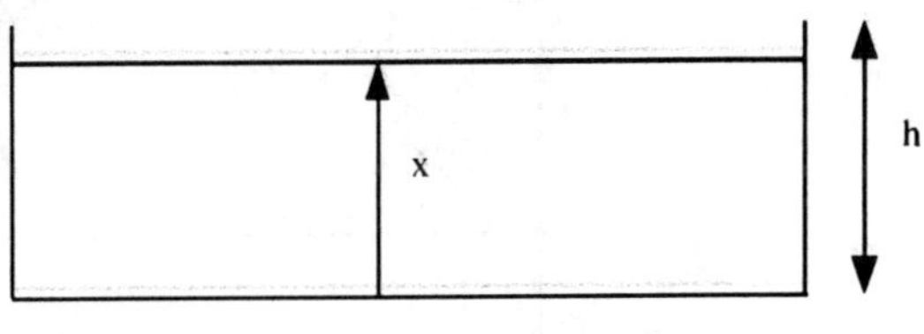

solve the equation $\dfrac{dx}{dt} = -k\sqrt{x}$ given $x = h$ when $t = 0$.

Solving the equation

File Load Utility file ode1.mth, **Declare Variables Domain**, enter each variable name and select **non-negative** and solve the equation using

DSOLVE1 $(k\sqrt{x}, 1, t, x, 0, h)$

Simplifying gives

$$2\sqrt{x} - 2\sqrt{h} + kt = 0 \tag{4.1}$$

solving algebraically for x gives, after removing the IF condition, $x = \dfrac{(2\sqrt{h} - kt)^2}{4}$.

The bath is empty at time T, when $x = 0$. Substituting these values gives $T = \dfrac{2\sqrt{h}}{k}$.

For a typical bath, suitable values are $h = 30$ cm, $k = 0.005\ m^{\frac{1}{2}} s^{-1}$ and then

$$x = \frac{(2\sqrt{0.3} - 0.005t)^2}{4} \qquad \text{and} \qquad T \approx 219 \text{ seconds}$$

Interpreting the solution

The graph of Figure 4.1 shows how the depth x of water varies with time t. The curve is a parabola with its minimum at time $t = T$ and is steepest at $t = 0$. This implies that, as might be expected, the fuller the bath is, the faster it is emptying.

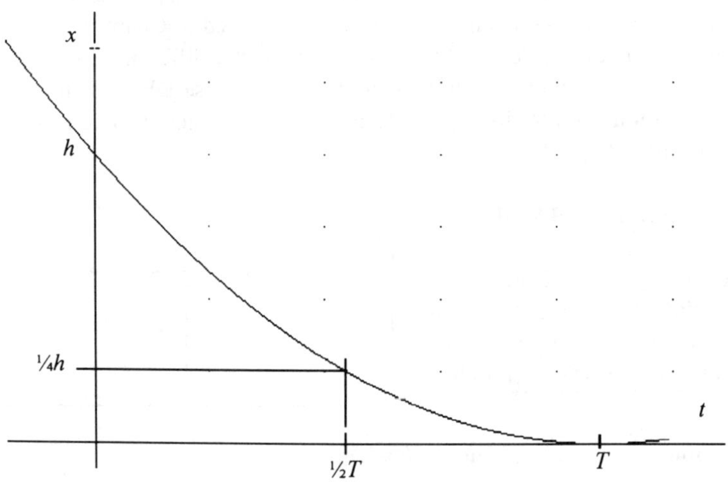

Figure 4.1 Depth x of water in bath against time t

(i) To find the depth x_1 of water when half the total time T has passed, substitute $t = \frac{1}{2}T = \dfrac{\sqrt{h}}{k}$ and obtain $x_1 = \frac{1}{4}h$.

In half the time three quarters of the water has gone.

(ii) To compare the times for the first and second halves of the bath to empty.
If $x = \frac{1}{2}h$ when $t = t_1$ then, from (4.1)

$$2\sqrt{\tfrac{1}{2}h} = -kt_1 + 2\sqrt{h}\text{, from which } t_1 = \frac{(2-\sqrt{2})\sqrt{h}}{k}$$

The time for the second half of the bath to empty is

$$t_2 = T - t_1 = \frac{2\sqrt{h}}{k} - \frac{(2-\sqrt{2})\sqrt{h}}{k} = \frac{\sqrt{2h}}{k}$$

The ratio α of the two times is

$$\alpha = \frac{t_2}{t_1} = \frac{\sqrt{2h}}{k} \times \frac{k}{(2-\sqrt{2})\sqrt{h}} = \sqrt{2} + 1 \approx 2.4$$

The second half takes about 2.4 times as long to empty as the first half. Hence it takes about 29% of the total time to empty the first half of the bath and 71% to empty the second half.

□

Example 4A

The first model of the bath problem assumed that the sides of the bath were vertical whereas it is usual for the bath to be wider at the top than at the bottom.

(i) Derive a revised model based on the assumption that a vertical cross section of the bath across its width is a trapezium. The ends of the bath remain vertical.

(ii) Find the time T for the bath to empty.

(iii) Predict the ratio of the time for the depth x to decrease from $\frac{1}{2}h$ to 0 with that for x to decrease from h to $\frac{1}{2}h$.

(iv) Obtain the predictions of the model for a typical bath for which a = 46 cm, b = 57 cm and h = 30 cm.

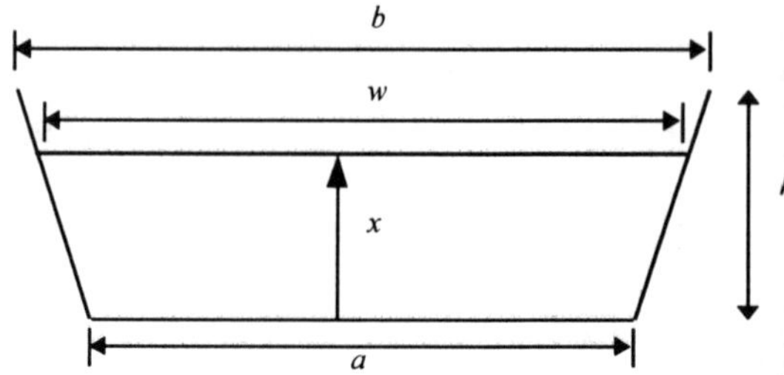

Solution

(i) Using the appropriate assumptions of the previous model the differential equation describing the situation is, as before,

$$-\frac{d}{dt}(Ax) = s\sqrt{2gx} \tag{4.2}$$

Whereas the cross sectional area was previously constant it will now vary with the depth x.

The width w of the bath at depth x is $w = a + \dfrac{(b-a)x}{h}$ and so the cross sectional area is

$$A = wl = \left[a + \frac{(b-a)x}{h}\right]l$$

where l is the length of the bath. Then, from (4.2),

$$-\frac{d}{dt}\left[alx + \frac{(b-a)lx^2}{h}\right] = s\sqrt{2gx}$$

This is our revised model of the problem.

(ii) Rearranging:

$$\left(\frac{a}{\sqrt{x}} + \frac{2(b-a)}{h}\sqrt{x}\right)\frac{dx}{dt} + c = 0\,, \quad \text{where } c = \frac{s\sqrt{2g}}{l}$$

To solve this equation given $x(0) = h$ use

$$\text{DSOLVE1}\left(c, \frac{a}{\sqrt{x}} + \frac{2(b-a)\sqrt{x}}{h}, t, x, 0, h\right)$$

Simplifying gives

$$\frac{4x^{3/2}(b-a)}{3h} + 2a\sqrt{x} - \frac{2a\sqrt{h} + 4b\sqrt{h} - 3ct}{3} = 0 \tag{4.3}$$

Substitute 0 for x, and T for t using **Simplify Substitute for Variables** and **Solve** for T to show that the time to empty the bath is $T = \dfrac{2\sqrt{h}(a+2b)}{3c}$.

(iii) Substituting $x = \frac{1}{2}h$, $t = t_1$ and **soLve** for t_1 to show that the time for the depth of water to decrease from h to $\frac{1}{2}h$ is $t_1 = \dfrac{\sqrt{2h}[(2\sqrt{2}-1)b - (2-\sqrt{2})a]}{3c}$. Similarly, the time t_2 for the depth to reduce from $\frac{1}{2}h$ to 0 is $t_2 = T - t_1 = \dfrac{\sqrt{2h}(2a+b)}{3c}$ and the ratio of these times is

$$\alpha = \frac{t_2}{t_1} = \frac{2a+b}{(2\sqrt{2}-1)b - (2-\sqrt{2})a}$$

Note that the value of α depends only on a and b and is independent of the depth h of water in the bath.

(iv) With $a = 46$ cm and $b = 57$ cm we find $\alpha \approx 1.93$. This means it will take about 34% of the total time for the depth to fall from h to $\frac{1}{2}h$, and 66% to fall from $\frac{1}{2}h$ to 0. Note that these are predictions for the height of water, not its volume. Since the cross section is a trapezium, when the depth of water has halved more than half of the water has escaped.

□

Exercise 4A

1. Water is pumped into a tank in the shape of a cone at a rate of $0.4\ \mathrm{m}^3$ per second. The cone has dimensions as shown.

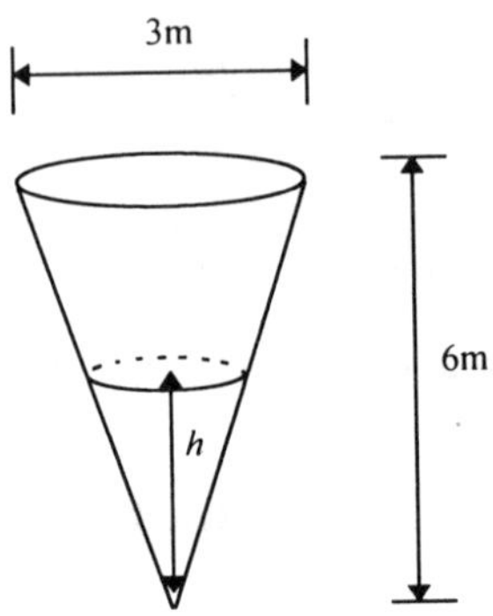

   (i) Show that the depth h of water at time t satisfies the differential equation $\pi h^2 \dfrac{dh}{dt} = 6.4$.

   (ii) Solve the differential equation for $h(t)$.

   (iii) How long does it take to fill the cone?

2. Water is contained in a tank that has a square base $2\mathrm{m} \times 2\mathrm{m}$ and height 4m. Initially the tank is full. Water leaves the tank through a hole at the bottom at a rate of $\sqrt{20h}$ litres per minute where h is the depth of water.

   (i) Show that the differential equation model for h is $\dfrac{dh}{dt} = -\sqrt{\dfrac{5h}{4}}$.

   (ii) Solve the differential equation for $h(t)$.

   (iii) How long does it take for the tank to empty?

3. A cylindrical tank has a circular cross section of radius 1 metre, a length of 4 metres and is initially full of oil. When the tap at the bottom of the tank is opened oil flows out at a rate of $100\sqrt{h}\ \mathrm{m}^3$ per minute, where h is the depth of oil.

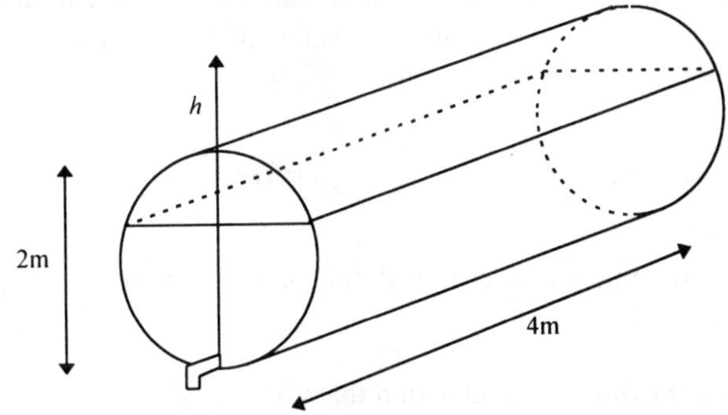

Show that the height h of oil at time t satisfies the equation

$$\frac{dh}{dt} = -\frac{2}{25\sqrt{2-h}}$$

and determine how long it takes to empty the tank.

4. A tank is filled with 400 litres of water containing 40 kg of salt. Brine containing 1.6 kg of salt per litre is pumped into the tank at a rate of 4 litres per minute. The mixture is stirred to ensure good mixing and leaves the tank at the rate of 4 litres per minute.

(i) Show that the concentration x kg of salt per litre at time t satisfies the equation $\frac{dx}{dt} = 6.4 - \frac{x}{100}$ and find the concentration, in kg per litre, of salt in the tank after 10 minutes.

(ii) A hole appears in the base of the tank so that an extra 1 litre per minute leaks from the tank. What will be the concentration of salt in the tank after a further 10 minutes?

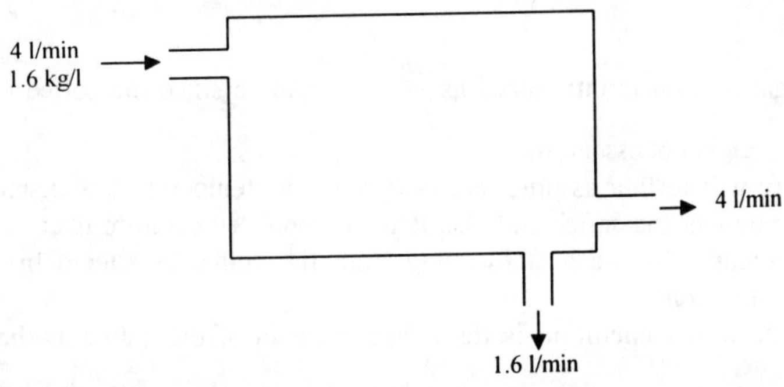

5. To control a cancer, an implant in a body releases a chemical into the blood in a periodic way, so that the rate of change of the level of the chemical is modelled by the differential equation

$$\frac{dx}{dt} = a - b\cos\frac{\pi t}{12} - kx \qquad x(0) = x_0$$

where x is the amount of chemical in the blood at time t and a, b, k are constants.

(i) Solve the differential equation for $x(t)$.

(ii) Sketch a graph for x for the case $a = b = 1$, $k = 2$ and $x_0 = 8$.

4.2 HEATING AND COOLING

Consider a freshly made cup of coffee on a dining room table. We may be interested in answering questions such as:

what will the temperature be after 10 minutes?
how long will it take for the temperature to drop so that it is cool enough to drink?
how long do I have to drink the coffee before it becomes too cold to enjoy?

The most important assumption to be made in setting up an initial model of the situation is to assume that Newton's law of cooling applies. This states that the rate of change of the temperature of the coffee is proportional to the difference between the temperature of the coffee and that of the room.

If y is the temperature of the coffee at time t and R is the room temperature then we obtain

$$\frac{dy}{dt} \propto y - R \quad \text{or} \quad \frac{dy}{dt} = -k(y-R) \tag{4.4}$$

(The negative sign is introduced as $\dfrac{dy}{dt}$ is negative since the coffee is cooling; this negative sign is not essential).

We will further assume that as it cools the temperature is instantaneously the same throughout the coffee and also that the room temperature is constant (so that R is not increased by the heat escaping from the coffee or altered by switching the central heating on).

The initial condition is the temperature y_0 of the coffee at the start; that is $y(0) = y_0$.

Solving the equation

The required command is DSOLVE1($ky - kR$, 1, t, y, 0, y_0). **Simplify** and **soLve** for y to give $y = e^{-kt}(y_0 - R) + R$.

(i) To find the temperature y_1 at time t_1, **Substitute** t_1 for t to obtain $y_1 = e^{-kt_1}(y_0 - R) + R$.

(ii) To find the time t_2 for the temperature to drop to y_2, **Substitute** y_2 for y, t_2 for t and **soLve** for t_2; this gives

$$t_2 = \frac{1}{k}\ln(y_0 - R) - \frac{1}{k}\ln(y_2 - R)$$

which is equivalent to $t_2 = \frac{1}{k}\ln\frac{y_0 - R}{y_2 - R}$.

Applying the solution

(iii) Consider the special case $R = 20°C$, $k = 0.04\ \text{min}^{-1}$, $y_0 = 100°C$. Then

$$y = 80e^{-0.04t} + 20$$

The temperature after 10 minutes is $y_1 = 80e^{-0.4} + 20 \approx 74°$. The time for the temperature to drop to 40° is

$$\frac{1}{0.04}\ln\left(\frac{100-20}{40-20}\right) = 25\ln 4 \approx 35 \text{ minutes}$$

The graph in Figure 4.2 shows how y varies with t. As time increases $e^{-kt} = e^{-0.04t}$ decreases towards 0 and y approaches the room temperature R as expected. The graph of y against t will be similar to that in Figure 4.2 whatever the numerical values of R, k and y_0.

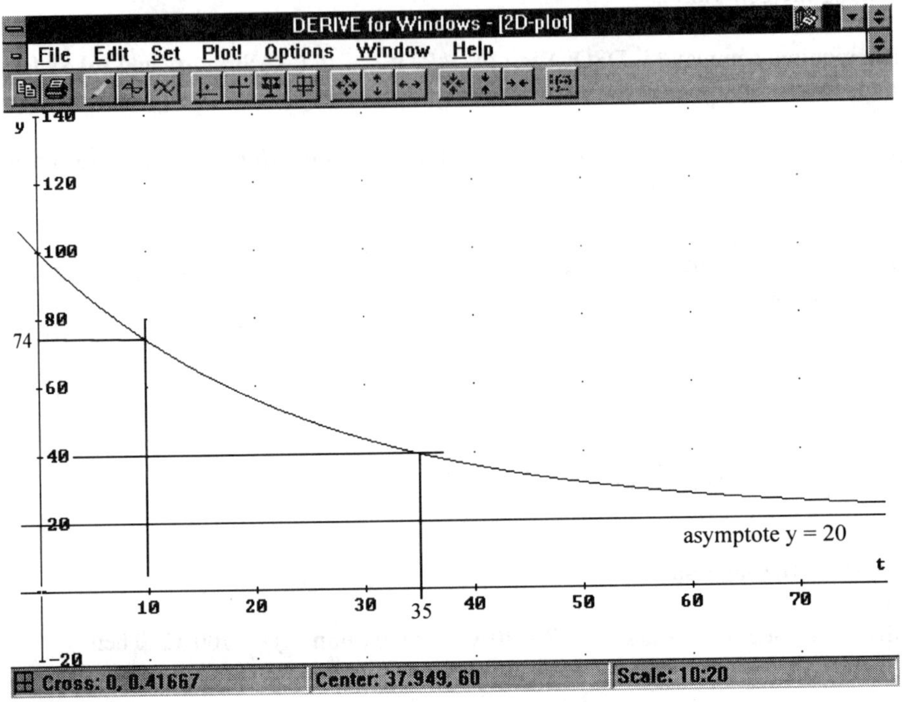

Figure 4.2 Graph of $y = 80e^{-0.04t} + 20$

Equation (4.4) can be adapted to other situations. Suppose, for example, that we are interested in how the temperature inside a house varies with time. We may then need to take into account

(i) a variation in the outside temperature R throughout the day and/or night. That is R need no longer be constant and $R = R(t)$;

(ii) the use of central heating in winter or air conditioning in summer. This will introduce a term $H(t)$ into the equation. $H(t)$ will be positive for central heating and negative for air conditioning.

Adjusting (4.4) to account for (i) and (ii):

$$\frac{dy}{dt} = -k[y - R(t)] + H(t)$$

Rearranging:

$$\frac{dy}{dt} + ky = kR(t) + H(t) \tag{4.5}$$

This is a first order linear equation with integrating factor $e^{\int k dt} = e^{kt}$. Then

$$e^{kt}\frac{dy}{dt} + ke^{kt}y = e^{kt}(kR(t) + H(t))$$

$$\frac{d}{dt}(ye^{kt}) = e^{kt}(kR(t) + H(t))$$

Integrating:

$$ye^{kt} = \int e^{kt}(kR(t) + H(t))dt + C$$

or $$y = e^{-kt}\left\{\int e^{kt}(kR(t) + H(t))dt + C\right\}$$

The form of the temperature variation y depends on the value of k as well as the expressions for $R(t)$ and $H(t)$. An example is given below.

Example 4B

For a time in the winter the temperature outside a house varies between 0°C at midnight and 8°C at midday. Investigate how the temperature inside the house varies under the following assumptions.

(i) The outside temperature varies sinusoidally with period 24 hours.

(ii) From midnight on a particular day, when the temperature outside and inside the house is zero, a constant heat source $H(t) = 1$ is supplied to the house.

(iii) The constant k has the value 0.1.

Solution

The function modelling the outside temperature is sinusoidal with mean value 4 and amplitude 4. If the minimum is zero, at midnight (t = 0), then it takes the form

$R(t) = 4 - 4\cos \omega t$. If the period T is 24 hours then $T = \dfrac{2\pi}{\omega} = 24$, so that $\omega = \dfrac{\pi}{12}$. Hence

$$R(t) = 4 - 4\cos\frac{\pi t}{12}$$

With $H(t) = 1$ and $k = 0.1$ equation (4.5) becomes

$$\frac{dy}{dt} + 0.1y = 1.4 - 0.4\cos\frac{\pi t}{12}$$

and the initial condition is $y(0) = 0$. **Simplifying**

$$\text{DSOLVE1}\left(0.1y - 1.4 + 0.4\cos\frac{\pi t}{12}, 1, t, y, 0, 0\right)$$

and then **Solving** for y with **Declare Algebra State Simplification Trigonometry** set to collect gives, to 2D,

$$y = -13.49e^{-0.1t} - 1.43\sin(0.26t + 0.36) + 14$$

The negative exponential quickly becomes very small and then y varies sinusoidally with mean 14°C and amplitude 1.43°C, that is between 12.57°C and 15.43°C, and period 24 hours. Note that the sinusoidal variation inside the building lags that outside by $\dfrac{1.2059}{\pi/12} \approx 4.6$ hours.

Figure 4.3 shows graphs of the inside and outside temperatures over a period of 4 days.

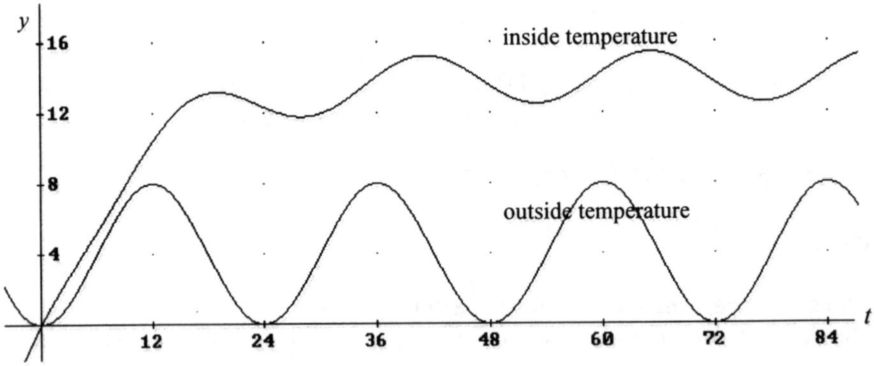

Figure 4.3 Graphs of temperature against time in Example 4B □

Exercise 4B

1. The temperature inside a garage with no means of heating satisfies Newton's law of cooling with constant $\frac{1}{3}\,\text{hr}^{-1}$. The air temperature outside the garage varies according to a sine wave with a minimum of 6°C at 4 am and a maximum of 20°C at 4 pm. Show that the temperature y at time t hours from midnight satisfies the equation $\dfrac{dy}{dt}+\dfrac{1}{3}y=\dfrac{13}{3}+\dfrac{7}{3}\sin\left(\dfrac{(t+14)\pi}{12}\right)$. Find the temperature in the garage as a function of time assuming that its temperature at midnight was 11°C. For how long in the day is the temperature above 18°C?

2. Stefan's law of radiation states that the rate of change of temperature of a body at T degrees Kelvin (0°C = 273K) in air at T_0 degrees Kelvin is given by

$$\frac{dT}{dt}=a(T_0^4-T^4)$$

where a is a positive constant.

(i) Solve this differential equation for $T(t)$.

(ii) Explain why Stefan's law and Newton's law are approximately the same model when T is close to T_0.

3. A motor under load generates heat at constant rate and radiates it at a rate proportional to the excess temperature T so that

$$\frac{dT}{dt}=10-kT$$

where t denotes time and k is a constant.

(i) Solve this differential equation for T given that T = 0°C initially and show that the ultimate rise in temperature is $(10/k)$°C.

(ii) After 10 minutes the rise in temperature is 50°C. Find the positive value of k.

4. In modelling the temperature inside an office, two assumptions are made:

   1. when the building is empty with the central heating turned off, the office cools according to Newton's law of cooling with proportionality constant k;

   2. during the day the people, lights and office machinery lead to a rate of increase of temperature $u(t)$;

   (i) Formulate a differential equation for the rate of change of temperature in the office.

   (ii) Find the temperature in an office for which $k = 0.5\text{hr}^{-1}$ during the summer months when the central heating is turned off. The outside temperature satisfies the equation

$$T_0 = 20 - 4\cos\frac{\pi t}{12}$$

   ($t = 0$ is 4 am) and $u = u_0$ (constant), $4 \le t \le 13$. Draw a graph for the room temperature $T(t)$ for various values of u_0. What is the maximum value of u if the temperature in the office is never to exceed 26°?

4.3 POPULATION MODELS

In Chapter 1 we discussed two models of population growth, the exponential model and the logistic model. Here we will solve the equations derived and show that they can sometimes describe real populations quite well. First of all we will look at graphs showing how some real populations have changed.

Certain of the data in these figures appear to suggest a constant population while others suggest a smooth change in population. For each graph think about the following points:

- What trends can you propose from the data?
- Do any populations appear constant?
- Do any populations appear to satisfy exponential growth?
- Which populations do you think will be difficult to describe mathematically? Why?

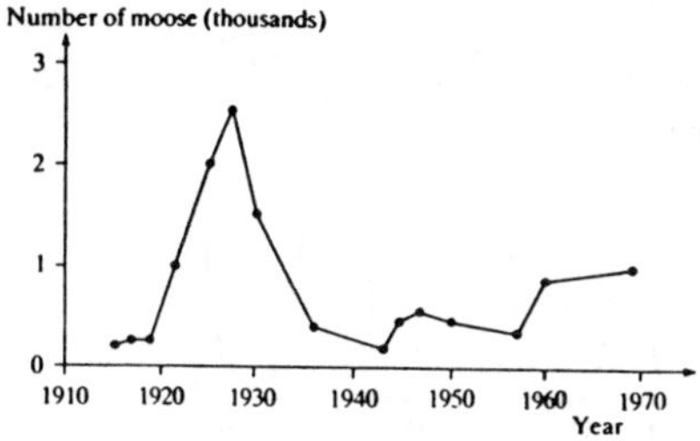

(a) Estimates of the population of moose on Isle Royale in Lake Superior. (Moose arrived on the island in about 1910.)

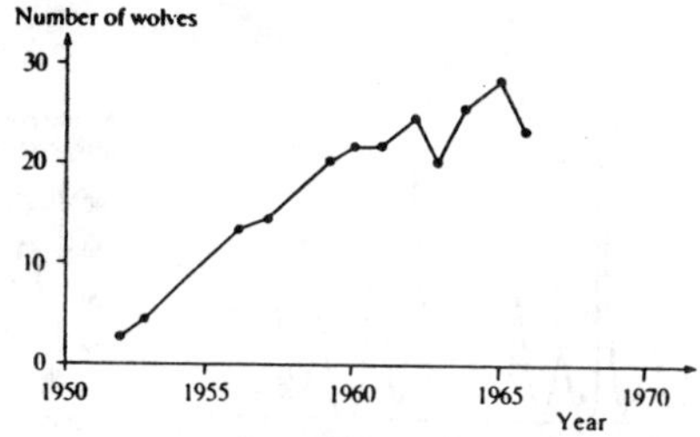

(b) Estimates of the number of wolves on Isle Royale. (Wolves arrived on the island in about 1948.)

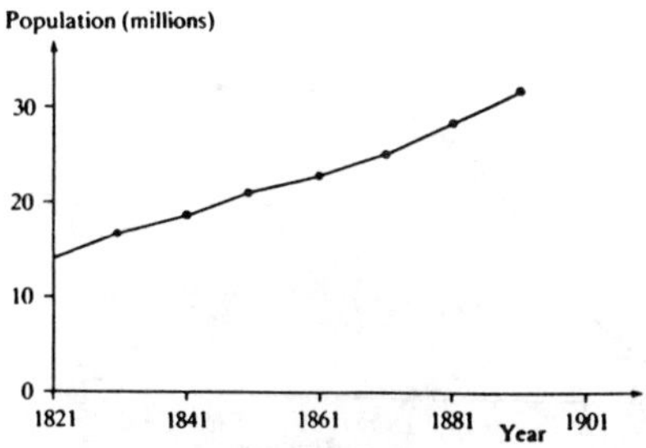

(c) The human population of Great Britain (1821-1891), taken to the nearest half million.

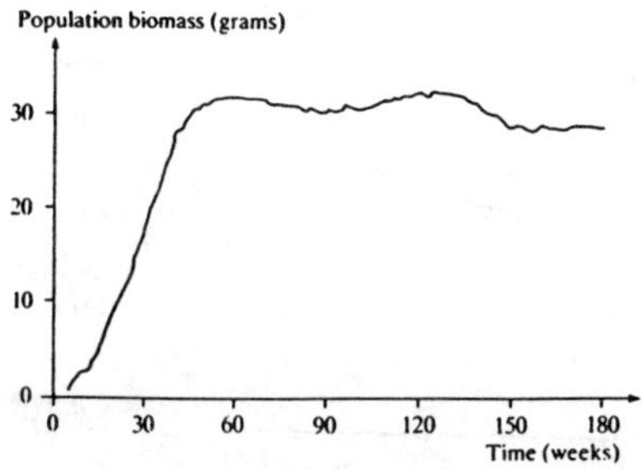

(d) The growth of a population of guppies in a laboratory aquarium (after Sulliman and Gulsell, 1958).

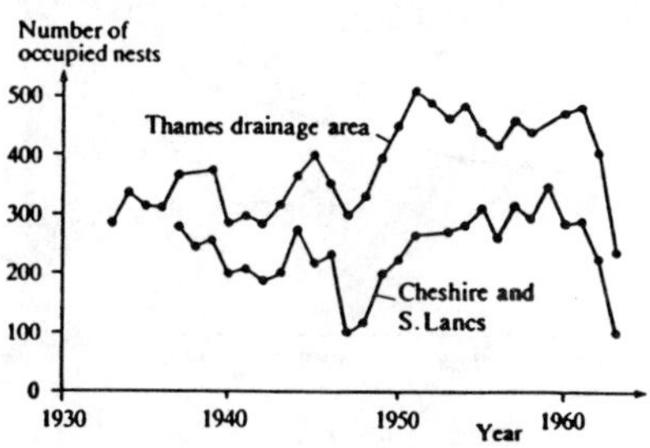

(e) Number of breeding pairs of heron (*Ardea anersa*) in two parts of England (1933-1963). Data from British Trust for Ornithology, analysed by J. Stafford (after Lack, 1966).

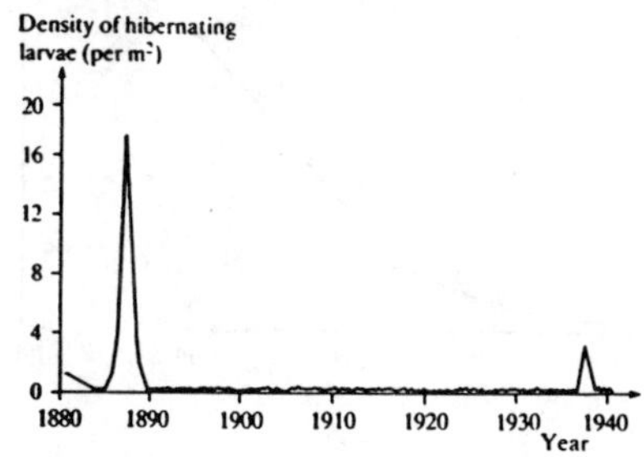

(f) The population density of a moth (*Dendrolimus*) in a forest at Letzlinger, Germany.

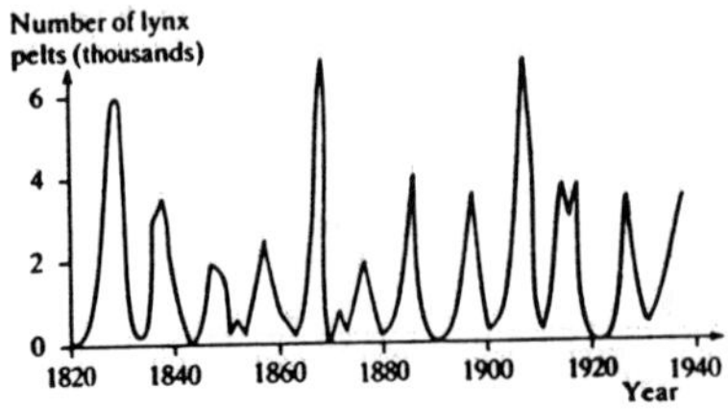

(g) Variations in the populations of lynx in the Hudson Bay area of Canada, estimated from the number of pelts sold to a company each year.

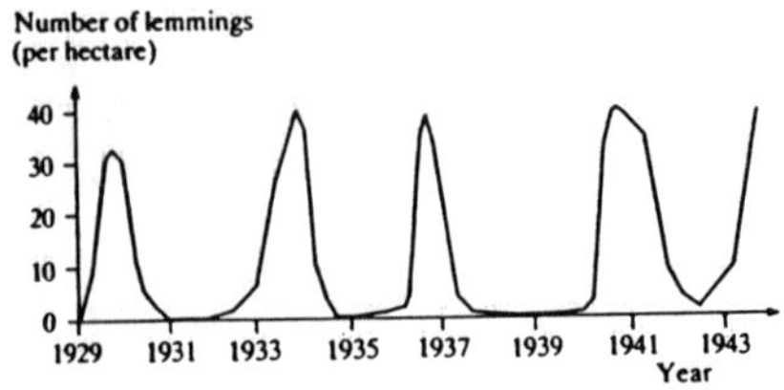

(h) The population density of lemmings in the area near Churchill, Canada.

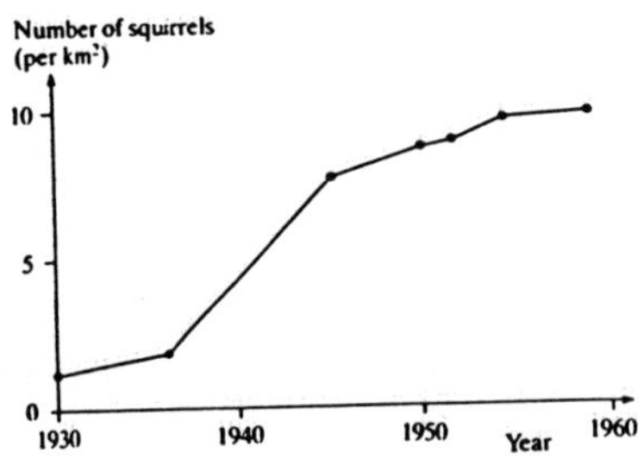

(l) The population density of the grey squirrel at selected sites in Great Britain

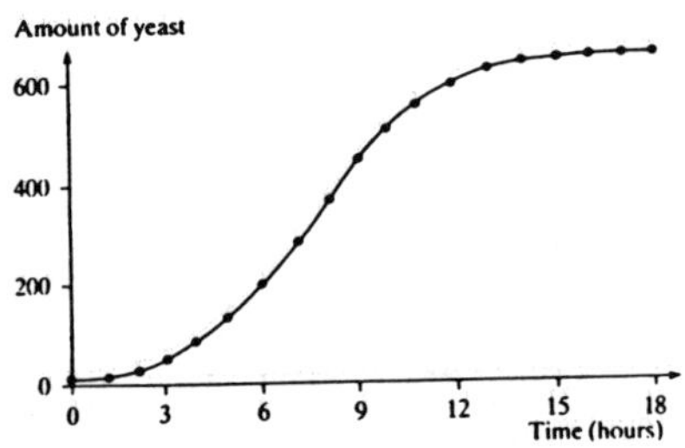

(j) Growth of a laboratory culture of yeast cells. Data from Carlson 1913 (after Pearl, 1927).

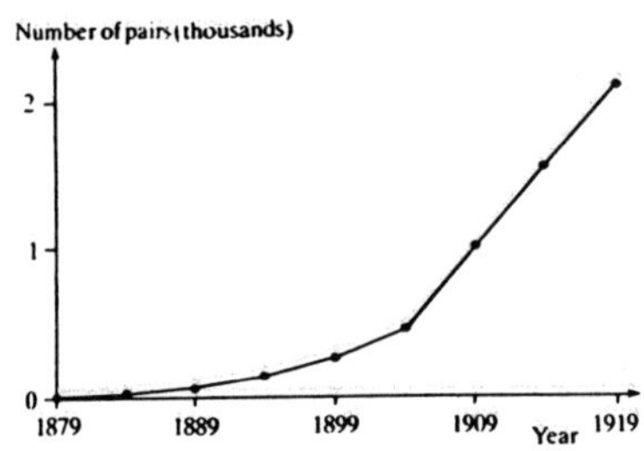

(k) The population of gannets at a colony at Cape St. Mary in the St. Lawrence, Canada.

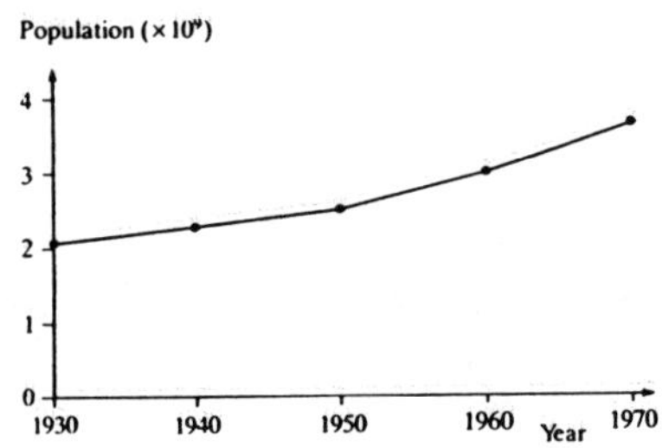

(l) The human population of the world (1930-1970).

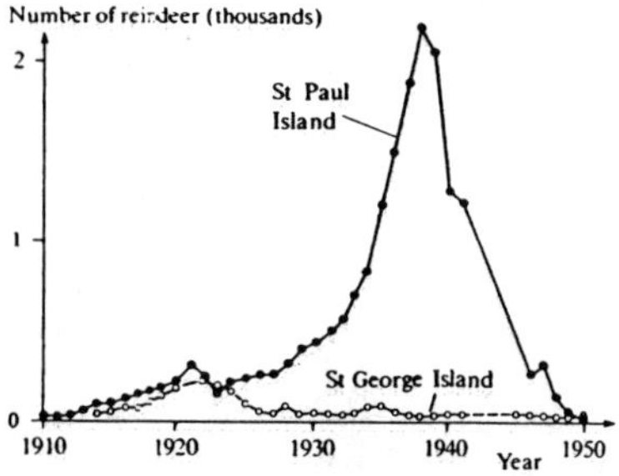

(m) Reindeer population on two of the Pribilof Islands in the Bering Sea, from 1911, when they were introduced, until 1950. (After Scheffer, 1951.)

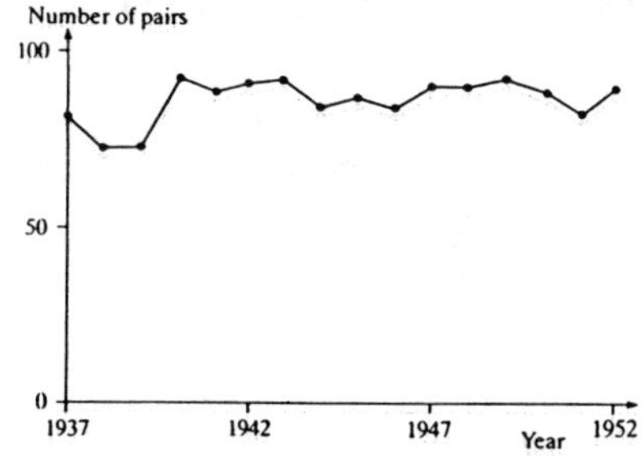

(n) Population of the yellow-eyed penguin (*Megadyptes antipodes*) in Dunedin, New Zealand.

We are interested in modelling populations and predicting what will happen in the future. Of the populations illustrated above:

1a appears to be decreasing towards zero;
2d and 2j seem to have reached a constant level;
2g follows regular patterns of increase and decrease;
2c, USA seem to be continually increasing.

The dangers of extrapolating to the future are demonstrated in 2m, where a prediction of the future population of reindeer on St Paul Island made in the late 1930's would almost certainly have been totally wrong.

Application 2A The Exponential Model

Under the assumptions of this model the population P at time t satisfies the equation $\dfrac{dP}{dt} = \alpha P$ (α constant). Separating the variables:

$$\frac{1}{P}dP = \alpha dt$$

Integrating:

$$\ln P = \alpha t + C \quad \text{or} \quad P = e^{\alpha t + C} = Ae^{\alpha t}$$

If the initial population is P_0, so that $P(0) = P_0$, then $A = P_0$ and

$$P = P_0 e^{\alpha t} \tag{4.6}$$

The predicted behaviour of the population depends on α.

If $\alpha > 0$ the population grows exponentially.
If $\alpha < 0$ the population decreases exponentially.
If $\alpha = 0$ the population remains constant.

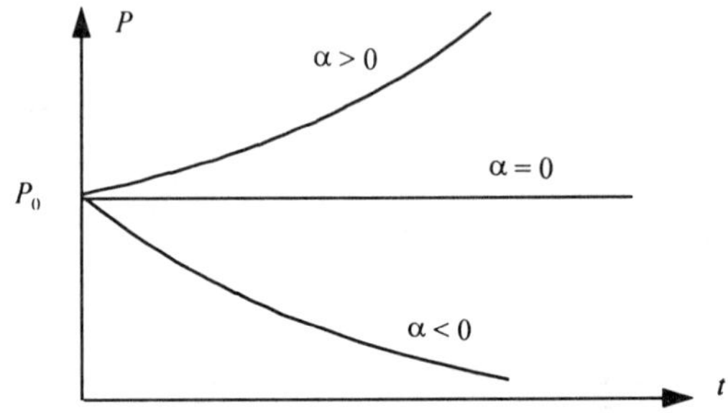

Example 4C

Censuses gave the population of the USA as 3.93 million in 1790 and 7.24 million in 1810. Use this data to predict the population of the USA in succeeding years and compare with the following, actual, values:

Year	1820	1840	1860	1880	1900	1930	1960	1990
Population	9.83	17.07	31.4	50.2	62.9	122.8	179.3	249.6

Solution

Taking $t = 0$ in 1790, so that $P(0) = P_0 = 3.93$, equation (4.6) becomes

$$P = 3.93e^{\alpha t}$$

Using $P = 7.24$ when $t = 20$: $7.24 = 3.93e^{20\alpha}$, from which $\alpha = 0.03055$. Hence our model is

$$P = 3.93e^{0.03055t}$$

We can now compare the predicted and actual populations:

Year	1820	1840	1860	1880
Actual population	9.64	17.07	31.4	50.2
Predicted population	9.83	18.10	33.4	61.4
Error	2%	6%	6%	22%

The model gives good predictions of the population up to 1860, where the error is a good deal less than 10%, but thereafter the error increases and the exponential model is no longer useful.

In selected years this century we find

Year	1900	1930	1960	1990
Actual population	62.9	122.8	179.3	249.6
Predicted population	113.2	283.0	707.7	1769.5
Error	80%	130%	295%	609%

The model is now quite useless, predicting a population in 1990 of almost 7 times its actual value!

□

Comparing real data with the Exponential Model

According to the exponential model the population P at time t is

$$P = P_0 e^{\alpha t}$$

Taking logs to the base e

$$\ln P = \ln(P_0 e^{\alpha t}) = \alpha t + \ln P_0$$

This is of the form $y = mx + c$ and so a graph of $\ln P$ (y) against t (x) will be linear with gradient α and intercept $\ln P_0$. Thus if, for a given population, a plot of $\ln P$ against t is approximately linear then the exponential model is a good one for the population.

Example 4D

The number of breeding pairs of collared doves was recorded.

Year	1955	1956	1957	1958	1959	1960	1961	1962	1963	1964
t = Year – 1955	0	1	2	3	4	5	6	7	8	9
Population P	4	16	45	100	205	675	1900	4650	10200	18855

Show that the data fits an exponential model and find its equation.

Solution

The graphs of P against t and $L = \ln P$ against t are shown in **Plot!**s 1 and 2 of Figure 4.4.

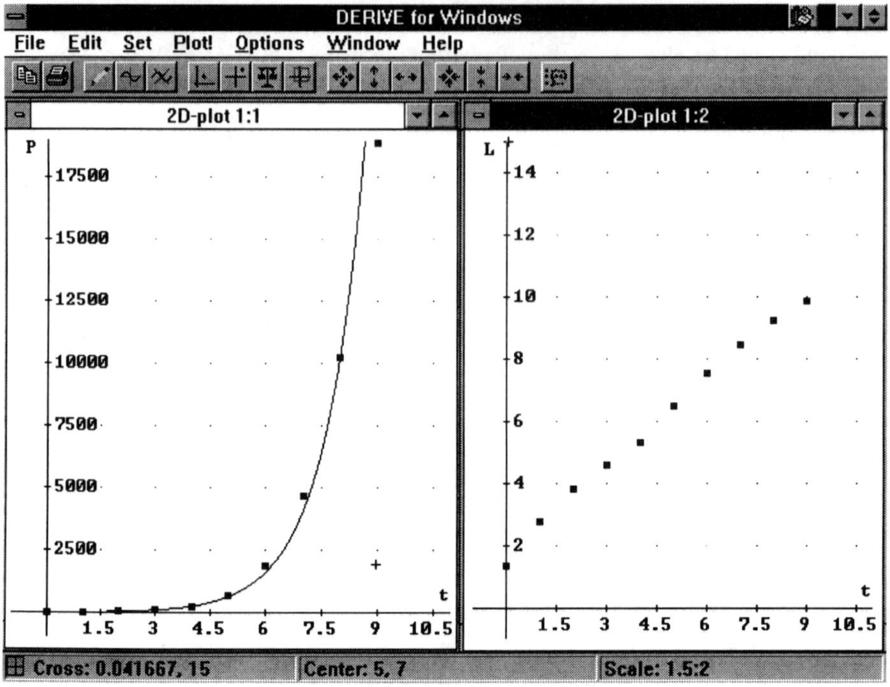

Figure 4.4 Graphs for Example 4D

The data is well approximated by a straight line, showing that the exponential model is a good fit. By taking two points on the line (or, better, using least squares) we find

slope of line = α = 0.93
intercept = $\ln P_0 = 1.8$, giving $P_0 = e^{1.8} \approx 6.05$

Hence the model is $P = 6.05e^{0.93t}$, where t is time in years from 1955.

Window 2 of Figure 4.4 shows the actual population values and the approximating function $P = 6.05e^{0.93t}$.

For an increasing population the exponential model predicts unlimited growth. While some populations correspond to this model fairly well for limited periods, the predicted long term behaviour, when the population becomes infinitely large, is not reasonable. Intuition says that there must come a time when crowding and a shortage of food and water coupled with increased disease will play a part and prevent further exponential increase. Application 2B in Chapter 1, the logistic model, tries to address these concerns.

Application 2B The Logistic Model

Under the assumptions of this model the population P at time t satisfies the equation $\frac{dP}{dt} = aP\left(1 - \frac{P}{M}\right)$ where a and M are constants and M is the equilibrium population level. (See Chapter 1, equation 1.3.)

This is a first order equation which can be solved analytically either by the separation of variables or as a Bernoulli equation. To take the hard work out of it we will use the DERIVE command

$$\text{DSOLVE1}\left(-aP\left(1 - \frac{P}{M}\right), 1, t, P, 0, P_0\right)$$

where P_0 is the initial population at $t = 0$.

After **Simplifying** and **Solving** for P this becomes

$$P = \frac{MP_0 \mathrm{e}^{at}}{P_0 \mathrm{e}^{at} + M - P_0} = \frac{M}{1 + \left(\frac{M}{P_0} - 1\right)\mathrm{e}^{-at}} \tag{4.7}$$

Figure 4.5 shows typical graphs of P against t in the case $M = 100$, $a = 0.1$. Plot 1 and Plot 2 show the effect of small (20) and larger (80) values of P_0; these graphs, and others, for $0 < P(0) < 100$ have the same basic shape, all converging to M as t increases.

Plot 3 shows that an initial population greater than the equilibrium level M (here $P_0 = 200$) decreases towards M.

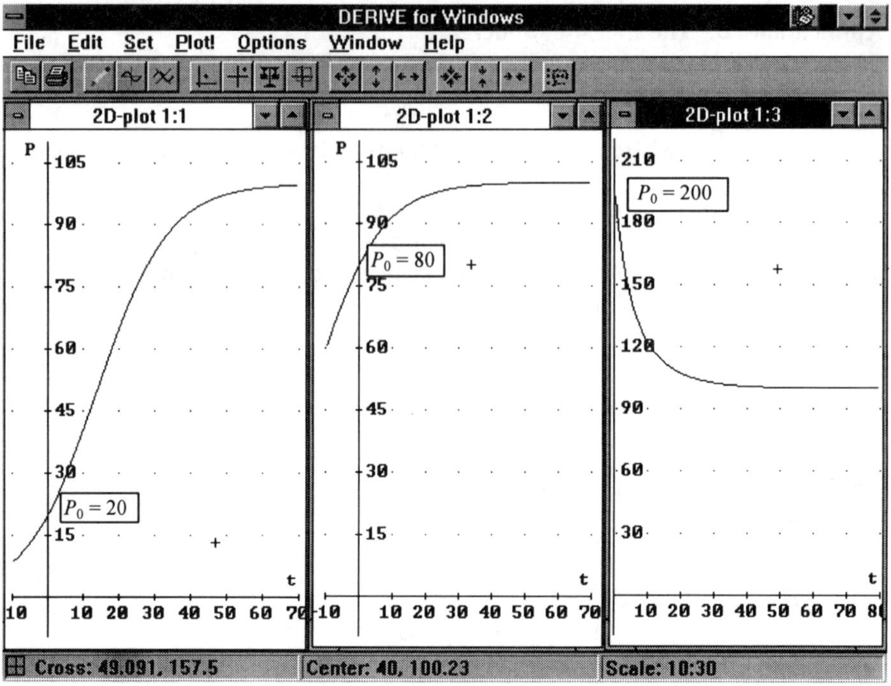

Figure 4.5 Typical solution curves for the logistic model

□

Comparing real data with the Logistic Model

To fit a logistic curve to a population we need to be able to estimate the parameters P_0, M and a. This can be done in a number of ways.

(i) Use the known population at three different times to obtain and solve three equations in P_0, M and a. (This does, however, use only some, not all, of the available data, possibly distorting the results.)

(ii) If the data follows the logistic model and the population is known at equal time intervals, so that $P(0) = P_0$, $P(t_1) = P_1$, $P(2t_1) = P_2$, $P(3t_1) = P_3$, … then a plot of $\dfrac{P_{n+1} - P_n}{P_n}$ against P_{n+1} will be approximately linear. Then M is the intercept on P_{n+1} axis. The intercept on the $\dfrac{P_{n+1} - P_n}{P_n}$ axis is $c = e^{a-1}$, from which $a = \ln(1 + c)$. (See Exercise 4C, problem 5.)

(iii) If we know or can estimate the equilibrium level M then it can be shown that for a population growing according to the logistic curve a graph of $y = \ln\left(\frac{M}{P} - 1\right)$ against t will be linear. The slope of the line will be $-a$ and the intercept on the y axis is $c = \ln\left(\frac{M}{P_0} - 1\right)$. (See Exercise 4C, problem 6.)

Example 4E

Fit a logistic curve to the following population data for the USA.

Year	1790	1830	1870
Population (millions)	3.93	12.87	39.82

Estimate the equilibrium population level M and compare the actual and predicted populations in 1910, 1950 and 1990.

Solution

With the time origin at 1790 the data is

Time t	0	40	80
Population P	3.93	12.87	39.82

As the initial population is $P_0 = 3.93$ the solution (4.7) becomes

$$P = \frac{M}{1 + \left(\frac{M}{3.93} - 1\right)e^{-at}}$$

Author the above **Expression**, then **Substitute** $P = 12.87$, $t = 40$ followed by $P = 39.82$, $t = 80$.

DERIVE will not successfully solve these equations as they stand so **Substitute** $z = e^{-40a}$ to produce a pair of equations in z. **soLve** one of the pair for M, substitute in the other and solve for z to give $z = 0.297517$. Back substituting then gives $a = 0.030307$ and $M = 351.942$. Hence, to 3S, the model of the population is

$$P = \frac{352}{1 + 88.6e^{-0.030t}}$$

The equilibrium population is about 352 million. This gives the following results:

Year	1910	1950	1990	2010
t = year – 1790	120	160	200	220
Actual population	92.0	151.2	246.6	?
Estimated population	105.6	207.8	291.7	316.3
Error	15%	37%	18%	

Figure 4.6 shows graphs of the actual and estimated populations.

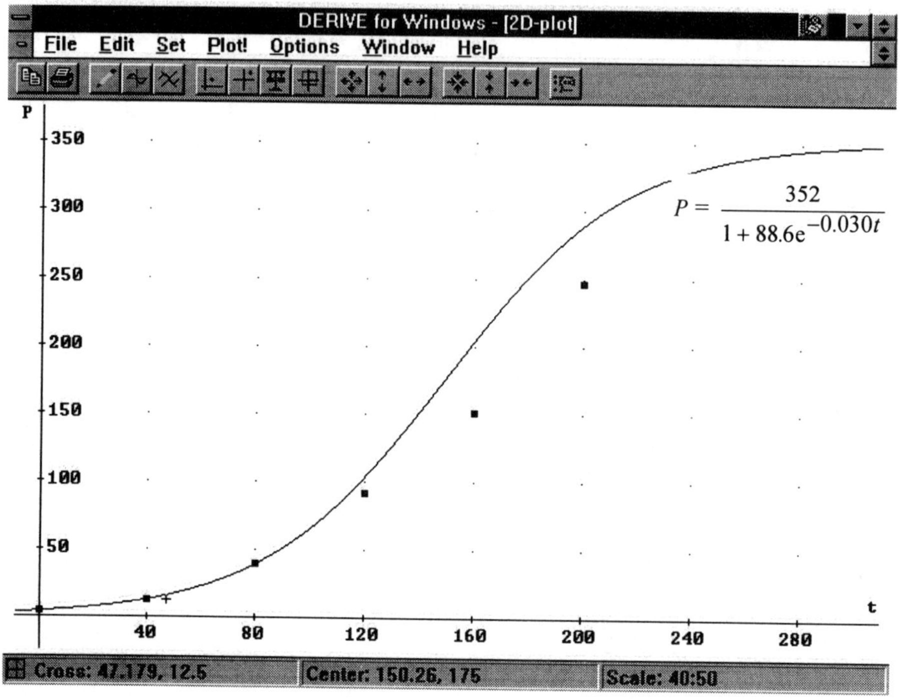

Figure 4.6 Estimated and actual populations in Example 4E

The above, and other, calculations show that the logistic model is a superior fit to the population than the exponential model, although in later years the model over estimates the actual population.

It should be noted that the above results are very dependent on which three years are chosen to estimate the parameters P_0, a and M, whereas other methods use all the data available.

□

Example 4F

The following table gives the growth of a yeast population (Carlson, 1913).

hours t	0	1	2	3	4	5	6	7	8	9	10	11	12	13	14	15	16	17	18
population P	9.6	18.3	29.0	47.2	71.1	119	175	257	351	441	513	560	595	629	641	651	656	660	662

(i) Estimate M, plot a graph of $\ln\left(\frac{M}{P}-1\right)$ against t and hence find an equation for P.

(ii) Plot $\frac{P_{n+1}-P_n}{P_n}$ against P_{n+1} and hence find an equation for P.

Solution

A graph of P against t is shown in Plot 1 of Figure 4.7.

(i) The population seems to have virtually reached its equilibrium level and a reasonable value to choose for M is $M = 665$. Tabulating the $\left[t, \ln\left(\frac{M}{P}-1\right)\right]$ pairs

Hours	0	1	2	3	4	5	6	7	8	
$\ln\left(\frac{665}{P}-1\right)$	4.233	3.565	3.088	2.572	2.123	1.523	1.050	0.462	–0.111	–(
Hours	10	11	12	13	14	15	16	17	18	
$\ln\left(\frac{665}{P}-1\right)$	–1.216	–1.674	–2.140	–2.861	–3.285	–3.839	–4.289	–4.883	–5.397	

A graph of $L = \ln\left(\frac{665}{P}-1\right)$ against t is shown in Plot 2 of Figure 4.7.

The points lie very close to a straight line and the logistic model is clearly a good one for this population.

From the graph the slope of the line is –0.52. This equals $-a$ and so $a = 0.52$ and with $M = 665$ the model is

$$P = \frac{665}{1+\left(\frac{665}{9.6}-1\right)e^{-0.52t}} = \frac{665}{1+68.27e^{-0.52t}}$$

(ii) A plot of $Q = \dfrac{P_{n+1} - P_n}{P_n}$ against P_n is shown in Plot 3 of Figure 4.7. The points lie reasonably close to a straight line, except that earlier points are rather scattered (small deviations from the logistic form are exaggerated by this method of plotting). The intercept on the vertical axis is about 0.7 so that $a = \ln 1.7 = 0.53$. The intercept on the horizontal axis is about 665, so that $M = 665$. Hence

$$P = \frac{665}{1 + 68.27\mathrm{e}^{-0.53t}}$$

which is similar to the equation in (i).

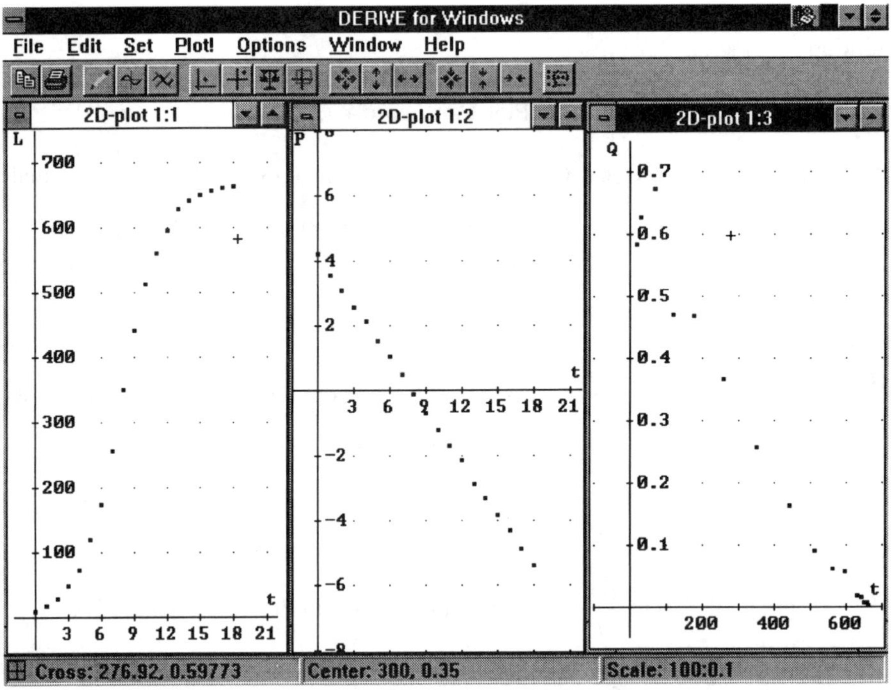

Figure 4.7 Graphs from Example 4F

□

Example 4G

The number of trout in a lake is approximately constant. Prior to this situation the population increased from the initially low level of 100. When the population was 1000 the proportionate birth rate was 55% per year and the proportionate death rate was 25% per year. When the population was 2000 the proportionate birth and death rates were 40% and 30% per year respectively.

A model of the population is based on the following assumptions:

(a) there is no migration and no fishing;
(b) the proportionate birth rate is a decreasing linear function of the population;
(c) the proportionate death rate is an increasing linear function of the population.

Show that a model based on these assumptions and the above data predicts that the population grows according to the logistic equation and find the current stable population size. [The investigation into the trout population is continued in Example 4H.]

Solution

(i) As there is no migration or fishing the population will only change due to births and deaths. Then

Proportionate growth rate	=	Proportionate birth rate	–	Proportionate death rate

$$\frac{1}{P}\frac{dP}{dt} = B - D \qquad (4.8)$$

From assumption (b), $B = a - bP$.
From assumption (c), $D = c + dP$.

Using the given data for births

$$0.55 = a - 1000b$$

$$0.25 = a - 2000b$$

with solution $a = 0.7$ and $b = 0.00015$ and so $B = 0.7 - 0.00015P$.

For deaths

$$0.25 = c + 1000d$$

$$0.30 = c + 2000d$$

with solution $c = 0.2$ and $d = 0.00005$ and so $D = 0.2 + 0.00005P$. Then from (4.8):

$$\frac{1}{P}\frac{dP}{dt} = (0.7 - 0.00015P) - (0.2 + 0.0005P)$$

$$\frac{1}{P}\frac{dP}{dt} = 0.5 - 0.0002P = 0.5\left(1 - \frac{P}{2500}\right)$$

This is the logistic equation with $a = 0.5$ and $M = 2500$ and has solution, using $P_0 = 100$, $P = \dfrac{2500}{1 + 24e^{-0.5t}}$.

The current constant population is $M = 2500$.

The graph of P against t is the left hand curve in Figure 4.8.

□

Modelling the Exploitation of an Animal Population

We shall assume that an unexploited population (of trout, say) will grow according to the logistic equation. That is, its population satisfies the equation $\frac{1}{P}\frac{dP}{dt} = aP\left(1 - \frac{P}{M}\right)$.

Suppose the population is now exploited at a rate C per unit time. (C is the rate at which the trout are caught.)

Models for the catch C

We will assume that

(i) C is constant

or (ii) C is proportional to the population, so that $C = kP$.

As an example we shall return to the trout population in Example 4G.

Example 4H

The trout population in the lake has now reached its equilibrium level of 2500. Fishing for trout is now to be allowed, but clearly the population must not be driven to extinction by overfishing. Determine the effect on the future population of two alternative fishing regimes:

(i) 200 trout can be caught each year
(ii) 15% of the trout can be caught each year.

It is assumed that the birth and (natural) death rates are as before. In each case estimate the time for the population to get to within 5% of any new equilibrium level.

Sketch the graph of P against t in each case.

Solution

When fishing is allowed

Proportionate growth rate	=	Proportionate birth rate	–	Proportionate death rate	–	Proportionate fishing rate

$$\frac{1}{P}\frac{dP}{dt} = (0.7 - 0.00015P) - (0.2 + 0.00005P) - \frac{C}{P}$$

$$\frac{1}{P}\frac{dP}{dt} = 0.5 - 0.0002P - \frac{C}{P}$$

(i) When C = 200 the equation becomes $\frac{dP}{dt} = 0.5P - 0.0002P^2 - 200$. With $P_0 = 2500$ use

DSOLVE1(200 + 0.0002P^2 – 0.5P, 1, t, P, 0, 2500)

Simplifying and **soLving** for P gives $P = \frac{500(16e^{0.3t} - 1)}{4e^{0.3t} - 1}$.

As t increases P approaches $\frac{500 \times 16}{4} = 2000$, which is the new equilibrium population level M_1.

P has to reduce to 2100 to be within 5% of M_1.

Substitute 2100 for P and **Solve** for t to show that it takes about 4.62 years for the population to get within 5% of its new equilibrium level of 2000.

(ii) When $C = 0.15P$ the equation becomes

$$\frac{1}{P}\frac{dP}{dt} = 0.5 - 0.0002P - 0.15 = 0.35 - 0.0002P$$
$$= 0.35\left(1 - \frac{P}{1750}\right)$$

This is another logistic equation with new equilibrium level M_2 = 1750 and solution

$$P = \frac{1750}{1 + \left(\frac{1750}{2500} - 1\right)e^{-0.35t}} = \frac{1750}{1 - 0.3e^{-0.35t}}$$

We wish to find t when $P = M_2 + 5\%\, M_2 = 1837.5$.
Substituting and solving for t we find that $t = 5.26$.
When catching 15% of the population each year it takes 5.26 years for the population to drop to within 5% of its new equilibrium level of 1750.

(iii) The graphs of P against t are shown in Figure 4.8.
The left hand graph is the solution of Example 4G while the right hand curves are from (i) and (ii) above.

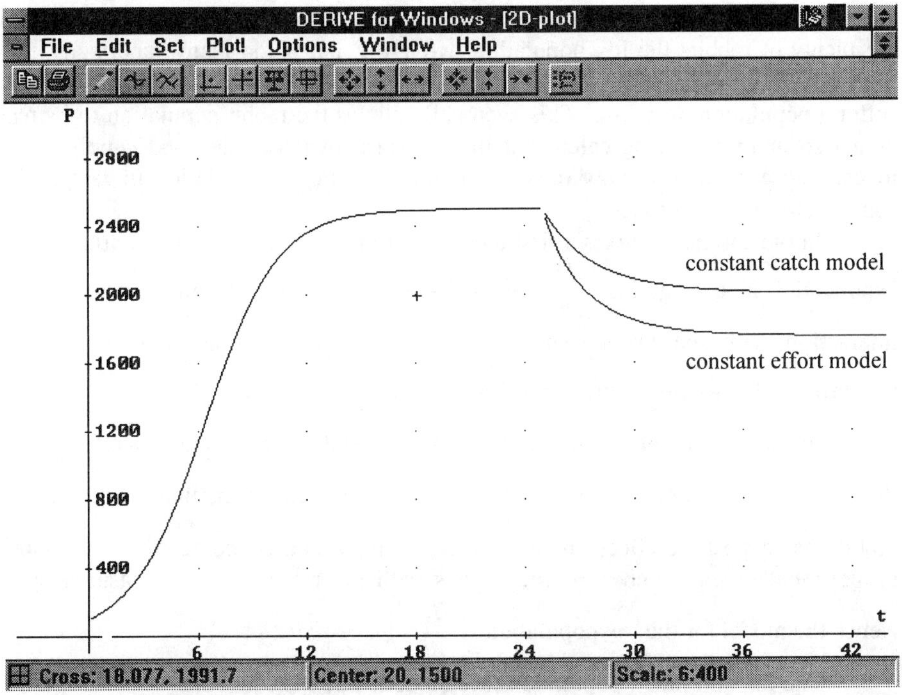

Figure 4.8 Graphs of P against t in Examples 4G and 4H

□

Interactions between populations

We used the exponential and logistic models to describe the growth of a single species, without reference to any other species. Here we will look at two populations which interact with each other. Examples are

(i) predator-prey systems in which one species (the prey) is the primary source of food for the other (the predators). For example, foxes eat rabbits, which live on vegetation;

(ii) systems where two species compete with each other for food and living space. An example is a lake stocked with trout and bass. Can they coexist, or will one of them dominate?

(iii) systems where species cooperate with each other. For example, bees use the nectar of a plant for food while simultaneously pollinating the plant.

Example 4I looks at a predator-prey system of foxes and rabbits. While there are plenty of rabbits the fox population will grow, but if too many rabbits are eaten their population will drop, and with it the food supply for the foxes; they then will suffer a population decrease. This eventually allows the rabbit population to increase as not so many are being eaten, but this, in turn, increases the food supply for the foxes, whose population now recovers, putting the rabbit population under pressure again. The cycle continues.

In the absence of foxes a first model for an increasing rabbit population x is the exponential model, giving $\frac{dx}{dt} = ax \quad (a > 0)$. This growth rate is reduced by interaction with the foxes (which eat them!); this reduction is assumed to be proportional to xy. Hence the rabbit equation is $\frac{dx}{dt} = ax - bxy \quad (b > 0)$.

In the absence of rabbits the foxes die out and the exponential model predicts that the fox population y satisfies $\frac{dy}{dt} = -cy \quad (c > 0)$. However, interaction with the rabbits has a positive effect on their growth rate, which is increased by an amount proportional to the number of interactions with the rabbits, again measured by xy. Hence the model for the fox population is $\frac{dy}{dt} = -cy + dxy \quad (d > 0)$.

The fox and rabbit populations are modelled by the equations

$$\frac{dx}{dt} = ax - bxy, \quad \frac{dy}{dt} = -cy + dxy \qquad (a,\ b,\ c,\ d > 0)$$

Example 4I illustrates the solution of these equations in a particular case.

Example 4I

Solve the equations $\frac{dx}{dt} = 4x - 0.02xy$, $\frac{dy}{dt} = -5y + 0.01xy$ given the initial conditions $x(0) = 1000$ (rabbits), $y(0) = 200$ (foxes).

Solution

The equations form a system of first order equations which will be solved numerically using the Runge-Kutta method, as described in Section 3.5 of Chapter 3. The DERIVE command is RK([4x – 0.02xy, –5y + 0.01xy], [t,x,y], [0,1000,200], 0.1, 60) which solves the equations using 60 steps of the classical Runge-Kutta method with step length 0.1. **Approximating** this gives a set of 60 sets of [t,x,y] values in line #2.

To plot values of y against x first use the command EXTRACT_2_COLUMNS(#2,2,3) to remove them from the [t,x,y] triples and then plot them; this graph is shown in Plot 1 of Figure 4.9. Plot 2 shows plots of x against t and y against t, these pairs of values being extracted from the $[t,x,y]$ triples in a similar way.

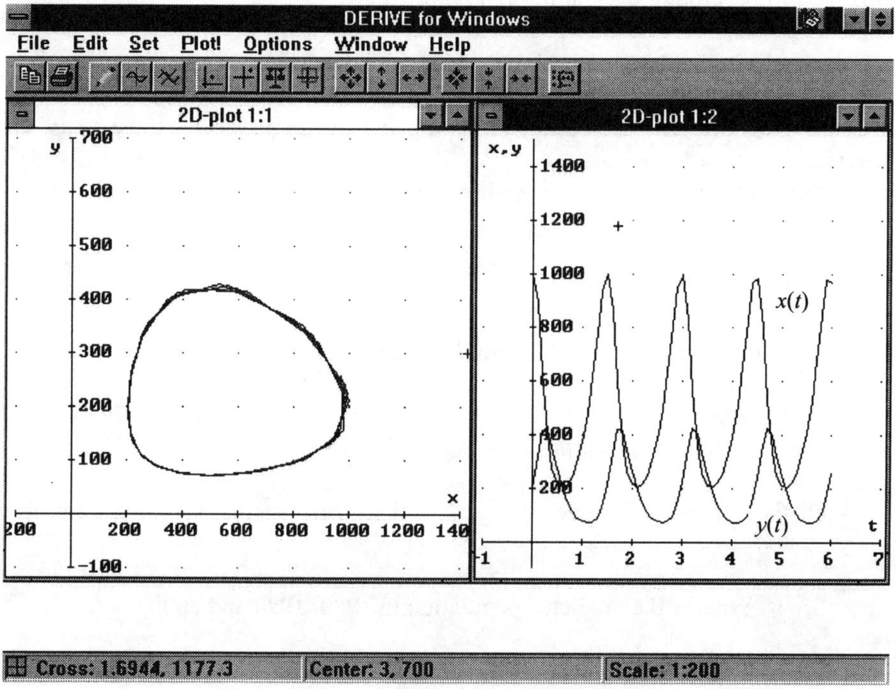

Figure 4.9 Solution to Example 4I

The graph of y against x in Plot 1 shows that the populations behave cyclically (the graph represents about 5 complete cycles). Plot 2 shows how the rabbits and foxes each vary with time. Each is periodic in nature, but with the rabbit population varying much more than the fox population. The changes in the fox population lag behind those of the rabbit population. Different initial conditions will give similar results to those above, but the equations also predict that given the right initial values the two populations will remain constant. This happens when $\frac{dx}{dt} = 0$ and $\frac{dy}{dt} = 0$ giving

$$4x - 0.02xy = 0 \quad \text{and} \quad -5y + 0.01xy = 0$$

from which the non zero solution is $x = 500$ and $y = 200$. Hence initial values $x(0) = 500$, $y(0) = 200$ will result in $x(t) = 500$, $y(t) = 200$ for all t. Use DERIVE to check this! The graph of y against x is called a phase plane diagram and the point

(500,200) an equilibrium point. The 'direction of flow' around the path is anti-clockwise.

Each cycle has four distinct regions, the boundaries of which are vertical and horizontal lines through the equilibrium point.

In A the foxes are increasing, forcing the rabbits to decrease.
In B both foxes and rabbits decrease.
In C the foxes are still decreasing becoming few in number, and the rabbits start recovering.
In D both rabbits and foxes are increasing.

□

Exercise 4C

1. The table gives the human population P of the world from 1930 to 1970.

Year	1930	1940	1950	1960	1970
Population (millions)	2070	2295	2485	2982	3635

Fit the exponential model $P = ae^{bt}$ to this data, where t measures time from 1930.

(i) What is the predicted population in 1980, 1990 and 2000?

(ii) The actual population in 1996 was about 5.8 billion. How does this compare with the predictions of the model?

(iii) When is the world population predicted to reach 10 billion?

2. The population of fish in a reservoir is affected by both fishing and restocking. The proportionate birth rate is constant at 60% per year and the proportionate death rate is constant at 65% per year. The reservoir is restocked at a constant rate of 4000 fish per year and the fishermen are allowed to catch 3500 fish per year.

(i) Using the above figures show that the differential equation modelling the fish population P at time t is

$$\frac{dP}{dt} = 500\left(1 - \frac{P}{10000}\right)$$

(ii) Without solving the differential equation describe the long term behaviour of the population in the two cases when the initial population is (a) 5000, (b) 15000.

(iii) Draw a direction field for the equation and confirm your answers to (ii).

(iv) Solve the differential equation to verify the results in (ii) and (iii).

3. This problem models the growth of sweet peas in the author's garden in Crapstone, Devon.

The following table gives data on the growth of sweet pea plants in the summer 1996. Two dozen plants were grown and their average heights were recorded every seven days.

week	1	2	3	4	5	6	7	8	9	10
height (cm)	14	26	47	69	93	121	146	162	177	182

The initial average height when measurements began was 8 cm.

(i) A model assumes that the rate of growth of the plants satisfies the logistic differential equation

$$\frac{dh}{dt} = ah\left(1 - \frac{h}{H}\right)$$

where H is the final (equilibrium) height of the sweet pea plants and a is a constant. Without solving this equation and by sketching a graph of the data, explain why it is an appropriate model.

(ii) Solve the differential equation. Use the data in the table at weeks 0, 5 and 10 to find the unknown constants in the solution.

(iii) Superimpose a graph of the logistic model onto a graph of the data. Comment on the appropriateness of the logistic equation as a model for plant growth in this case.

4. A colony of a rare species of duck was established to provide breeding stock. The population grew from an initial value of 500 but has now stabilised. During the period the population was growing studies at two different times produced the following results. When the population was 2000 the proportionate birth rate was 60% per year and the proportionate death rate

15% per year. When the population was 4000 the proportionate birth and death rates were 40% and 25% per year respectively.

A model of the population is based on the assumptions that there is no migration or exploitation and that the proportionate birth and death rates are linear functions of the population.

(i) Show that the model of the population P at time t, based on the above assumptions and data, satisfies the equation

$$\frac{dP}{dt} = 0.75P - 0.00015P^2$$

and deduce the stable duck population.

(ii) It is now proposed to remove a number of ducks each year in order to supplement other colonies. Deduce the effect on the future population of each of the strategies

(a) removing 500 ducks each year,
(b) removing 20% of the ducks each year.

The birth and death rates remain as before.

(iii) Sketch the graph of P against t in each case, from the establishment of the colony.

(iv) Investigate how the population varies as the number or percentage of ducks removed each year is varied. What is the maximum number, or percentage, of ducks that can be removed for the colony to remain economically viable, with at least 3000 ducks?

5. Population data is available at equal intervals so that the population after n intervals of time is P_n. If P_n is given by the logistic equation then $P_n = \dfrac{M}{1 + ke^{-an}}$, where $k = \dfrac{M}{P_0} - 1$ is a constant.

(i) Apply the above expression after n and $(n + 1)$ intervals, when the population is P_n and P_{n+1} respectively and show, by division, that

$$e^a\left(\frac{M}{P_{n+1}} - 1\right) = \frac{M}{P_n} - 1$$

(ii) Show that the above equation can be rearranged to

$$\frac{P_{n+1} - P_n}{P_n} = \left(\frac{1-e^a}{M}\right)P_{n+1} + (e^a - 1)$$

(iii) Deduce that a graph of $y = \dfrac{P_{n+1} - P_n}{P_n}$ against $x = P_{n+1}$ is linear and that it intersects the x axis at M and the y axis at $e^a - 1$.

(iv) Plot a straight line graph for the sweet pea data in question 3 and hence estimate a and the maximum height H of the sweet peas.

6. (i) The solution of the logistic equation is $P = \dfrac{M}{1+ke^{-at}}$, where $k = \dfrac{M}{P_0} - 1$. Show that this equation can be rearranged into the form

$$\ln\left(\frac{M}{P} - 1\right) = -at + \ln k$$

and deduce that a graph of $\ln\left(\dfrac{M}{P} - 1\right)$ against t will be linear.

(ii) For the sweet pea data in question 3 take $H = 191$, plot a graph of $\ln\left(\dfrac{H}{h} - 1\right)$ against t and confirm that the logistic model is a good one for this data.

7. A more general model for population growth of a species is given by the differential equation

$$\frac{dP}{dt} = aP - bP^r, \qquad P(0) = P_0$$

where a, b and $r > 1$ are constants.

If $a = 6$, $b = 2$ and $P_0 = 1$ find the solution of this model for $0 \le t \le 2$ for each of the cases $r = 1.5$, $r = 3$ and $r = 4$.

Sketch a graph of $P(t)$ against t in each case and compare your results with the logistic model.

8. Another equation for modelling animal population growth is the **Gompertz** equation

$$\frac{dP}{dt} = aP(b - \ln P)$$

where a and b are constants.

(a) Solve the differential equation for P and deduce, from the equation and its solution, that the equilibrium population is $M = e^b$.

(b) Sketch some of the solution curves for different values of a and b.

9. Trout and bass compete with each other for food and living space. A model of the trout (x) and bass (y) populations is based on the following assumptions:

1. In the absence of one species, the other will grow according to the exponential model.

2. The intensity of competition between trout and bass is proportional to the number of possible interactions between the species, as measured by xy.

Show that the above assumptions lead to the equation

$$\frac{dx}{dt} = ax - bxy, \quad \frac{dy}{dt} = cy - dxy \qquad (a,b,c,d > 0).$$

For what initial populations are the trout and bass populations predicted to remain constant?

(a) Consider the particular case

$$\frac{dx}{dt} = 0.4x - 0.0002xy, \quad \frac{dy}{dt} = 0.2y - 0.0002xy$$

Investigate the long term behaviour of x and y given the initial conditions (i) $x(0) = 500$, $y(0) = 700$, (ii) $x(0) = 200$, $y(0) = 700$, (iii) $x(0) = 1000$, $y(0) = 2000$.

Under what initial conditions will trout (or bass) eventually dominate?

(b) Investigate how x and y change for different values of a, b, c and d.

(c) Investigate the effect of fishing on the two populations.

4.4 MECHANICS

Mechanics is the study of the motion of objects and the effects of forces acting on them. For example

(i) when a ball is thrown in the air, how high will it go?

(ii) when an athlete throws a javelin, at what angle should it be released for maximum distance?

(iii) when a rocket takes off from earth can it escape from the pull of the earth's gravity?

Modelling mechanics problems, of which the above are examples, is based on Newton's laws of motion. These are:

1. If there is no resultant external force acting on a body then it will move with constant velocity. (That is, along a straight line with constant speed.)

2. The rate of change of momentum (momentum = mass × velocity) of a body is equal to the resultant force acting on the body. That is,

$$\frac{d}{dt}\text{(momentum)} = \text{forces acting}$$

$$\frac{d}{dt}(mv) = F \tag{4.9}$$

If the mass m is constant then this reduces to the more familiar form $m\dfrac{dv}{dt} = F$

(mass × acceleration = forces acting).

3. When one body interacts with another the force of the first body on the second is equal in magnitude, but opposite in direction, to the force of the second body on the first.

The Newtonian model is a good one in every day situations but breaks down for bodies

(i) which are very small, when quantum mechanics must be used

and (ii) moving very quickly (close to the speed of light) when relativity must be used.

Units and Notation

Quantities are measured in SI units. They are:

Quantity	Unit	Abbreviation
distance	metre	m
time	second	s or sec
mass	kilogram	kg
velocity	metres per second	ms^{-1} or m/s
acceleration	metres per second per second	ms^{-2} or $\mathrm{m/s}^2$
force	newton	N

For objects close to the earth's surface the acceleration due to gravity is $g \approx 9.81\mathrm{ms}^{-2}$, and the gravitational force acting on a body of mass m is $F = mg$.

If x is the displacement of a body at time t then its

velocity (the rate of change of distance) is $v = \dfrac{dx}{dt}$

acceleration (the rate of change of velocity) is $a = \dfrac{dv}{dt} = \dfrac{d^2x}{dt^2} = v\dfrac{dv}{dx}$

Newton's second law (N2) is
$$m\frac{dv}{dt} = F \tag{4.10}$$

which can also be expressed as
$$mv\frac{dv}{dx} = F \tag{4.11}$$

$$\text{or } m\frac{d^2x}{dt^2} = F$$

In this section we will look at situations where a body is moving in a straight line and for which the application of Newton's second law leads to a first order differential equation of the form (4.10) or (4.11). In general we shall use (4.10) when time is involved in the problem and use (4.11) when connections between speed and distance only are required.

Procedure for solving problems

1. Draw a diagram showing

   (i) the positive x axis (normally the direction in which the body is moving),
   (ii) arrows giving the directions of all the forces acting on the body.

2. Write down the magnitude of each force.

3. Apply Newton's second law.

4. Solve the differential equation using the techniques of Chapters 2 and 3 and/or DERIVE.

In all examples the body under consideration will be modelled as a particle; that is, it has negligible size and no internal structure.

In Example 4J we will consider motion under gravity where there is no resistive force, but such a force is included in Examples 4K to 4N.

Example 4J

A ball is thrown vertically upwards with speed u from a point A which is at height h above ground level. Assuming that there is no resistance to motion

(i) use N2 with $a = \dfrac{dv}{dt}$ and find the time the ball takes to reach its highest point, and find this height,

(ii) use N2 with $a = v\dfrac{dv}{dx}$ to find the greatest height reached by the ball and compare with (i),

(iii) find the time taken for the ball to complete its flight and hit the ground, and also the speed with which the ball hits the ground.

Solution

As the ball is moving upwards choose a positive x axis in this direction with the origin at ground level.

The only force acting is that due to gravity, with magnitude $F = mg$.

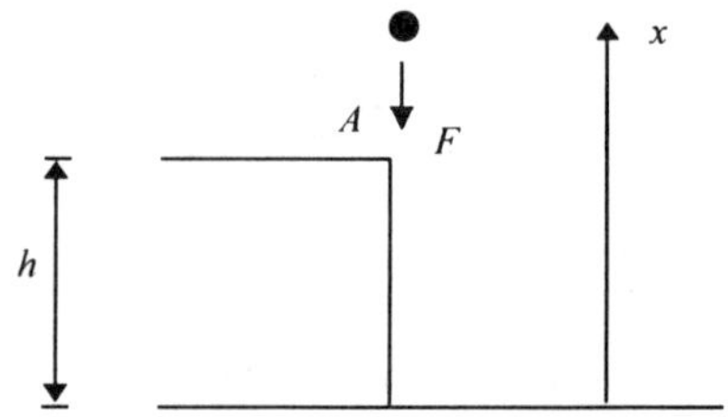

(i) Applying N2: $m\dfrac{dv}{dt} = -mg$, or $\dfrac{dv}{dt} = -g$ (4.12)

Integrating: $v = -gt + A$

Applying the initial condition $v = u$ at $t = 0$ gives $A = u$ and then

$$v = u - gt \tag{4.13}$$

At the highest point the speed of the ball is instantaneously zero. Putting $v = 0$ we find that the time t_1 to the highest point is $t_1 = \dfrac{u}{g}$. To find the greatest height put $v = \dfrac{dx}{dt}$ in equation (4.13) and integrate again. Then

$$x = ut - \frac{1}{2}gt^2 + B \tag{4.14}$$

The initial condition is $x(0) = h$, so that $B = h$ and

$$x = ut - \frac{1}{2}gt^2 + h \tag{4.15}$$

The greatest height, x_1, when $t = t_1 = \dfrac{u}{g}$ is

$$x_1 = u\left(\frac{u}{g}\right) - \frac{1}{2}g\left(\frac{u}{g}\right)^2 + h = \frac{u^2}{2g} + h$$

(ii) Using $v\dfrac{dv}{dx}$ for acceleration in place of $\dfrac{dv}{dt}$, (4.12) becomes

$$v\frac{dv}{dx} = -g \tag{4.16}$$

To solve this equation using DERIVE use the command

DSOLVE1(g, v, x, v, h, u)

Simplifying, **Solving** for x, putting $v = 0$ to find the greatest height and **Expanding** for h gives

$$x = h + \frac{u^2}{2g}, \qquad \text{as in (i).}$$

Notice that if the greatest height only was required, using $v\dfrac{dv}{dx}$ for acceleration would be quicker than using $\dfrac{dv}{dt}$.

(iii) In the absence of any resisting force the equations (4.12) and (4.16) apply when the ball is falling back towards the ground. The ball hits the ground at time t_2, when $x = 0$. **Substitute** 0 for x in (4.15) and **Solve** for t to give the two solutions $\dfrac{u \pm \sqrt{2gh + u^2}}{g}$. As $t_2 > 0$ we require the positive root and so

$$t_2 = \frac{u + \sqrt{2gh + u^2}}{g}.$$

To find the speed v_2 with which the ball hits the ground either use (4.13) with $t = t_2$ or (4.15) with $x = 0$. (4.13) gives $v_2 = u - (u + \sqrt{u^2 + 2gh}) = -\sqrt{u^2 + 2gh}$. Remember that this is the velocity of the ball. Its speed is $\sqrt{u^2 + 2gh}$ and the minus sign indicates it is travelling in the negative x direction (that is, downwards).

□

Models of Resistance

A particle moving in air, water or some other fluid will experience a force, such as air resistance, which slows it down. The force directly opposes the direction of motion (the velocity) and is called **drag**. Drag can have a beneficial effect as in parachuting where the parachute ensures a soft(ish) landing, or may be something to be minimised, such as in motor racing where manufacturers go to great lengths to reduce its effect.

Intuition suggests that the magnitude of a drag force increases with speed. For example, air resistance has little effect on a cyclist at low speeds but becomes more noticeable as the speed increases. The air, or fluid, in which the body is moving also has an effect. While jogging in air is quite easy, it is certainly more difficult under water and virtually impossible in treacle.

Experiments show that the magnitude R of the drag force is often proportional to some power of the speed v. That is $R \propto v^n$ or $R = kv^n$. If the particle is moving through air at low speeds or through a viscous fluid such as treacle the magnitude of the drag force can be assumed to be proportional to the speed. That is $n = 1$ and $R = kv$.

For many every day situations such as the motion of various types of balls (golf, cricket, football, ...) in air the quadratic approximation is more appropriate. That is $n = 2$ and $R = kv^2$. In other situations another value of n or a different form for R (such as $R = cv + kv^2$) is more appropriate.

If $n = 1$ or 2, so that $R = kv$ or $R = kv^2$ one dimensional, problems can be solved analytically using separation of variables or the integrating factor method, while for other values of n the numerical techniques of Chapter 3 must be used.

Example 4K

This example looks at Application 3 in Chapter 1. When a ball is thrown vertically upwards we wish to find how high the ball will go and the time this takes.

Solution

The mathematical problem is to solve the equation of motion of the ball, $m\dfrac{dv}{dt} = -kv - mg$ given $v = u$ at $t = 0$ in order to find the greatest height reached by the ball and the time it takes to reach this height. The equation can be solved either by separating the variables or by using the integrating factor method. We will use the former.

The equation is

$$m\frac{dv}{dt} = -(kv + mg)$$

To solve this equation use **Declare Variable Domain** to define *m, g, k, u* and *v* to be non negative and choose **Declare Algebra State** select **Simplification** and set **Logarithms** to **Collect**.

Simplify DSOLVE1($kv + mg,m,t,v,0,u$) and **soLve** for t to give

$$t = -\frac{m}{k}\ln\frac{gm+kv}{gm+ku} \tag{4.17}$$

Find the time t_1 to the highest point by substituting $v = 0$, so that

$$t_1 = -\frac{m}{k}\ln\frac{gm}{gm+u} = \frac{m}{k}\ln\left(\frac{ku}{gm}+1\right)$$

To find the greatest height first **Solve** (4.17) for v to give

$$v = \frac{e^{-\frac{kt}{m}}(gm+ku)}{k} - \frac{gm}{k}$$

Next replace v by $\frac{dx}{dt}$ and integrate with the condition $x(0) = 0$. This can be achieved by DSOLVE1$\left(\frac{gm}{k} - \frac{e^{-\frac{kt}{m}}(gm+ku)}{k},1,t,x,0,0\right)$. **Simplifying** and **Solving** for x gives

$$x = -\frac{m}{k^2}e^{-kt}(gm+ku) - \frac{m}{k^2}(g(kt-m)-ku)$$

The greatest height x_1 occurs when $t = t_1$. **Substitute** this value for t, **Simplify** and **Solve** for x to obtain $x_1 = \frac{mu}{k} - \frac{gm^2}{k^2}\ln\frac{gm+ku}{gm}$.

In a modelling situation we would now continue the modelling loop by obtaining data for the parameters m, k and u, compute x_1 and t_1 and compare them with the results of experiments.

□

Example 4L

A ball of mass 0.4kg is thrown vertically upwards from ground level at $25ms^{-1}$ and experiences a resistive force of magnitude $R = 0.004v^2$ when its speed is v. Taking $g = 10ms^{-2}$ find the greatest height reached by the ball.

Solution

The two forces acting on the ball are its weight $mg = 0.4 \times 10 = 4$N and the air resistance $R = 0.004v^2$, both acting downwards. With a positive x axis pointing upwards from ground level N2 gives

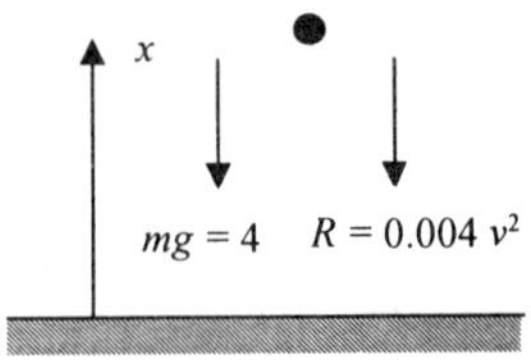

$$ma = -mg - R$$

$$0.4a = 4 - 0.004v^2$$

Using $a = v\dfrac{dv}{dx}$ and simplifying this becomes $100v\dfrac{dv}{dx} = -(1000 + v^2)$. The initial condition is $v(0) = 25$. Using DSOLVE1($1000 + v^2, 100v, x, v, 0, 25$), **Simplifying** and **Solving** for x gives $x = 150\ln 5 - 50\ln\left(\dfrac{v^2 + 1000}{13}\right)$. To find the greatest height x_1 **Manage Substitute** 0 for v. This shows that $x_1 = 24.3$ to 1D.

□

Downward Motion

When a particle is falling the air resistance and weight act in opposite directions. Applying N2 with a positive x axis downwards:

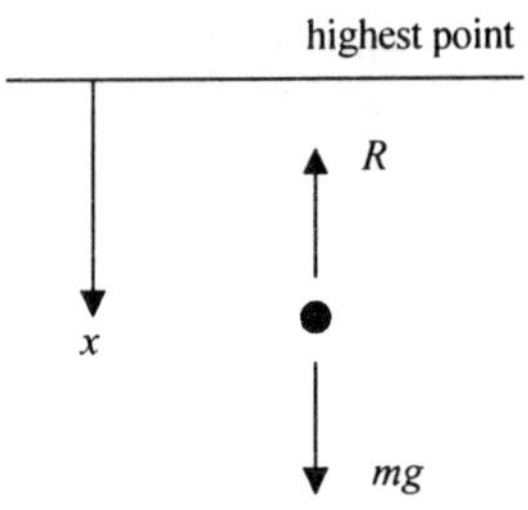

$$m\frac{dv}{dt} = mg - R$$

(i) If $R = kv$ then $m\dfrac{dv}{dt} = mg - kv$.

(ii) If $R = kv^2$ then $m\dfrac{dv}{dt} = mg - kv^2$.

If the particle starts from rest, so that $v(0) = 0$, then $R = 0$ initially, but as the particle picks up speed R increases (and the acceleration $\dfrac{dv}{dt}$ decreases). If R increases to mg the two forces balance, the acceleration is zero and the particle falls at a constant speed, called the **terminal speed** v_T.

In (i) $v_T = \frac{mg}{k}$.

In (ii) $v_T = \sqrt{\frac{mg}{k}}$.

More generally, if $R = kv^n$ then $v_T = \left(\frac{mg}{k}\right)^{\frac{1}{n}}$.

Mathematically a particle will approach, but never reach, its terminal speed.

Example 4M

A particle of mass 0.2kg is dropped, from rest, from the top of a building 60 metres high. During its fall the particle experiences a resistive force $0.1v$ newtons. Taking $g = 10\text{ms}^{-2}$ find

(i) the time the particle takes to reach the ground
(ii) the speed with which the particle hits the ground
(iii) the proportion of its terminal speed the particle has reached when it hits the ground.

Solution

(i) The forces acting on the particle are its weight $mg = 0.2 \times 10 = 2$ newtons acting downwards and the air resistance $0.1v$ newtons acting upwards. Choose the positive x axis to be down with the origin at the top of the building.

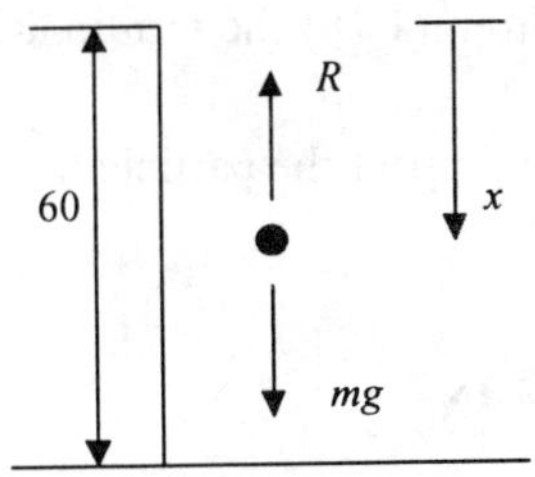

Applying N2: $0.2\frac{dv}{dt} = 2 - 0.1v$ (4.18)

The initial condition is $v(0) = 0$.
Declare Variable v to be real in the interval $0 \le v \le 20$ and use

DSOLVE1(0.1v – 2,0.2,t,v,0,0)

Simplify and **Solve** for v to obtain

$v = 20 - 20e^{-t/2}$ (4.19)

To find x in terms of t use DSOLVE1($20e^{-t/2}$ – 20,1,t,x,0,0). **Solve** for x, **Substitute** 60 for x and **Simplify** to

$$-40e^{-t/2} - 20t + 100 = 0 \tag{4.20}$$

To solve this equation numerically using the Newton-Raphson method define the function

$$\text{NEWTON}(f,t,t_0,n) := \text{ITERATES}\left(t - \frac{f}{\text{DIF}(f,t)}, t, t_0, n\right)$$

This will produce n iterations, starting from t_0, for the equation $f(t) = 0$. We require a value for t_0 for equation (4.20). This could be done by guesswork, but it is preferable to look at what would happen in the absence of air resistance. Calculation shows the time to the ground would then be $\sqrt{12} \approx 3.46$ seconds. As the time is likely to be larger when air resistance is included a sensible initial value is $t_0 = 4$.

The required command is NEWTON($-40e^{-t/2}$ – $20t$ + 100,t,4,5). The five iterations show that, to 2 decimal places, the time $t_1 = 4.82$ seconds.

(ii) Using (4.19) the speed of the particle in impact is $v(4.82) \approx 18.2\text{ms}^{-1}$.

(iii) From (4.18) the terminal speed of the particle when $\frac{dv}{dt} = 0$, is 20ms^{-1} and so on impact the particle has achieved $\frac{18.2}{20} \times 100 = 91\%$ of this.

□

Example 4N

A particle of mass 1kg is projected vertically upwards from ground level with an initial speed of 40ms^{-1}. Assuming that $g = 10\text{ms}^{-2}$ and that the magnitude of the air resistance force is $R = 0.01v^{1.5}$ find

(i) the time the particle takes to reach its greatest height and this height,
(ii) the total time of flight and the speed with which the particle hits the ground.
(iii) Plot graphs showing how the speed of the particle and its height above the ground vary with time throughout the flight.

Solution

(i) Applying N2: $m\dfrac{dv}{dt} = -mg - R$

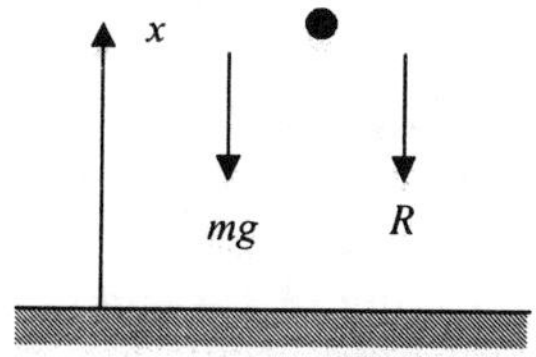

$$\frac{dv}{dt} = -10 - 0.01v^{1.5}$$

We can find the height x from $v = \dfrac{dx}{dt}$. The initial conditions are $x(0) = 0$ and $v(0) = 40$. Hence we must solve the pair of first order equations

$$\frac{dx}{dt} = v, \quad \frac{dv}{dt} = -10 - 0.01v^{1.5} \quad \text{given } x(0) = 0,\ v(0) = 40.$$

Although DERIVE can find analytical solutions to these equations they are very cumbersome and not at all concise, and so it is sensible to solve them numerically using the Runge Kutta function RK located in the utility file ODE_APPR. The command is of the form

$$\text{RK}([v,-10 - 0.01v^{1.5}],\ [t,x,v],\ [0,0,40],\ h,\ n) \tag{4.20}$$

and will produce a set of n $[t,x,v]$ triples, where t is incremented by h each time. The highest point is identified by searching for $v = 0$.

Experimentation with different values of h and n shows that, to 1 decimal place, v becomes zero after 3.6 seconds, when $x = 70.1$. Thus the particle reaches a height of 70.1 metres in 3.6 seconds.

(ii) For the downward motion the equation becomes

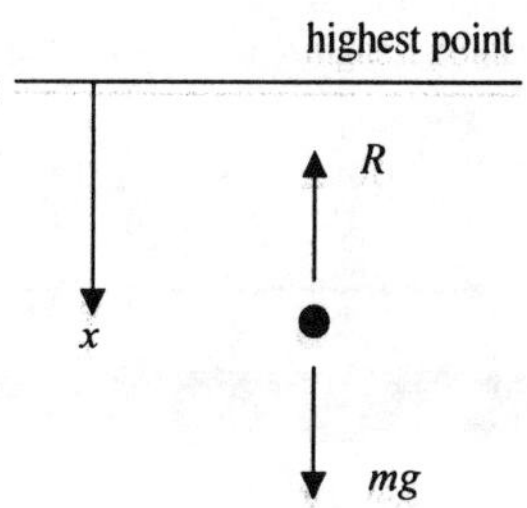

$$m\frac{dv}{dt} = mg - R, \quad \text{or} \quad \frac{dv}{dt} = 10 - 0.01v^{1.5}$$

We must now solve the pair

$$\frac{dx}{dt} = v, \quad \frac{dv}{dt} = 10 - 0.01v^{1.5} \quad \text{given } x(0) = 0,\ v(0) = 0.$$

To find the time to the ground and the speed on impact we must find t and v when $x = 70.1$. The command is

$$\text{RK}([v,10 - 0.01v^{1.5}],\ [t,x,v],\ [0,0,0],\ 0.1,\ 40) \tag{4.21}$$

and shows that, correct to 1D, the time to the ground is 3.9 seconds and the particle hits the ground at about 35ms^{-1}. The total time of flight is about 7.5 seconds.

(iii) The graphs of speed and height above the ground against time are shown in Figure 4.10. The RK functions produce sets of $[t,x,v]$ triples and the EXTRACT_2_COLUMNS command must be used to obtain $[t,v]$ and $[t,x]$ pairs. Solving (4.20) and extracting and plotting these pairs gives the graphs for $0 \leq t \leq 3.6$. To obtain the graphs for $t > 3.6$ the $[t,v]$ and $[t,x]$ pairs obtained from the solution to (4.21) have to be amended by increasing the t values by 3.6 and replacing each x value by $70.1 - x$ (to get the height above the ground rather than from the highest point). How this was done is left to the reader.

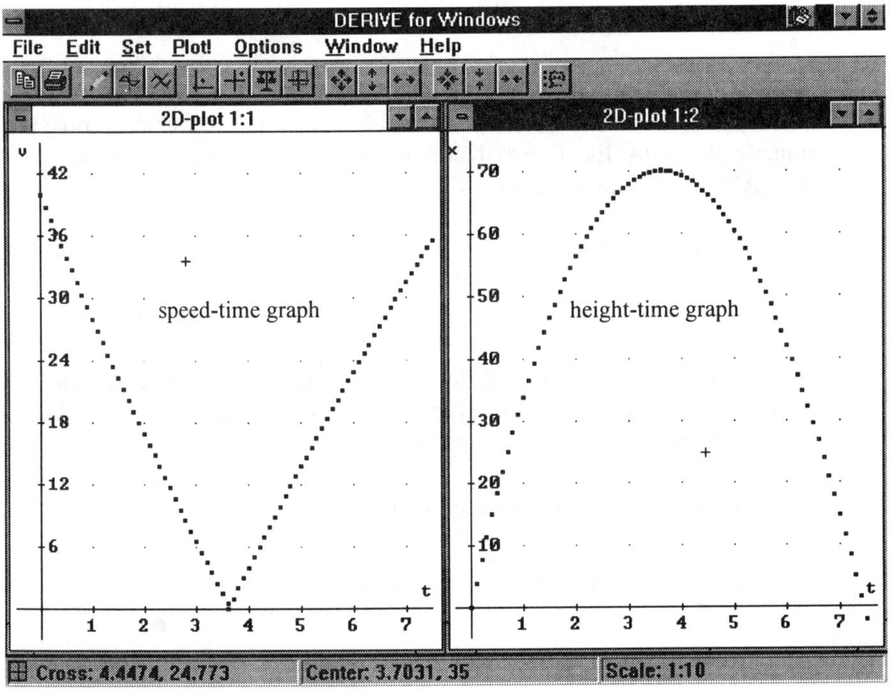

Figure 4.10 Speed-Time and Height-Time graphs in Example 4N

□

The motion of a particle in two dimensions, with and without air resistance will be considered in Chapter 6.

Exercise 4D

In each problem take $g = 9.8\text{ms}^{-2}$.

1. A parachutist with his equipment has a mass of 80kg. He drops vertically and opens his parachute having reached a speed of 40ms^{-1}. The parachute then exerts a resistive force of magnitude $320\sqrt{v}$ where v is the speed of the parachutist (in ms^{-1}).

   (a) Show that when the parachute is open the equation of motion of the parachutist is

   $$v\frac{dv}{dx} = 9.8 - 4\sqrt{v} \quad \text{and} \quad v(0) = 40$$

   (b) Find the minimum height above the ground that the parachute should be opened so that the parachutist reaches the ground with a speed of less than 7ms^{-1}.

2. Emergency food and medicine are to be dropped from a helicopter to a village cut off by snowstorms in Lapland. The parcels fall under gravity and experience air resistance of magnitude $0.02mv^2$ where m is the mass of a parcel and v is its speed. The parcels can withstand an impact at a maximum speed of 20ms^{-1}.

   (a) Find the maximum height from which the parcels can be dropped.

   (b) If they are dropped from the maximum height, find how long each parcel takes to drop.

   (c) What assumptions have you made in modelling the problem?

3. A stone of mass 1kg is dropped through a liquid which provides a resisting force proportional to the square root of the magnitude of its velocity. The initial speed of the stone is 5ms^{-1} and the terminal speed is 12ms^{-1}.

   (a) Formulate the differential equation of motion of the stone.

   (b) Solve the differential equation and draw a graph of speed of the stone against time.

   (c) Find how long it takes for the speed to be within 10% of the terminal speed.

4. A boy catapults a toy parachute of mass 0.5kg vertically upwards from ground level at 30ms^{-1}. At its highest point the parachute opens and it floats back to the ground. On the upward motion the force due to air resistance has magnitude $0.2v$, where v is the speed of the parachute, while on the downward motion the resistive force has magnitude $0.2v^2$.

   (i) Find the greatest height reached by the parachute and the time taken.

   (ii) Show that, on the downward path, the parachute is within 1% of its terminal speed after falling 5 metres and, assuming it falls at terminal speed from then on, find the time of the downward path and speed on impact with the ground.

   (iii) Plot graphs to show how the speed and height above ground of the parachute will vary with time throughout the flight.

5. Heat seeking missiles have a guidance system that enables the missile to follow an object that emits heat. For example, an anti-aircraft missile will lock on to the heat of the exhaust from a jet engine and move so that the missile is always pointing toward its source, the aircraft.

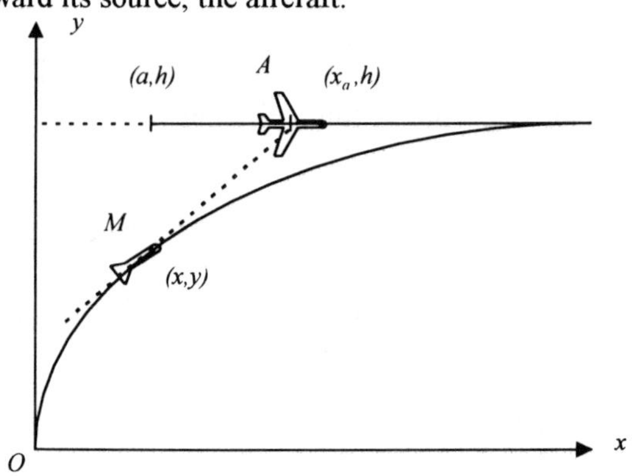

A missile is launched at an aircraft that is flying horizontally with constant speed v at a constant height of h km. At the instant the missile is launched the initial position of the aircraft has co-ordinates (a,h) relative to the launch pad. The missile travels with constant speed u (where $u > v$). At time t, the missile has position (x,y).

   (i) Explain why the slope of the tangent at M is $\dfrac{\dot{y}}{\dot{x}}$ and why

$$\dot{x}^2 + \dot{y}^2 = u^2 \qquad (1)$$

(ii) Show that t seconds after the launch of the missile the aircraft is at the point $A(vt + a,h)$.

(iii) Write down the equation of the tangent at M, make the point A lie on the tangent and show that

$$\dot{x}(h-y) = \dot{y}(a+vt-x) \qquad (2)$$

(iv) Solve (1) and (2) to obtain

$$\dot{x} = \frac{u(a+vt-x)}{\sqrt{(a+vt-x)^2+(h-y)^2}}, \quad \dot{y} = \frac{u(h-y)}{\sqrt{(a+vt-x)^2+(h-y)^2}}$$

(v) Solve the equations to find where and when the missile strikes the aircraft, given

(a) $a = 0$, $h = 2500$, $v = 200$, $u = 1000$

(b) $a = 2000$, $h = 5000$, $v = 400$, $u = 1500$.

Assume a hit when the missile gets within 50 metres of the aircraft.

6. A parachutist jumps from a plane that is travelling horizontally at a constant speed of 200ms^{-1}. Before opening his parachute the air resistance on the parachutist is $0.14m\mathbf{v}$ where m is the mass and $\mathbf{v}$ is the velocity.

(a) Show that the horizontal and vertical velocity components u and v at any time satisfy the equations

$$\frac{du}{dt} = -0.14u \qquad u(0) = 200$$
$$\frac{dv}{dt} = 9.8 - 0.14v \qquad v(0) = 0$$

(b) Solve the equations for u and v and find the speed of the parachutist ($\sqrt{u^2+v^2}$) as a function of time t.

(c) By graphing the speed against time, describe the motion of the parachutist.

5

Analytical Solution of Second Order Linear Differential Equations

5.1 CLASSIFICATION OF SECOND ORDER DIFFERENTIAL EQUATIONS

Any second order differential equation is a function of the independent variable x, the dependent variable y and the first two derivatives y' and y'' of y with respect to x. It is of the form

$$F(x,y,y',y'') = 0 \qquad \text{for some function } F.$$

For example

$$(x^3 - 1)y'' + 3x(y')^4 + y^2 = x^2$$

is a second order differential equation.

A **linear** second order equation can be written in the form

$$a(x)y'' + b(x)y' + c(x)y = f(x) \tag{5.1}$$

For example, the equation $x^2y'' - (2x + 1)y' + 4y = \sin x$ is linear but $xy'' + y'y^3 - y = x$ is not.

A linear equation with zero right hand side, of the form

$$a(x)y'' + b(x)y' + c(x)y = 0$$

is called **homogeneous**, whereas if $f(x)$ is not zero the equation is **non-homogeneous**.

Finally, if $a(x)$, $b(x)$ and $c(x)$ are constants we obtain the non-homogenous linear equation

$$ay'' + by' + cy = f(x) \tag{5.2}$$

or, when $f(x) = 0$, the homogeneous linear equation is

$$ay'' + by' + cy = 0 \tag{5.3}$$

Example 5A

Classify the following equations as linear or non-linear.

(i) $\sin x y'' + \sqrt{x}\, y' + 2y = e^x$ (ii) $2y'' - 4y' + 9y = 0$

(iii) $y'' + 2y^2 = \sin x$ (iv) $xy'' + x^2y' - x^3y = \cos x$

(v) $y'' - 3y' + 2y = 4x - 1$ (vi) $xy'' + yy' = 1$

Solution

An equation is linear if it can be put in the form (5.1), where the terms involving y, y' and y'' do not occur in products, raised to powers or in functions such as $\sin y$. This applies to (i), (ii), (iv) and (v) above, which are all linear. (iii) is not linear because of the term y^2 while (vi) is not because of the product yy'.

□

Example 5B

Which of the linear equations in Example 5A are homogeneous, and which are constant coefficient?

Solution

Equation	Homogeneous?	Constant coefficient?
(i)	No	No
(ii)	Yes	Yes
(iv)	No	No
(v)	No	Yes

□

5.2 HOMOGENEOUS LINEAR EQUATIONS WITH CONSTANT COEFFICIENTS

These are equations of the form

$$ay'' + by' + cy = 0 \qquad (5.3)$$

where a, b and c are constants.

Equations of this type have two independent solutions, as the following example demonstrates.

Example 5C

Verify that $y = e^x$ and $y = e^{2x}$ are both solutions of the equation

$$y'' - 3y' + 2y = 0$$

Solution

If $y = e^x$ then $y' = e^x$ and $y'' = e^x$. Then

$$y'' - 3y' + 2y = e^x - 3e^x + 2e^x = 0$$

as required. Similarly, if $y = e^{2x}$ then $y' = 2e^{2x}$ and $y'' = 4e^{2x}$. Then

$$y'' - 3y' + 2y = 4e^{2x} - 6e^{2x} + 2e^{2x} = 0$$

as required. This verifies that $y = e^x$ and $y = e^{2x}$ are both solutions of $y'' - 3y' + 2y = 0$.

□

5.2.1 The general solution of $ay'' + by' + cy = 0$

In general, if $y = y_1(x)$ and $y = y_2(x)$ are both solutions of (5.3) then any linear combination of y_1 and y_2 is also a solution of (5.3), and is the general solution. Let

$$y = Ay_1(x) + By_2(x) \qquad \text{(a linear combination of } y_1 \text{ and } y_2\text{)}$$

where A and B are arbitrary constants. Then

$$\begin{aligned}
ay''+by'+cy &= a\frac{d^2}{dx^2}[Ay_1+By_2]+b\frac{d}{dx}[Ay_1+By_2]+c[Ay_1+By_2] \\
&= aA\frac{d^2y_1}{dx^2}+aB\frac{d^2y_2}{dx^2}+bA\frac{dy_1}{dx}+bB\frac{dy_2}{dx}+cAy_1+cBy_2 \\
&= A\left(a\frac{d^2y_1}{dx^2}+b\frac{dy_1}{dx}+cy_1\right)+B\left(a\frac{d^2y_2}{dx^2}+b\frac{dy_2}{dx}+cy_2\right) \\
&= A\times 0+B\times 0 \qquad \text{as } y_1 \text{ and } y_2 \text{ are solutions of (5.3)} \\
&= 0
\end{aligned}$$

Hence, if y_1 and y_2 are solutions of (5.3) then the general solution is

$$y = Ay_1(x) + By_2(x)$$

Example 5D

Write down the general solution of $y'' - 3y' + 2y = 0$.

Solution

From the results of the preceding section and Example 5C, the general solution of the equation is a linear combination of e^x and e^{2x}. This is

$$y = Ae^x + Be^{2x}$$

□

We wish to find the solution of equation (5.3) for any a, b and c. A similar first order equation is $ay' + by = 0$ and it can easily be shown that its solution is

$$y = Ae^{-\frac{b}{a}x}$$

This and the result of Example 5D suggest that exponential functions appear in the solution of (5.3).

Let us see whether $y = e^{mx}$ can be a solution of (5.3). Differentiating, $y' = me^{mx}$ and $y'' = m^2e^{mx}$. Then

$$\begin{aligned}
ay''+by'+cy &= am^2e^{mx}+bme^{mx}+ce^{mx} \\
&= e^{mx}(am^2+bm+c)
\end{aligned}$$

For this to equal zero, as required, $am^2 + bm + c = 0$ as $e^{mx} > 0$ for all real x. This equation is called the **auxiliary equation** (AE) and has roots $m_1 = \dfrac{-b+\sqrt{b^2-4ac}}{2a}$ and $m_2 = \dfrac{-b-\sqrt{b^2-4ac}}{2a}$. Provided m takes one of the two values m_1 or m_2 we have found solutions to (5.3). Then $y_1(x) = e^{m_1 x}$ and $y_2(x) = e^{m_2 x}$ are both solutions of (5.3). The previous analysis shows that a linear combination of solutions is also a solution and it follows that the general solution of (5.3) is

$$y = Ae^{m_1 x} + Be^{m_2 x}$$

Example 5E

Find the general solution of $2y'' - 5y' + 2y = 0$.

Solution

The auxiliary equation (AE) is $2m^2 - 5m + 2 = 0$. Factorising

$$(2m - 1)(m - 2) = 0$$

giving $m = \frac{1}{2}, 2$. Hence the general solution is

$$y = Ae^{\frac{1}{2}x} + Be^{2x}$$

□

The auxiliary equation $am^2 + bm + c = 0$ is a quadratic equation whose roots can be one of three types.

Case A The AE can have distinct roots, as in Example 5E. In general this occurs when $b^2 > 4ac$.

Case B The AE can have equal roots, when $b^2 = 4ac$.
For example $m^2 - 4m + 4 = 0$ factorizes to $(m - 2)(m - 2) = 0$ and so the roots are $m = 2$ and 2.

Case C If $b^2 < 4ac$ the number under the root will be negative and so the roots will be complex.
For example the roots of $m^2 - 2m + 2 = 0$ are

$$m = \frac{2 \pm \sqrt{(-2)^2 - 4 \times 1 \times 2}}{2 \times 1} = \frac{2 \pm \sqrt{-4}}{2} = \frac{2 \pm 2i}{2} = 1 \pm i$$

and we obtain two roots $1 + i$ and $1 - i$, which are complex conjugates.

The form of the general solution of (5.3) will be different in each of the three cases A, B and C.

Case A Auxiliary equation with distinct real roots (when $b^2 > 4ac$)

$$m_1 = \frac{-b + \sqrt{b^2 - 4ac}}{2a} \quad \text{and} \quad m_2 = \frac{-b - \sqrt{b^2 - 4ac}}{2a}$$

As we have seen, the GS is $y = Ae^{m_1 x} + Be^{m_2 x}$.

Case B Auxiliary equation with equal roots m (when $b^2 = 4ac$). The solution is then

$$y = Ae^{mx} + Be^{mx} = (A + B)e^{mx} = Ce^{mx}$$

and only one solution has been found.

Suppose that the complete solution is $y = e^{mx} \times g(x)$, for some function $g(x)$. Then

$$y' = e^{mx} \times g'(x) + me^{mx} \times g(x) = e^{mx}(g' + mg)$$

$$y'' = e^{mx}(g'' + mg') + me^{mx}(g' + mg)$$

$$= e^{mx}(g'' + 2mg' + m^2 g)$$

Substituting into the left hand side of (5.3):

$$\begin{aligned} ay''+by'+cy &= ae^{mx}(g''+2mg+m^2 g) + be^{mx}(g'+mg) + ce^{mx} g \\ &= e^{mx}(ag''+2amg'+am^2 g + bg'+bmg + cg) \\ &= e^{mx}[ag''+(2am+b)g'+(am^2 + bm + c)g] \\ &= ae^{mx} g'' \end{aligned}$$

since $2am + b = 0$ as, for equal roots, $m = -\dfrac{b}{2a}$, and $am^2 + bm + c = 0$ is the AE. Hence for $y = e^{mx}g(x)$ to be a solution of (5.3) we require $g''(x) = 0$.

Integrating: $g'(x) = B$

Integrating: $g(x) = A + Bx$

Hence the solution is $y = e^{mx}(A + Bx)$.

Case C Auxiliary equation with complex roots (when $b^2 < 4ac$). The roots are

$$m = \frac{-b \pm \sqrt{b^2 - 4ac}}{2a} = \frac{-b \pm i\sqrt{4ac - b^2}}{2a}$$
$$= -\frac{b}{2a} \pm i\frac{\sqrt{4ac - b^2}}{2a} = \alpha \pm i\beta$$

(where $\alpha = -\dfrac{b}{2a}$ and $\beta = \dfrac{\sqrt{4ac - b^2}}{2a}$).

The general solution is then

$$y = Ce^{(\alpha+\beta i)} + De^{(\alpha-\beta i)x}$$
$$= Ce^{\alpha x}e^{i\beta x} + De^{\alpha x}e^{-i\beta x}$$
$$= e^{\alpha x}(Ce^{i\beta x} + De^{-i\beta x})$$
$$= e^{\alpha x}[C(\cos\beta x + i\sin\beta x) + D(\cos(-\beta x) + i\sin(-\beta x)]$$
(using Euler's formula)
$$= e^{\alpha x}[C(\cos\beta x + i\sin\beta x) + D(\cos\beta x - i\sin\beta x)]$$
(using $\cos(-\theta) = \cos\theta$ and $\sin(-\theta) = -\sin\theta$)
$$= e^{\alpha x}[(C + D)\cos\beta x + i(C - D)\sin\beta x]$$
$$= e^{\alpha x}(A\cos\beta x + B\sin\beta x) \quad \text{(where } A = C + D \text{ and } B = i(C - D))$$

Hence, when $b^2 < 4ac$, the general solution is

$$y = e^{\alpha x}(A\cos\beta x + B\sin\beta x)$$

This can also be put in the form

$$y = Re^{\alpha x}\sin(\beta x + \theta) \quad \text{or} \quad y = Se^{\alpha x}\cos(\beta x + \phi)$$

if required.

Summary

To solve $ay'' + by' + cy = 0$.

1. Write down the AE $am^2 + bm + c = 0$.

2. Solve the AE (3 cases):

   (i) Case A. AE has two distinct real roots m_1 and m_2.
   The GS is $y = Ae^{m_1 x} + Be^{m_2 x}$.

   (ii) Case B. AE has equal roots m, m.
   The GS is $y = e^{mx}(A + Bx)$.

   (iii) Case C. AE has complex roots $\alpha \pm i\beta$.
   The GS is $y = e^{\alpha x}(A\cos \beta x + B\sin \beta x)$.

Example 5F

Find the general solution of

(i) $y'' + 6y' + 8y = 0$ (ii) $y'' - 6y' + 9y = 0$

(iii) $y'' + 2y' + 10y = 0$ (iv) $y'' + 4y = 0$.

Solution

(i) AE is $m^2 + 6m + 8 = 0$, giving $(m + 4)(m + 2) = 0$, from which $m = -4, -2$.
The roots are real and distinct (Case A).
The general solution is $y = Ae^{-4x} + Be^{-2x}$.

(ii) AE is $m^2 - 6m + 9 = 0$, from which $(m - 3)(m - 3) = 0$, giving $m = 3, 3$.
The roots are equal (Case B).
The general solution is $y = e^{3x}(A + Bx)$.

(iii) AE is $m^2 + 2m + 10 = 0$.

Then $m = \dfrac{-2 \pm \sqrt{4-40}}{2} = \dfrac{-2 \pm 6i}{2} = -1 \pm 3i$.

The roots are complex (Case C).

The general solution is $y = e^{-x}(A\cos 3x + B\sin 3x)$

(iv) AE is $m^2 + 4 = 0$, from which $m = 0 \pm 2i$ (Case C).

The general solution is $y = e^{0x}(A\cos 2x + B\sin 2x)$ or $y = A\cos 2x + B\sin 2x$.

□

Example 5G

Solve $y'' + 5y' + 4y = 0$ given $y(0) = 1$ and $y'(0) = 2$.

Solution

Factorising the AE $m^2 + 5m + 4 = 0$ gives $(m + 1)(m + 4) = 0$ with roots $m = -1, -4$ (Case A). The GS is

$$y = Ae^{-x} + Be^{-4x}$$

and then

$$y' = -Ae^{-x} - 4Be^{-4x}$$

Using $y(0) = 1$: $1 = A + B$.
Using $y'(0) = 2$: $2 = -A - 4B$.

Solving simultaneously, we find $A = 2$, $B = -1$ and so the required particular solution is

$$y = 2e^{-x} - e^{-4x}.$$

□

5.2.2 Sketching solution curves to $ay'' + by' + cy = 0$

A sketch often provides a good qualitative solution of a differential equation.

The form of the solution curve depends on the initial conditions and the values of a, b and c.

(i) If $b = 0$ the equation becomes $ay'' + cy = 0$, or $y'' + \frac{c}{a} y = 0$.

(a) If $\frac{c}{a} > 0$ let $\omega^2 = \frac{a}{c}$ so that the equation becomes $y'' + \omega^2 y = 0$. (This is the equation of simple harmonic motion.)

The auxiliary equation is $m^2 + \omega^2 = 0$, from which $m = 0 \pm i\omega$ and the equation has general solution $y = A\cos \omega x + B\sin \omega x$. This can also be put in the form $y = C\cos(\omega x + \phi)$ or $y = C\sin(\omega x + \theta)$. y oscillates sinusoidally with x.

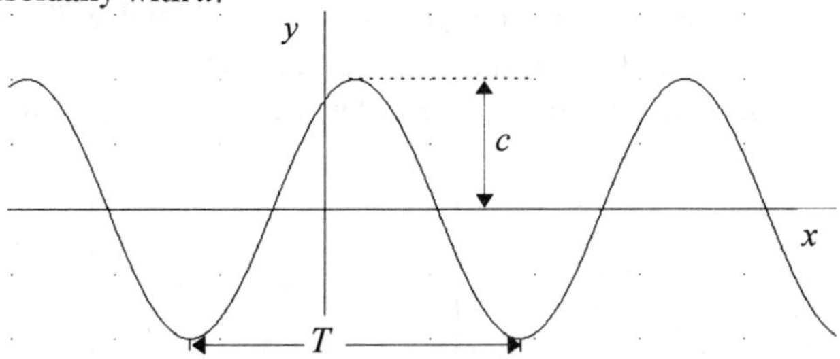

ω is the **angular frequency**. $C = \sqrt{A^2 + B^2}$ is the **amplitude**.
ϕ is the **phase**. $T = \frac{2\pi}{\omega}$ is the **period** of the oscillations.

(b) If $\frac{c}{a} < 0$ let $\omega^2 = -\frac{c}{a}$ so that the equation becomes $y'' - \omega^2 y = 0$. The auxiliary equation is $m^2 - \omega^2 = 0$, giving $m = \pm\omega$ and general solution $y = Ae^{\omega x} + Be^{-\omega x}$. Unless the initial conditions make $A = 0$, y grows exponentially in magnitude with x. That is, as $x \to \infty$, $y \to \infty$ or $-\infty$. A sketch of the solutions will resemble:

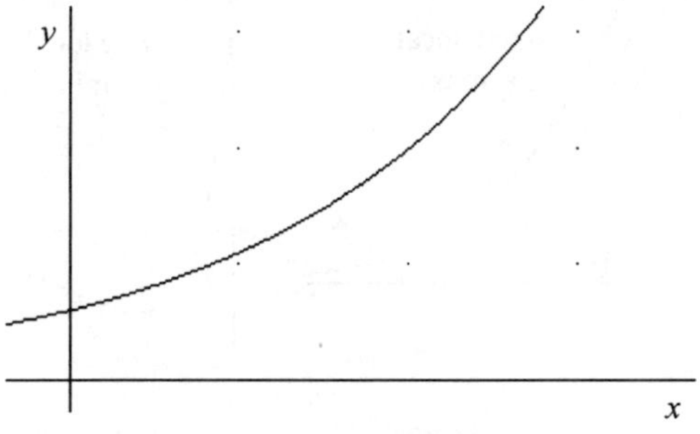

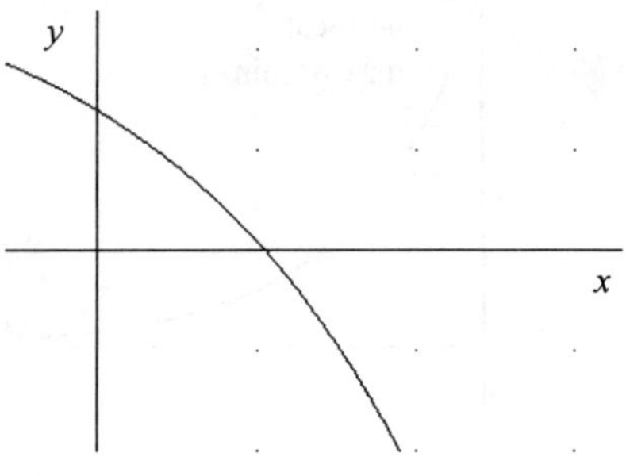

(ii) If $b^2 > 4ac$ then the auxiliary equation has real roots (which could be equal) given by

$$m_1 = \frac{-b+\sqrt{b^2-4ac}}{2a} \quad \text{and} \quad m_2 = \frac{-b-\sqrt{b^2-4ac}}{2a}$$

Depending on the relative signs and magnitudes of a, b and c, the roots m_1 and m_2 could be either (a) both positive or (b) both negative or (c) one positive and one negative.

(a) If m_1 and m_2 are both positive then $e^{m_1 x}$ and $e^{m_2 x} \to \infty$ as $x \to \infty$ and so, depending on the initial conditions, $y \to \infty$ or $-\infty$ as $x \to \infty$. The graph of y against x for $x \geq 0$ will resemble one of those below.

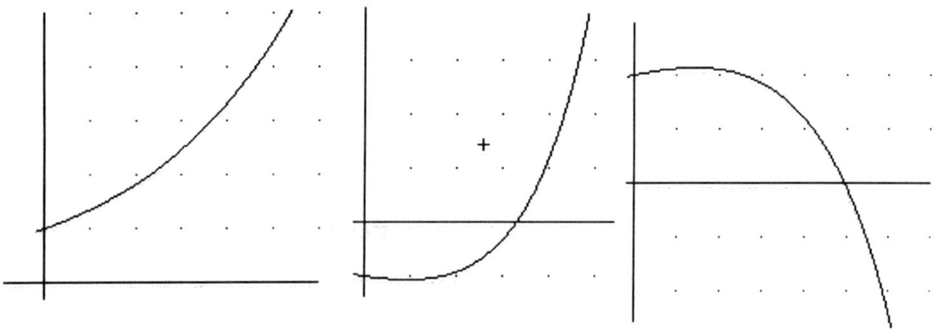

(b) If m_1 and m_2 are both negative $e^{m_1 x}$ and $e^{m_2 x} \to 0$ as $x \to \infty$. Hence $y \to 0$ as $x \to \infty$ whatever the initial conditions, as illustrated below.

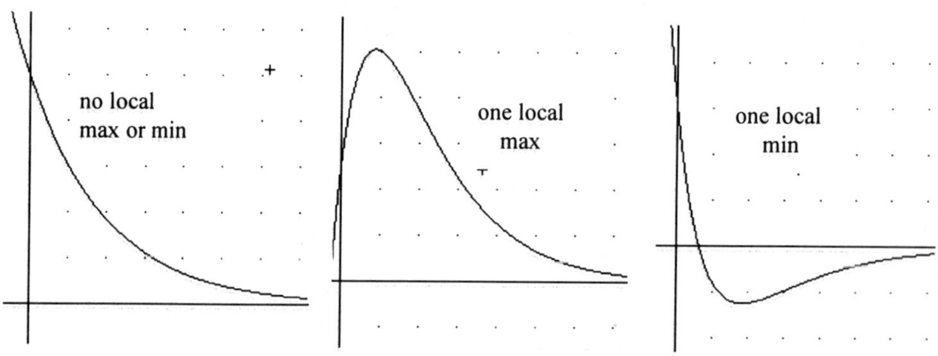

(c) If the roots are of opposite sign suppose $m_1 > 0$ and $m_2 < 0$. Then, as $x \to \infty$, $e^{m_1 x} \to \infty$ while $e^{m_2 x} \to 0$ and the solution $y = Ae^{m_1 x} + Be^{m_2 x}$ will increase in magnitude as $x \to \infty$ and the graph will resemble one of those in part (a). (The only exception to this would be if the initial conditions were such that $A = 0$, in which case $y \to 0$ as $x \to \infty$; the graph of y would then resemble one of those in part (b)).

(iii) If $b^2 < 4ac$ the auxiliary equation has complex roots $m = \dfrac{-b \pm i\sqrt{4ac-b^2}}{2a}$ and the general solution is

$$y = \mathrm{e}^{-\frac{bx}{2a}}\left(A\cos\frac{\sqrt{4ac-b^2}}{2a}x + B\sin\frac{\sqrt{4ac-b^2}}{2a}x\right)$$

The sine and cosine terms indicate that y oscillates as x increases. The amplitude of the oscillations is governed by $\mathrm{e}^{-\frac{b}{2a}x}$.

(a) If a and b have the same sign $\mathrm{e}^{-\frac{b}{2a}x}$ decreases as x increases and so $y \to 0$ as $x \to \infty$. The sinusoidal terms indicate that y oscillates, and the graph of y exhibits damped vibrations.

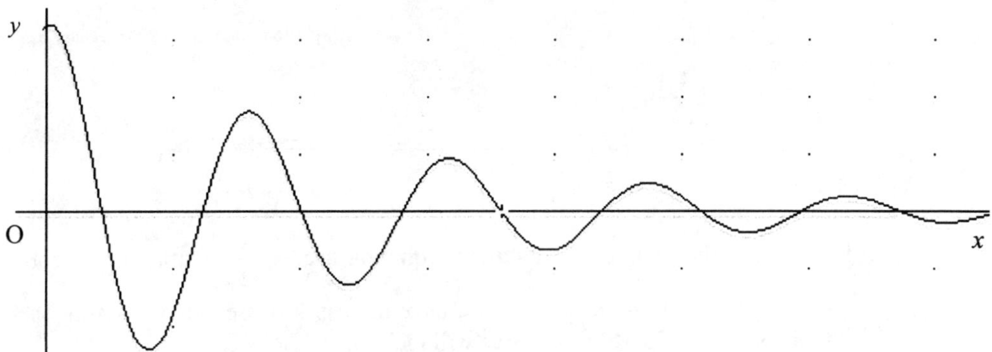

(b) If a and b have opposite signs then $\mathrm{e}^{-\frac{b}{2a}x}$ increases with x and the oscillations will have increasing magnitude.

Example 5H

Without solving the equations explicitly determine the form of the solution curve of each of the following equations and sketch their graphs for $x \geq 0$.

(i) $y'' + 2y' + 7y = 0, \quad y(0) = 4, y'(0) = 1$

(ii) $y'' - 6y' + 3y = 0, \quad y(0) = 1, y'(0) = 0$

(iii) $y'' + 9y' + 4y = 0, \quad y(0) = 2, y'(0) = 1$

Solution

(i) Here $b^2 - 4ac = 2^2 - 4 \times 1 \times 7 = -24 < 0$ and, as a and b have the same sign, we have case (iii) (a). Hence y oscillates with decreasing amplitude.

The condition $y(0) = 4$ shows where the graph crosses the y axis, while $y'(0) = 1$ gives the gradient of the curve at this point.

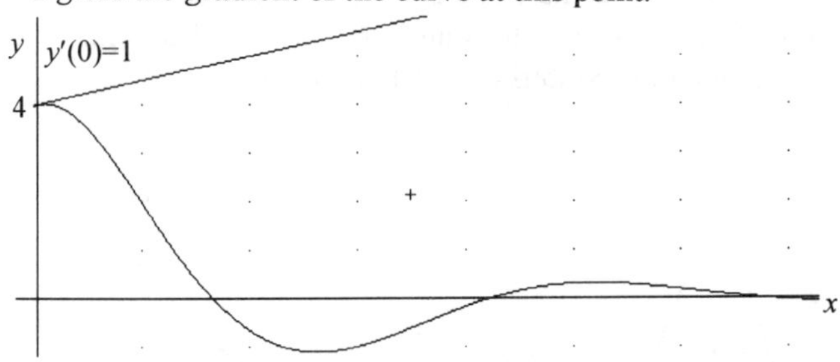

(ii) In this case the roots of the auxiliary equation are $\dfrac{6 \pm \sqrt{24}}{2}$ and so are both positive. Hence $y \to \infty$ or $y \to -\infty$ as x increases (Case (b)(i)). Using the initial conditions the solution curve will look like

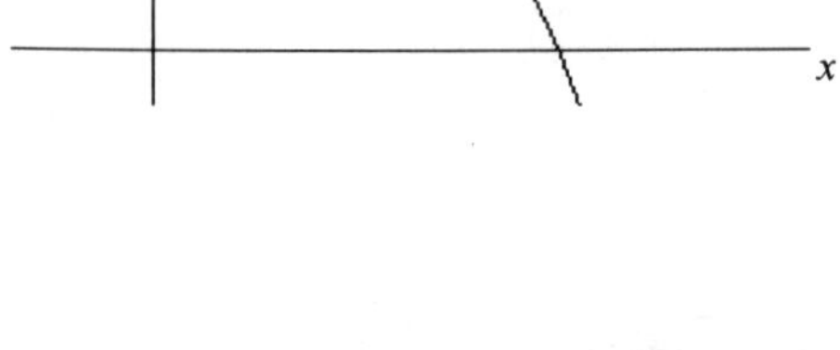

(iii) The roots of the auxiliary equation are $\dfrac{-9 \pm \sqrt{65}}{2}$ and are both negative. It follows that $y \to 0$ as $x \to \infty$ (Case (b)(ii)). The initial conditions give the intercept and slope at $x = 0$.

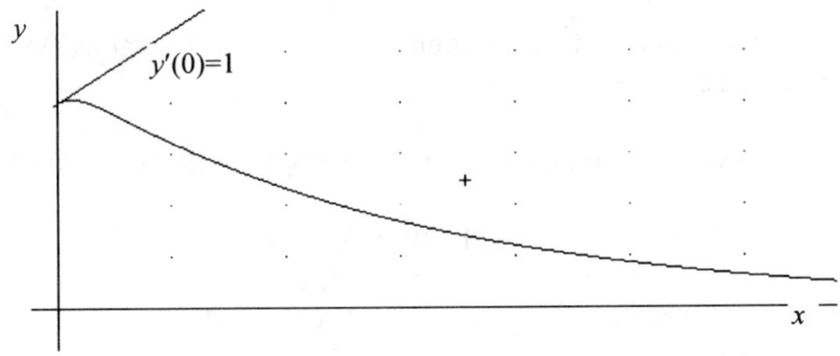

□

Exercise 5A

1. Find the general solution of

(i)	$y'' - y' - 2y = 0$	(ii)	$y'' + 4y' + 4y = 0$
(iii)	$y'' + 2y' + 5y = 0$	(iv)	$4y'' + 12y' + 9y = 0$
(v)	$6y'' + 5y' - 4y = 0$	(vi)	$9y'' - 18y' + 10y = 0$
(vii)	$y'' - 4y = 0$	(viii)	$y'' + 4y = 0$
(ix)	$y'' + 4y' = 0$	(x)	$25y'' - 20y' + 4y = 0$
(xi)	$8y'' - 6y' - 35y = 0$	(xii)	$9y'' - 12y' + 8y = 0$

2. Find the particular solution of

(i) 1 (viii) given $y(0) = 0$, $y'(0) = 2$
(ii) 1 (vi) given $y(0) = 0$, $y'(0) = 1$
(iii) 1 (i) given $y(0) = 4$, $y'(0) = -1$
(iv) 1 (iii) given $y(0) = 1$, $y'(0) = 3$
(v) 1 (ix) given $y(0) = 1$, $y'(0) = 4$
(vi) 1 (v) given $y(0) = 2$ and $y \to 0$ as $x \to \infty$.
(vii) 1 (vii) given $y(0) = 3$, $y'(0) = 2$ and show that it can be expressed in the form $y = 3\cosh 2x + \sinh 2x$.

In each case sketch the graph of the solution y for $x \geq 0$.

3. Solve the equation $y'' + ky' + 4y = 0$, $y(0) = 0$, $y'(0) = 3$ when k takes the value (i) 0, (ii) 2, (iii) 4, (iv) 5 and sketch the graph of the solution in each case.

4. Construct the differential equation whose solution is

(i) $y = Ae^{x} + Be^{-5x}$ (ii) $y = e^{4x}(A\cos 3x + B\sin 3x)$

(iii) $y = e^{4x}(A + Bx)$

5. Find the value of a so that the solution of $y'' - y' - 2y = 0$, $y(0) = a$, $y'(0) = 3$ approaches zero as x increases.

6. Without solving the equations, sketch the graphs of the solutions of

(i) $y'' - 4y' + 9y = 0$ with $y(0) = 0$, $y'(0) = 2$
(ii) $y'' - 8y' + 4y = 0$ with $y(0) = -1$, $y'(0) = -1$
(iii) $y'' + y' + y = 0$ with $y(0) = 1$, $y'(0) = -1$
(iv) $y'' + 6y' + y = 0$ with $y(0) = -1$, $y'(0) = 5$

7. Select in each case the graph showing the solution of the following equations with the initial equations $y(0) = 1$, $y'(0) = 0$.

(i) $y'' + 4y' + 5y = 0$ (ii) $y'' - 4y' + 5y = 0$
(iii) $y'' + 9y = 0$ (iv) $y'' - 9y = 0$
(v) $y'' - 6y' + 16y = 0$ (vi) $y'' + 8y' + y = 0$

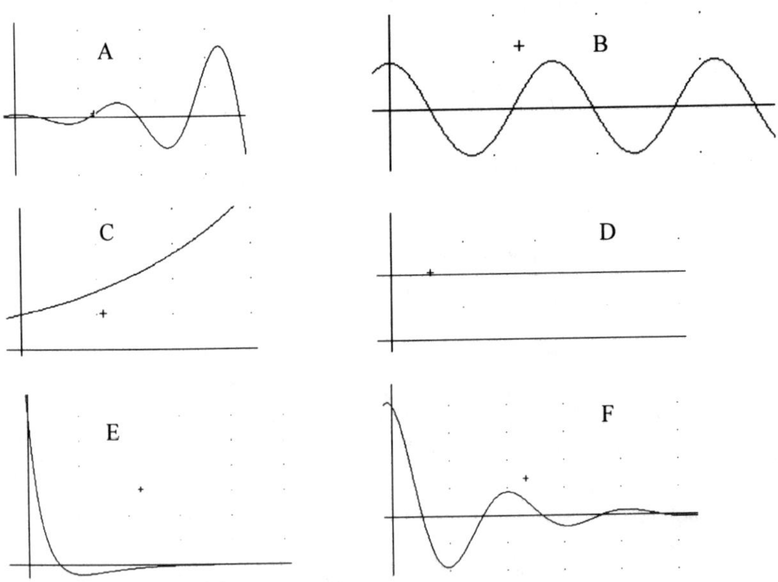

8. The equation $\frac{d^2y}{dx^2}+n^2y=0$, where $n^2=\frac{P}{EI}$, gives the deflection, y, of a beam due to a force P acting along its length at each end. The ends of the beam remain on the x axis (i.e. $y = 0$ at $x = 0$ and $x = l$). Show that the smallest force P that will hold the beam in a deflected position is $P=\frac{EI\pi^2}{l^2}$.

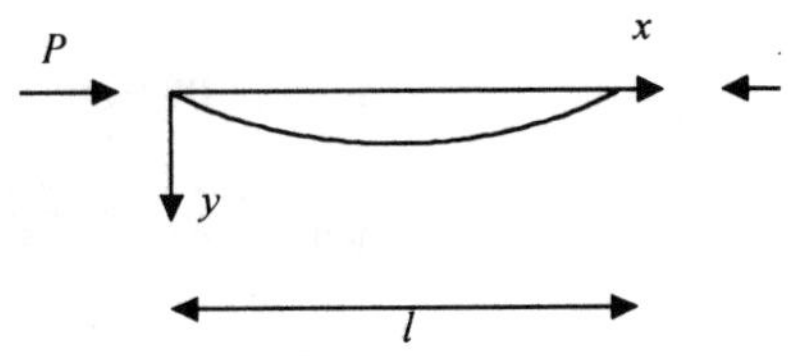

DERIVE ACTIVITY 5A

This activity introduces the DERIVE library functions for solving second order differential equations of the form

$$\frac{d^2y}{dx^2}+p(x)\frac{dy}{dx}+q(x)y=r(x) \quad \text{or} \quad y''+p(x)y'+q(x)y=r(x)$$

Notice that in its standard form the coefficient of y'' is 1 and the coefficients p, q and r can be functions of x. In this activity p and q will be constants and the differential equations will be homogeneous so that $r(x) = 0$.

File Load the **Utility** file ode2.mth.

The library function DSOLVE2(p,q,r,x,a,b) will give the general solution with arbitrary constants a and b. If a and b are omitted, DERIVE uses the constants of integration c_1 and c_2.

(A) (i) To solve the differential equation

$$y''-2.5y'+y=0$$

Author DSOLVE2(−2.5,1,0,x,a,b) and **Simplify**.

(ii) **Author** DSOLVE2(−2.5,1,0,x) and **Simplify**.

Compare your answers to parts (i) and (ii)

```
#1:  DSOLVE2(-2.5, 1, 0, x, a, b)

            2·x          0.5·x
#2:  a·ê        + b·ê

#3:  DSOLVE2(-2.5, 1, 0, x)

             2·x           0.5·x
#4:  c1·ê        + c2·ê
```

Figure 5.1 Using DSOLVE2 to solve differential equations

(B) When using DSOLVE2 it is important to write the equation in the standard form with the coefficient of y'' being 1. Use DSOLVE2 to find the general solution of the following differential equations.

(i) $3y'' + 5y' + 2y = 0$
(ii) $y'' - 6y' + 9y = 0$
(iii) $y'' + 2y' + 10y = 0$
(iv) $4y'' + 9y = 0$

When initial conditions are given in a problem these can be input directly into the DERIVE library function DSOLVE2_IV(p,q,r,x,x_0,y_0,v_0) to obtain a particular solution. x_0 and y_0 are the initial values of x and y and v_0 is the initial value of y' i.e. $y_0 = y(x_0)$ and $v_0 = y'(x_0)$.

(C) (i) To solve the problem

$$y'' + 5y' + 4y = 0$$

given $y(0) = 1$ and $y'(0) = 2$

Author DSOLVE2_IV(5,4,0,x,0,1,2) and **Simplify**.

(ii) Now plot the graph of the particular solution. Figure 5.2 shows the outcome of this activity. In the graphics window **Scale** has been set to x:1, y:0.5 with **Centre** at (1.5,0.5).

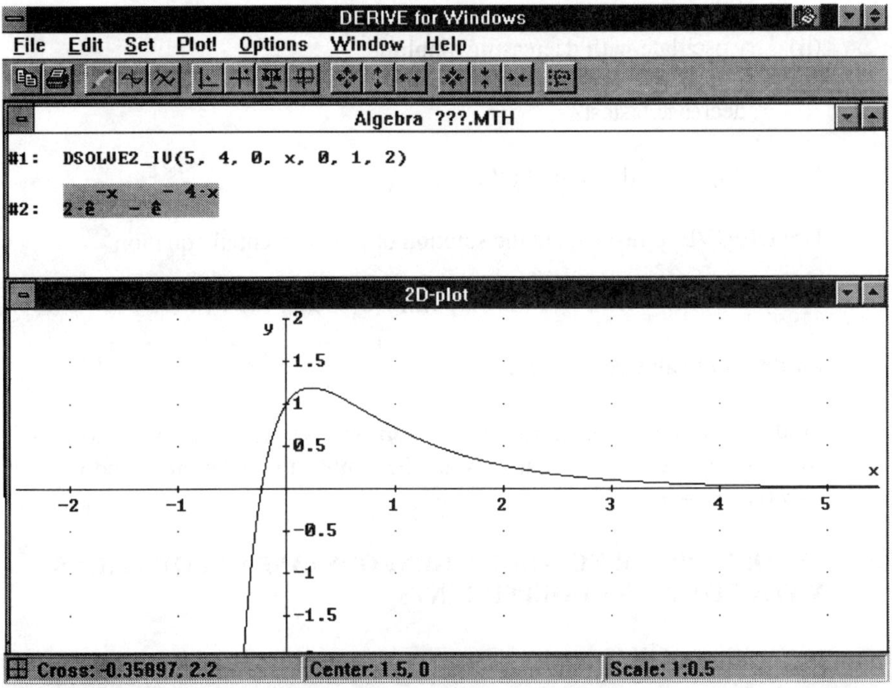

Figure 5.2 Solving and graphing the particular solution of an initial value problem

(D) Solve each of the following problems using DERIVE and **Plot!** a graph of the solution in each case. From the graph describe the type of solution giving the period and amplitude for those which represent oscillations.

(i)	$y'' + 4y = 0$	$y(0) = 0, y'(0) = 2$
(ii)	$y'' - 8y' + 4y = 0$	$y(0) = -1, y'(0) = 1$
(iii)	$y'' - 4y' + 9y = 0$	$y(0) = 0, y'(0) = 2$
(iv)	$y'' + y' + y = 0$	$y(0) = 1, y'(0) = -1$

(E) Use DERIVE to investigate the solution of the differential equation

$$y'' + py' + 16y = 0$$

for values of $p \geq 0$.

In particular when will

(i) y oscillate sinusoidally,
(ii) y oscillate with decreasing amplitude,
(iii) y decrease towards zero exponentially? For what value of p does y decrease fastest?

What happens to the solution if $p < 0$?

(F) Use DERIVE to investigate the solution of the differential equation

$$y'' + 3y' - 4y = 0 \qquad \text{with } y(0) = y_0,\ y'(0) = v_0$$

for different values of y_0 and v_0.

Find initial conditions so that (i) $y \to \infty$ as $x \to \infty$, (ii) $y \to -\infty$ as $x \to \infty$, (iii) $y \to 0$ as $x \to \infty$. What must be the connection between y_0 and v_0 so that $y \to 0$ as $x \to \infty$?

5.3 SECOND ORDER NON-HOMOGENEOUS LINEAR EQUATIONS WITH CONSTANT COEFFICIENTS

These are equations of the form

$$ay'' + by' + cy = f(x) \tag{5.4}$$

5.3.1 The form of the solution

We are interested in solving (5.4) for $f(x)$ in the form

(i) polynomial — example: $x^2 + 2x - 1$

(ii) e^{ax} — example: e^{4x}

(iii) $A\sin \alpha x$ and/or $B\cos \alpha x$ — example: $5\cos 3x$

(iv) combinations of (i), (ii) and (iii) — example: x^2e^x or $3 + 2\sin 5x$

We first find the general solution of $ay'' + by' + cy = 0$. Let this be $y = y_1(x)$

$y_1(x)$ is called the **Complementary Function** (CF)

Consider $y = y_1(x) + y_2(x)$ ($y_2(x)$ unspecified). Substitute in equation (5.4)

$$ay''+by'+cy = a\frac{d^2}{dx^2}[y_1+y_2]+b\frac{d}{dx}[y_1+y_2]+c[y_1+y_2]$$

$$= a\frac{d^2y_1}{dx^2}+a\frac{d^2y_2}{dx^2}+b\frac{dy_1}{dx}+b\frac{dy_2}{dx}+cy_1+cy_2$$

$$= \left(a\frac{d^2y_1}{dx^2}+b\frac{dy_1}{dx}+cy_1\right)+\left(a\frac{d^2y_2}{dx^2}+b\frac{dy_2}{dx}+cy_2\right)$$

$$= a\frac{d^2y_2}{dx^2}+b\frac{dy_2}{dx}+cy_2$$

(since y_1 is the CF and so satisfies $ay'' + by' + cy = 0$)

Thus $y = y_1 + y_2$ is the solution of (5.4) if

$$ay_2'' + by_2' + cy_2 = f(x)$$

That is if y_2(x) is a particular solution, or **particular integral**, (PI) of (5.4).

Thus the GS of		$ay'' + by' + cy = f(x)$
takes the form		y = CF + PI
We must		
	(i)	solve $ay'' + by' + cy = 0$ to find the CF
	(ii)	find a PI of $ay'' + by' + cy = f(x)$
	(iii)	add (i) and (ii) to find the GS

5.3.2 Finding Particular Integrals

We must find a function y that satisfies $ay'' + by' + cy = f(x)$. Consider, for example, the equation $y'' - 6y' + 9y = 8e^{2x}$; we must find a y which, when substituted into the left hand side, simplifies to $8e^{2x}$. Bearing in mind that the exponential function differentiates to itself the required y must look something like $8e^{2x}$ as there is no possibility of a y involving a polynomial or sinusoid or different exponential such as e^{5x} turning the left hand side into $8e^{2x}$. Thus the technique is to assume a form for the particular integral which is similar to $f(x)$; in the above example we would assume the particular integral takes the form $y = ae^{2x}$.

(i) $f(x)$ a constant

Assume the particular integral (PI) is $y = c$, substitute into the equation and determine c.

Example 5I

Find a PI of $y'' - 3y' + 2y = 8$.

Solution

Assume the PI is $y = c$. Then $y' = y'' = 0$ and the left hand side of the equation becomes $2c$. This equals 8 if $c = 4$ and so a PI is $y = 4$.

□

(ii) $f(x)$ a polynomial

Assume the particular integral is the most general polynomial of the same degree.

Examples:	$f(x)$	assumed form of PI
	$2x - 1$	$ax + b$
	$x^2 - x + 3$	$ax^2 + bx + c$
	x^3	$ax^3 + bx^2 + cx + d$

Example 5J

Find a PI of $2y'' - 5y' + 2y = 4x + 2$.

Solution

Assume the PI is $y = ax + b$. Then $y' = a$ and $y'' = 0$. Substituting in the equation

$$2 \times 0 - 5a + 2(ax + b) = 4x + 2$$

Rearranging:

$$2ax + (2b - 5a) = 4x + 2$$

For equivalence we require $2a = 4$, giving $a = 2$ and $2b - 5a = 2$, giving $b = 6$. Hence a PI is $y = 2x + 6$.

□

Example 5K

Find a PI of $y'' + 2y' + 10y = 50x^2$.

Solution

As the right hand side is a quadratic, assume that the PI is $y = ax^2 + bx + c$. Then $y' = 2ax + b$ and $y'' = 2a$. Substituting into the equation:

$$2a + 2(2ax + b) + 10(ax^2 + bx + c) = 50x^2$$

$$10ax^2 + (4a + 10b)x + (2a + 2b + 10c) = 50x^2$$

For equality we require

$10a = 50,$ giving $a = 5$

$4a + 10b = 0,$ from which $b = -2$

and $2a + 2b + 10c = 0,$ giving $c = -\frac{3}{5}$

Hence a PI is

$$y = 5x^2 - 2x - \frac{3}{5}$$

□

(iii) $f(x)$ an exponential

Assume the particular integral is an exponential function of the same form. Examples:

$f(x)$	assumed form of PI
e^{3x}	ae^{3x}
$6e^{-x}$	ae^{-x}

Example 5L

Find a PI of $y'' - 6y' + 9y = 100e^{-2x}$.

Solution

Assume $y = ae^{-2x}$. Then $y' = -2ae^{-2x}$ and $y'' = 4ae^{-2x}$. Substituting

$$4ae^{-2x} + 12ae^{-2x} + 9ae^{-2x} = 100e^{-2x}$$

giving $25a = 100$, so that $a = 4$ and a particular integral is

$$y = 4e^{-2x}$$

□

(iv) $f(x)$ sinusoidal

$f(x)$ will be of the form $c \cos \alpha x$ or $d \sin \alpha x$ or $c \cos \alpha x + d \sin \alpha x$.

Method 1

Assume the particular integral is $y = a \cos \alpha x + b \sin \alpha x$. Examples:

$f(x)$	assumed form of PI
$2 \sin x$	$a \cos x + b \sin x$
$10 \cos 3x$	$a \cos 3x + b \sin 3x$
$5 \sin 2x - 4 \cos 2x$	$a \cos 2x + b \sin 2x$

Example 5M

Find a PI of $y'' - y' - 2y = 20 \sin 2x$.

Solution

Assume $y = a \cos 2x + b \sin 2x$. Then

$$y' = -2a \sin 2x + 2b \cos 2x \quad \text{and} \quad y'' = -4a \cos 2x - 4b \sin 2x$$

Substituting:

$$(-4a\cos2x - 4b\sin2x) - (-2a\sin2x + 2b\cos2x) - 2(a\cos2x + b\sin2x) = 20\sin2x$$

$$(-6a - 2b)\cos2x + (2a - 6b)\sin2x = 20 \sin2x$$

For equality the coefficients of $\cos 2x$, and of $\sin 2x$, on each side of the equation must be the same. Then

$$-6a - 2b = 0 \quad \text{and} \quad 2a - 6b = 20$$

Solving these equations simultaneously gives $a = 1$ and $b = -3$ and so a PI is $y = \cos 2x - 3\sin 2x$.

□

Method 2

This approach makes use of Euler's formula $e^{i\theta} = \cos\theta + i\sin\theta$, from which $\cos\theta = \text{Re}(e^{i\theta})$. Multiplying Euler's formula by $-i$ gives $-ie^{i\theta} = -i\cos\theta + \sin\theta$, from which $\sin\theta = \text{Re}(-ie^{i\theta})$. Hence a sine or cosine term can be expressed as the real part of a complex number. For example

$$\cos 2x = \text{Re}(e^{2ix})$$

$$3\sin 4x = \text{Re}(-3ie^{4ix})$$

$$4\cos x + 5\sin x = \text{Re}(4e^{ix}) + \text{Re}(-5ie^{ix}) = \text{Re}[(4-5i)e^{ix}]$$

In general

$$f(x) = c\cos\alpha x + d\sin\alpha x = \text{Re}[(c-di)e^{i\alpha x}]$$

Assume the particular integral is of the form $y = \text{Re}[ze^{i\alpha x}]$ substitute into the equation to find z, and then find y.

Example 5N

Find a PI of $y'' - y' - 2y = 20\sin 2x$ in the form (i) $a\cos 2x + b\sin 2x$ and (ii) $r\cos(2x+\alpha)$.

Solution

Rewriting the right hand side, the equation becomes

$$y'' - y' - 2y = \text{Re}(-20ie^{2ix})$$

Assume $y = \text{Re}(ze^{2ix})$. Then $y' = \text{Re}(2ize^{2ix})$ and $y'' = \text{Re}(-4ze^{2ix})$. Substituting

$$\text{Re}(-4ze^{2ix}) - \text{Re}(2ize^{2ix}) - 2\text{Re}(ze^{2ix}) = \text{Re}(-20ie^{2ix})$$

$$\text{Re}[(-6-2i)ze^{2ix}] = \text{Re}(-20ie^{2ix})$$

For equality we require $(-6-2i)z = -20i$ from which

$$z = \frac{-20i}{-6-2i} = \frac{10i}{3+i} = \frac{10i(3-i)}{(3+i)(3-i)} = \frac{10(1+3i)}{10} = 1+3i$$

(i) Hence a PI is

$$y = \mathrm{Re}(ze^{2ix})$$

$$y = \mathrm{Re}[(1 + 3i)(\cos 2x + i\sin 2x)]$$

$$y = \cos 2x - 3\sin 2x, \qquad \text{as in Example 5M.}$$

(ii) To express the particular integral in the form $r\cos(2x + \alpha)$ we must first put $z = 1 + 3i$ in exponential form.

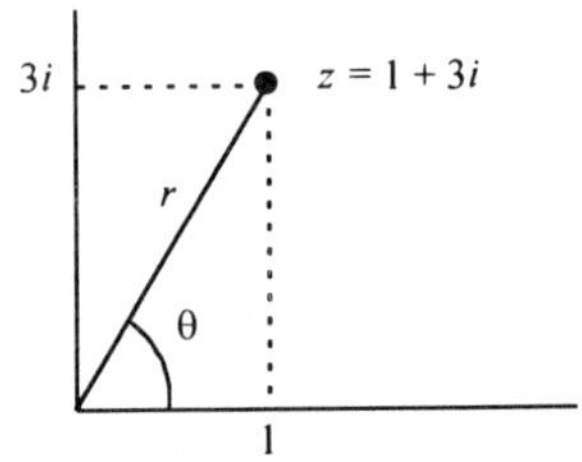

Then $r = \sqrt{1^2 + 3^2} = \sqrt{10}$

and $\theta = \tan^{-1}3 = 71.6°$ or 1.25 radians.

Hence $z = 1 + 3i = \sqrt{10}e^{1.25i}$

The PI is

$$y = \mathrm{Re}(ze^{2ix})$$
$$y = \mathrm{Re}(\sqrt{10}e^{1.25i} \cdot e^{2ix})$$
$$y = \mathrm{Re}(\sqrt{10}e^{(2x+1.25)i})$$
$$y = \sqrt{10}\cos(2x + 1.25)$$

or $y = \sqrt{10}\cos(2x + 71.6°)$

□

(v) $f(x)$ the sum of two different types of functions

The assumed form for the particular integral is the sum of the assumed forms for the particular integrals corresponding to the separate terms. Examples:

$f(x)$	assumed form of PI
$2x + e^x$	$y = ax + b + ce^x$
$\begin{cases} e^{2x} - \sin x \\ e^{2x} - \mathrm{Re}(-ie^{ix}) \end{cases}$	$\begin{cases} y = ae^{2x} + b\cos x + c\sin x \\ y = ae^{2x} + \mathrm{Re}[ze^{ix}] \end{cases}$

Example 5O

Find a PI of $y'' + 6y' + 8y = 4x + 3e^{-x}$.

Solution

Assume $y = ax + b + ce^{-x}$. Then $y' = a - ce^{-x}$ and $y'' = ce^{-x}$. Substituting and collecting terms

$$8ax + (6a + 8b) + 3ce^{-x} = 4x + 3e^{-x}$$

For equality

$$8a = 4, \quad \text{giving } a = \tfrac{1}{2}$$

$$6a + 8b = 0, \quad \text{giving } b = -\tfrac{3}{8}$$

$$3c = 3, \quad \text{giving } c = 1.$$

Hence a PI is

$$y = \tfrac{1}{2}x - \tfrac{3}{8} + e^{-x}$$

□

(vi) $f(x)$ the product of two functions

The assumed form of the particular integral is the product of the assumed forms for the particular integrals corresponding to the separate terms. Note that the product of two constants can be replaced as a single constant. For example the particular integral for xe^x would be the product of that for x, $(ax + b)$, with that for e^x (ce^x). That is

$$y = (ax + b)ce^x$$

This can be rewritten as

$$y = (acx + bc)e^x = (px + q)e^x$$

where $p = ac$ and $q = bc$. p and q can be replaced with a and b if preferred. Examples:

$f(x)$	assumed for of PI
xe^x	$(ax + b)e^x$
$e^x \sin 5x$	$e^x(a\cos 5x + b\sin 5x)$
$x \cos 2x$	$a\cos 3x + b\sin 3x + ax\cos 3x + dx \sin 3x$

Example 5P

Find a PI of $y'' - 2y' + 2y = 25xe^{-x}$.

Solution

Assume $y = (ax + b)e^{-x}$. Then

$$y' = -(ax + b)e^{-x} + ae^{-x} \quad \text{and} \quad y'' = (ax + b)e^{-x} - ae^{-x} - ae^{-x}$$

Substituting and collecting terms

$$5axe^{-x} + (5b - 4a)e^{-x} = 25xe^{-x}$$

from which $5a = 25$, giving $a = 5$ and $5b - 4a = 0$, given $b = 4$. Hence a PI is

$$y = (5x + 4)e^{-x}$$

□

Cases of Failure

Occasionally the function suggested in (i) to (vi) above fails to provide a particular integral. Consider, for example, the equation

$$y'' - 3y' + 2y = 6e^x$$

To find the particular integral we would try $y = ae^x$, from which $y' = ae^x$ and $y'' = ae^x$. The left hand side of the equation then becomes

$$y'' - 3y' + 2y = ae^x - 3ae^x + 2ae^x = 0$$

and so can never equal the right hand side $6e^x$.

To see why this happens consider the associated homogeneous equation $y'' - 3y' + 2y = 0$. This equation was solved in Example 5D and has general solution $y = Ae^x + Be^{2x}$. We also showed in Example 5C that $y = e^x$ and $y = e^{2x}$ were solutions of $y'' - 3y' + 2y = 0$. The trouble with our guess $y = ae^x$ is that it contains e^x, which is a solution of the associated homogeneous equation (and part of the complementary function for this equation); this is why its substitution into the left hand side of the equation gives zero. We are in a case of failure situation when the right hand side of the equation is the same as one of the terms in the complementary function.

Let us amend our choice of particular integral to $y = g(x)e^x$ and we will find $g(x)$ so that the equation $y'' - 3y' + 2y = 6e^x$ is satisfied. If $y = g(x)e^x$, or $y = ge^x$ for brevity, then

$$y' = ge^x + g'e^x \quad \text{and} \quad y'' = (ge^x + g'e^x) + (g'e^x + g''e^x) = ge^x + 2g'e^x + g''e^x$$

Substituting into the equation

$$(ge^x + 2g'e^x + g''e^x) - 3(ge^x + g'e^x) + 2ge^x = 6e^x$$

$$(g'' - g')e^x = 6e^x$$

For equality we require $g'' - g' = 6$. If we make the substitution $z = g'$, so that $z' = g''$, the equation becomes $z' - z = 6$, which is a first order linear equation that can be solved by the integrating factor method. Multiplying throughout by the integrating factor $e^{\int -1dx} = e^{-x}$, we find

$$e^{-x}z' - e^{-x}z = 6e^{-x}$$

or $$\frac{d}{dx}(e^{-x}z) = 6e^{-x}$$

Integrating:

$$e^{-x}z = -6e^{-x} + A$$

$$z = -6 + Ae^x$$

or $$g' = -6 + Ae^x$$

Integrating

$$g = -6x + Ae^x + B$$

and so the particular integral is

$$y = g(x)e^x = (-6x + Ae^x + B)e^x$$

or $$y = -6xe^x + Ae^{2x} + Be^x$$

We need to find **any** particular integral and so, putting $A = B = 0$, a particular integral is

$$y = -6xe^x$$

Note that whereas our original guess at the PI contained e^x, we now know it contains xe^x.

We do not need to go through an analysis like that above for every example. When the function on the right hand side (here e^x) is part of the complementary function (here $Ae^x + Be^{2x}$) the usual guess at the particular integral will fail and the new guess is found by multiplying the previous one by x. That is

new assumed PI = x.old assumed PI

Example 5Q

Find a PI of $y'' - 3y' + 2y = 6e^x$.

Solution

Since e^x is part of the complementary function assume the particular integral takes the form $y = axe^x$. Then

$$y' = axe^x + ae^x \quad \text{and} \quad y'' = axe^x + ae^x + ae^x = axe^x + 2ae^x$$

Substituting in the equation

$$(axe^x + 2ae^x) - 3(axe^x + ae^x) + 2axe^x = 6e^x$$

which simplifies to

$$-ae^x = 6e^x$$

giving $a = -6$. Hence a PI is

$$y = -6xe^x, \text{ as above.}$$

□

Example 5R

Find a PI of $y'' + 4y = 8\sin 2x$.

Solution

We would normally assume a particular integral of the form $y = a\cos 2x + b\sin 2x$. This will fail as the right hand side, $\sin 2x$ is part of the complementary function, which is $y = A\cos 2x + B\sin 2x$ (see Example 5 F (iv)).

Multiplying by x, assume $y = ax\cos 2x + bx\sin 2x$.

Then $y' = -2ax\sin 2x + a\cos 2x + 2bx\cos 2x + b\sin 2x$

and $y'' = -4ax\cos 2x - 2a\sin 2x - 2a\sin 2x - 4bx\sin 2x + 2b\cos 2x + 2b\cos 2x$

$$= -4ax\cos 2x - 4a\sin 2x - 4bx\sin 2x + 4b\cos 2x$$

Substituting into the equation

$$(-4ax\cos 2x - 4a\sin 2x - 4bx\sin 2x + 4b\cos 2x) + 4(ax\cos 2x + bx\sin 2x) = 8\sin 2x$$

$$-4a\sin 2x + 4b\cos 2x = 8\sin 2x$$

from which $-4a = 8$, giving $a = -2$ and $4b = 0$, or $b = 0$. Hence a PI is

$$y = -2x\cos 2x$$

□

Example 5S

Are the following cases of failure? Give the form of the particular integral for each.

(i) $y'' + 4y = 2x + 1$ (ii) $y'' + 4y = 3\cos 2x$ (iii) $y'' + 4y = e^x\sin 2x$

(iv) $y'' - 3y' + 2y = 10e^{2x}$ (v) $y'' - 6y + 9y = 5e^{3x}$

Solution

(i) We saw in Example 5R that the complementary function is

$$y = A\cos 2x + B\sin 2x$$

Since $2x + 1$ is not one of these terms this is **not** a case of failure and the particular integral takes the form $y = ax + b$.

(ii) As in (i) the CF is $y = A\cos 2x + B\sin 2x$. As $\cos 2x$ is one of these terms this **is** a case of failure and the particular integral takes the form

$$y = ax\cos 2x + bx\sin 2x$$

(iii) As in (i) the CF is $y = A\cos 2x + B\sin 2x$. As $e^x\sin 2x$ is not one of these terms this is **not** a case of failure and the particular integral takes the form

$$y = e^x(a\cos 2x + b\sin 2x)$$

(iv) From Example 5D the complementary function is $y = Ae^x + Be^{2x}$. As e^{2x} is one of these terms this **is** a case of failure and the particular integral takes the form

$$y = axe^{2x}.$$

(v) From Example 5F (ii) the complementary function is $y = e^{3x}(Ax + B)$. As e^{3x} is one of the terms in the CF this **is** a case of failure. Multiplying by x, the PI becomes $y = axe^{3x}$, but as this is the other term in the CF this also fails. Multiplying by x again, the particular integral takes the form

$$y = ax^2e^{3x}$$

□

Exercise 5B

1. Find particular integrals for the following equations.

(i) $y'' - 4y = 100$ (ii) $y'' - y' - 2y = 2 + 5x$

(iii) $y'' + 2y' + 5y = 1 + x + x^2$

(iv) $25y'' - 20y' + 4y = 4x^3 - 20x^2 - 150x - 200$

(v) $y'' + 4y' = e^x$ (vi) $9y'' - 18y' + 10y = 20e^{2x}$

(vii) $8y'' - 6y' - 35y = e^{3x}$ (viii) $9y'' - 12y' + 8y = 10e^x$

(ix) $y'' + 4y' + 4y = 50\cos x$ (x) $y'' - y' - 2y = 4\cos 2x - 3\sin 2x$

(xi) $4y'' + 12y' + 9y = 12\cos 3x - 9\sin 3x$

(xii) $6y'' + 5y' - 4y = 221\sin 2x$

2. (i) Use the results of question 1 to find a particular integral for

$$y'' - y' - 2y = 2 + 5x + 4\cos 2x - 3\sin 2x$$

(ii) Find the particular integrals for

(a) $y'' + 2y' - 3y = 8e^{-x} - 9x$ (b) $y'' + 4y = 4e^{2x} + 6\cos x$

(c) $y'' - 9y = 25xe^{2x}$ (d) $y'' + y' = 5e^{2x}\sin x + 15e^{2x}\cos x$

3. Show that the initial guess of the form of a particular integral in each of the following equations fails. Find a particular integral in each case.

(i) $y'' + 4y' = 12$ (ii) $y'' - y' - 2y = 15e^{2x}$

(iii) $y'' - y' - 2y = 6e^{-x}$ (iv) $y'' + 4y = 12\sin 2x$

(v) $y'' + 4y' + 4y = 6e^{-2x}$

4. Show that the complementary function of the equation $y'' - 3y' + 2y = f(x)$ is

$$y = Ae^{x} + Be^{2x}$$

Give the **form** of the particular integral for each $f(x)$ given. (Do **not** find the particular integral explicitly.)

(i) $f(x) = 10x$ (ii) $f(x) = 2\sin 3x$ (iii) $f(x) = 4e^{x}$

(iv) $f(x) = e^{-x} + e^{2x}$ (v) $f(x) = e^{2x}\cos x$ (vi) $f(x) = 7xe^{x}$

5.3.3 The general solution of $ay'' + by' + cy = f(x)$

We saw in Section 5.3.1 that the general solution (GS) is found by adding the complementary function (CF), which is the solution of $ay'' + by' + cy = 0$, to the particular integral (PI), which is any solution of $ay'' + by' + cy = f(x)$. That is, the GS is

$$y = \text{CF} + \text{PI}$$

Particular solutions are found by applying initial or boundary conditions to the general solution.

Example 5T

Use the results of previous examples in Sections 5.3.1 and 5.3.2 to write down the general solution of

(i) $y'' - 3y' + 2y = 8$ (ii) $2y'' - 5y' + 2y = 4x + 2$

(iii) $y'' - 6y' + 9y = 100e^{-2x}$ (iv) $y'' - y' - 2y = 20\sin 2x$

(v) $y'' + 6y' + 8y = 4x + 3e^{-x}$ (vi) $y'' + 4y = 8\sin 2x$

Solution

(i) The CF and PI are, from Examples 5D and 5I respectively, $y = Ae^{x} + Be^{2x}$ and $y = 4$. Hence the general solution is $y = Ae^{x} + Be^{2x} + 4$.

(ii) From Example 5E and 5J the CF and PI are $y = Ae^{\frac{1}{2}x} + Be^{2x}$ and $y = 2x + 6$, giving the general solution $y = Ae^{\frac{1}{2}x} + Be^{2x} + 2x + 6$. Similarly:

	Equation	CF	PI	GS
(iii)	$y'' - 6y' + 9y = 100e^{-2x}$	$e^{3x}(A + Bx)$	$4e^{-2x}$	$y = e^{3x}(A + Bx) + 4e^{-2x}$
(iv)	$y'' - y' - 2y = 20\sin 2x$	$Ae^{-x} + Be^{2x}$	$\cos 2x - 3\sin 2x$	$y = Ae^{-x} + Be^{2x} + \cos 2x - 2\sin 2x$
(v)	$y'' + 6y' + 8y = 4x + 3e^{-x}$	$Ae^{-4x} + Be^{-2x}$	$\frac{1}{2}x - \frac{3}{8} + e^{-x}$	$y = Ae^{-4x} + Be^{-2x} + \frac{1}{2}x - \frac{3}{8} + e^{-x}$
(vi)	$y'' + 4y = 8\sin 2x$	$A\cos 2x + B\sin 2x$	$-2x\cos 2x$	$y = A\cos 2x + B\sin 2x - 2x\cos 2x$

□

5.3.4 Particular solutions of $ay'' + by' + cy = f(x)$

The general solution contains two arbitrary constants and so two conditions are required to evaluate them.

Example 5U

Find the particular solution of

(i) $y'' + 6y' + 8y = 4x + 3e^{-x}$ given $y(0) = 0$, $y'(0) = 4$

(ii) $y'' + 4y = 8\sin 2x$ given $y(0) = 1$, $y'(0) = 2$.

Solution

(i) The general solution is, from Example 5T (v),

$$y = Ae^{-4x} + Be^{-2x} + \tfrac{1}{2}x - \tfrac{3}{8} + e^{-x}$$

Differentiating

$$y' = -4Ae^{-4x} - 2Be^{-2x} + \tfrac{1}{2} - e^{-x}$$

Using $y(0) = 0$: $A + B - \tfrac{3}{8} + 1 = 0$, so that $A + B = -\tfrac{5}{8}$.
Using $y'(0) = 4$: $-4A - 2B + \tfrac{1}{2} - 1 = 4$, so that $-4A - 2B = \tfrac{9}{2}$.
Solving these equations simultaneously gives $A = -\tfrac{13}{8}$, $B = 1$ and so the required particular solution is

$$y = e^{-2x} - \tfrac{13}{8}e^{-4x} + \tfrac{1}{2}x - \tfrac{3}{8} + e^{-x}$$

(ii) From Example 5T (vi) the general solution is $y = A\cos 2x + B\sin 2x - 2x\cos 2x$ from which

$$y' = -2A\sin 2x + 2B\cos 2x + 4x\sin 2x - 2\cos 2x$$

Using $y(0) = 1$: $1 = A \times 1 + B \times 0 - 2 \times 0 \times 1$, giving $A = 1$.
Using $y'(0) = 2$: $2 = -2A \times 0 + 2B \times 1 - 2 \times 1 + 4 \times 0 \times 0$, giving $B = 2$.

Hence the required solution is $y = \cos 2x + 2\sin 2x - 2x\cos 2x$.

□

Example 5V

Solve the equation $y'' - 3y' = 6 + 10\cos x + 10\sin x$ given $y(0) = 1$ and $y'(0) = 0$.

Solution

The auxiliary equation is $m^2 - 3m = 0$, giving $m = 0, 3$ so that the complementary function is $y = A + Be^{3x}$.

The PI will be of the form $y = ax + b\cos x + c\sin x$. (The first guess with respect to the 6 on the right hand side will fail as one of the terms in the complementary function is a constant.) Then

$$y' = a - b\sin x + c\cos x \quad \text{and} \quad y'' = -b\cos x - c\sin x$$

Substituting in the equation and collecting terms gives

$$(-b - 3c)\cos x + (3b - c)\sin x - 3a = 6 + 10\cos x + 10\sin x$$

For equality we require

$$\left.\begin{aligned} -b - 3c &= 10 \\ 3b - c &= 10 \end{aligned}\right\} \text{ with solution } b = 2,\ c = -4$$

$$-3a = 6, \text{ so that } a = -2$$

Hence the PI is

$$y = -2x + 2\cos x - 4\sin x$$

and the equation has general solution

$$y = A + Be^{3x} - 2x + 2\cos x - 4\sin x$$

Using $y(0) = 1$ gives $A + B = -1$. Differentiating the GS

$$y' = 3Be^{3x} - 2 - 2\sin x - 4\cos x$$

Using $y'(0) = 0$ gives $3B - 6 = 0$, from which $B = 2$ and then $A = -3$. The particular solution is

$$y = -3 + 2e^{3x} - 2x + 2\cos x - 4\sin x$$

Note that as the equation did not contain a term in y it could have been integrated directly and the constant of integration evaluated using the initial conditions. The resulting equation can be solved using the integrating factor method. You may like to verify that this approach gives the same answer as before.

□

Exercises 5C

1. Use the solutions to question 1 of Exercise 5A and questions 1, 3 of Exercise 5B to find the general solution of

(i)	$y'' - y' - 2y = 2 + 5x$	(ii)	$y'' - y' - 2y = 4\cos 2x - 3\sin 2x$
(iii)	$9y'' - 18y' + 10y = 20e^{2x}$	(iv)	$y'' - 4y = 100$
(v)	$6y'' + 5y' - 4y = 221\sin 2x$	(vi)	$9y'' - 12y' + 8y = 10e^{x}$
(vii)	$y'' + 2y' + 5y = 1 + x + x^2$	(viii)	$4y'' + 12y' + 9y = 12\cos 3x - 9\sin 3x$
(ix)	$y'' + 4y' = e^{x}$	(x)	$y'' + 4y = 4e^{2x} + 6\cos x$
(xi)	$y'' + 4y = 12\sin 2x$	(xii)	$y'' + 4y' + 4y = 6e^{-2x}$

2. Use the solutions to question 1 to find particular solutions of

(i) $y'' - y' - 2y = 2 + 5x$ given $y(0) = 0$ and $y'(0) = 0$

(ii) $y'' - 4y = 100$ given $y(0) = 1$ and $y'(0) = 0$

(iii) $y'' - y' - 2y = 4\cos 2x - 3\sin 2x$ given $y(0) = 2$ and $y'(0) = \frac{3}{4}$

(iv) $9y'' - 12y' - 8y = 10e^{x}$ given $y(0) = 1$ and $y'(0) = 2$

3. Solve

(i) $y'' - 3y' + 2y = 5\sin x$ given $y(0) = 1$, $y'(0) = 2$

(ii) $y'' - 2y' + 5y = 40e^{-x}$ given $y(0) = 0$, $y'(0) = -2$

(iii) $4y'' - 12y' + 9y = 169\cos x$ given $y(0) = 3$, $y'(0) = -9$

(iv) $y'' - 8y' + 15y = 18xe^{2x}$ given $y(0) = -2$, $y'(0) = 0$

(v) $y'' + y' - 6y = 5e^{2x} - 6e^{3x} - 12x - 22$ given $y(0) = -1$, $y'(0) = 2$

4. A long strut of length l, uniform cross section, flexural rigidity EI and negligible weight is designed to support a load P which acts along its length. The deflection, y, at distance x from the left hand end satisfies the equation $\frac{d^2y}{dx^2}+n^2y=\frac{G}{EI}$ where G is the constant couple holding the ends of the strut horizontal and $n^2=\frac{P}{EI}$. Solve the equations and apply the conditions $y = 0$ at $x = 0$ and $\frac{dy}{dx}=0$ at $x = 0$ to show that $y=\frac{G}{P}(1-\cos nx)$. Deduce that if y is to satisfy the condition $y = 0$ at $x = l$ the smallest possible value of P is $\frac{4\pi^2EI}{l^2}$. (This is Euler's crippling load for the strut).

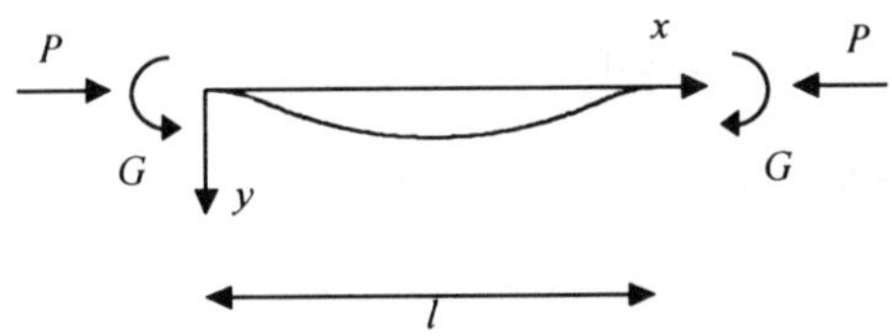

5. A light horizontal cantilever is of length l. At the free end there is a vertical load W and a horizontal compressive force P. Taking the origin at the free end the equation for the deflection, y, is

$$EI\frac{d^2y}{dx^2}+Py=-Wx \quad \text{or} \quad \frac{d^2y}{dx^2}+n^2y=-\frac{Wx}{EI}$$

where $n^2=\frac{P}{EI}$. Use the conditions $y = 0$ at $x = 0$ and $\frac{dy}{dx}=0$ at $x = l$ to show that the height of the clamp above the free end is $y=\frac{W}{nP}(\tan(nl)-nl)$.

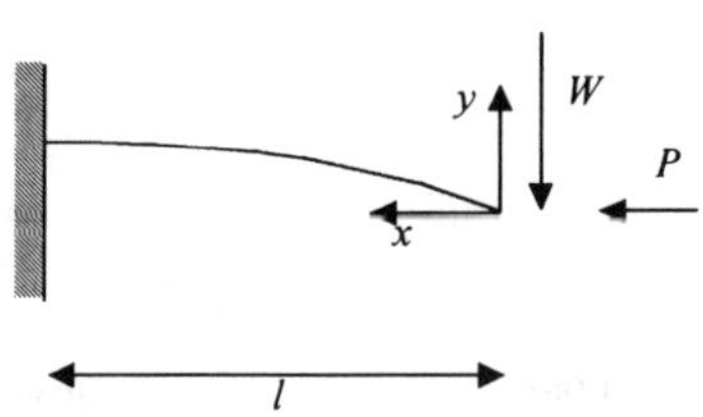

6. An inductance L and a capacitance C are connected in series to a battery of constant voltage E_0. The charge, q, on the capacitor plate satisfies the equation $\frac{d^2q}{dt^2}+\omega^2 q=\frac{E_0}{L}$, where $\omega^2=\frac{1}{LC}$. The current i is related to q by $i=\frac{dq}{dt}$. Apply the conditions $q = 0$ and $i = 0$ at $t = 0$ to show that $q = E_0C(1 - \cos\omega t)$ and $i = \omega E_0C\sin\omega t$. Show further that when the current reverses (next becomes zero) the capacitor is charged to a voltage $2E_0C$.

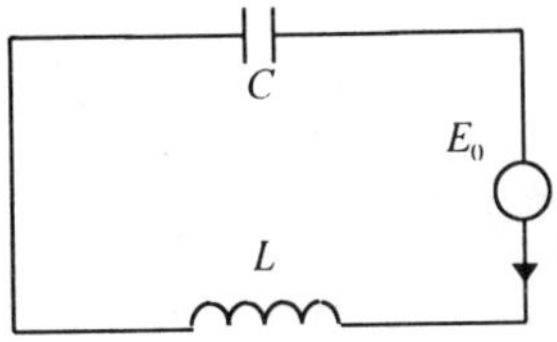

7. In an oscillator circuit the differential equation for the charge q on the capacitor C at time t is $L\frac{d^2q}{dt^2}+\frac{q}{C}=E(t)$. If the applied e.m.f. $E(t)$ has the form $E_0\cos pt$, find q and the current i $\left(=\frac{dq}{dt}\right)$ if both q and i are zero initially. (Assume $p \neq \omega$, where $\omega^2=\frac{1}{LC}$.)

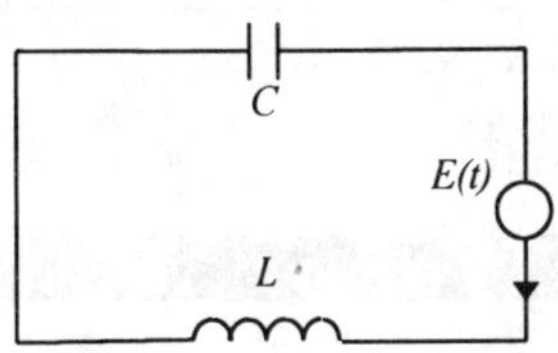

DERIVE ACTIVITY 5B

The aim of this activity is to investigate the use of the utility functions DSOLVE2 and DSOLVE2_IV for inhomogeneous differential equations.

File Load the **Utility** file ODE2.MTH.

(A) To solve the differential equation

$$y'' - 2.5y' + y = 2x + 1$$

Author DSOLVE2(–2.5, 1, 2*x* + 1, *x*, *a*, *b*) and **Simplify**. From the solution identify the complementary function and particular integral. Figure 5.3 shows a DERIVE screen dump for this activity.

```
#1:  DSOLVE2(-2.5, 1, 2·x + 1, x, a, b)

           2·x        0.5·x
#2:  a·ê      + b·ê        + 2·(x + 3)
```

complementary function   particular integral

Figure 5.3 Using DERIVE for non-homogeneous equations

(B) To solve the initial value problem

$$y'' - y' - 2y = 2 + 5x \qquad \text{with } y(0) = 1, y'(0) = -1$$

Author DSOLVE2_IV(–1, –2, 2 + 5x, x, 0, 1, –1) and **Simplify**. **Plot!** the particular solution for this problem. Figure 5.4 shows a screen dump for this activity. The graph of the particular shows a minimum value which occurs at the point (0.26,0.86) and the solution then grows unboundedly as x increases.

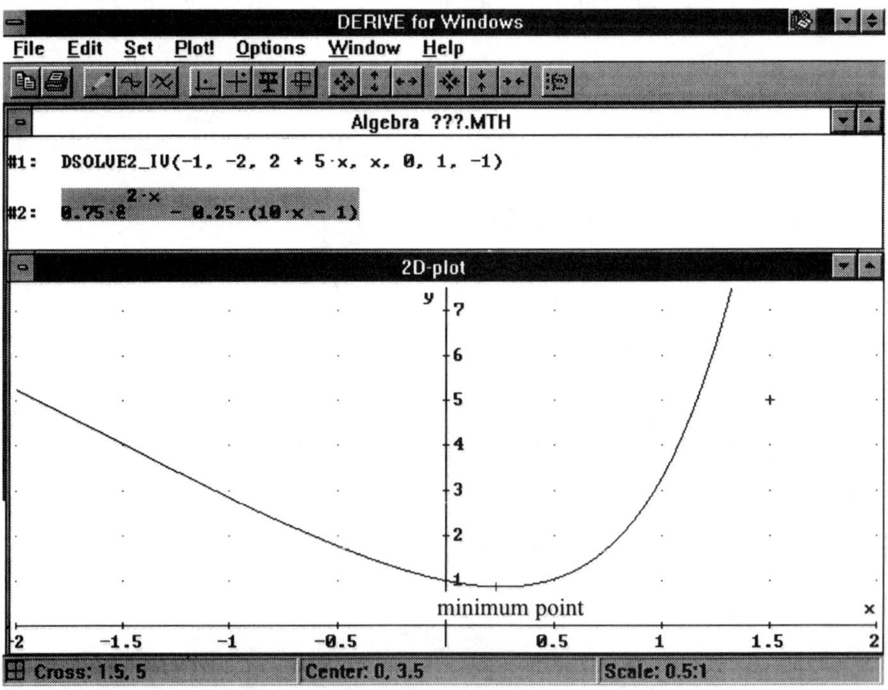

Figure 5.4 Solving and graphing the solution of an initial value problem of an inhomogeneous differential equation

(C) Use DERIVE to solve and graph the solution of $y'' + 7y' + 10y = 10\sin x$ given $y(0) = 2$, $y'(0) = 3$. Draw graphs of the complementary function and particular integral separately. How do different initial conditions affect the long term behaviour of y? What happens when the right hand side of the equation is changed to $20 + 5e^{-x}$ or $x\cos x$? What must be true about the original equation so that after a short time the solution is effectively just the particular integral?

(D) Use DERIVE to investigate the solutions of the equation $y'' + ay' + y = 1$, $y(0) = 0$ and $y'(0) = 0$, for different values of a.

For each choice of a draw a graph of the solution for $x \geq 0$ and deduce what happens to the solution as x increases. What conditions can be placed on the parameter a for

(i) the solution to oscillate forever
(ii) the solution to grow every larger
(iii) a non-zero steady state solution to occur?

(E) Solve each of the following problems using DERIVE. Draw a graph of each solution and hence interpret the solution.

(i) $y'' - 3y' + 2y = 5\sin x$ given $y(0) = 1$, $y'(0) = 2$

(ii) $y'' - 2y' + 5y = 40e^{-x}$ given $y(0) = 0$, $y'(0) = -2$

(iii) $4y'' - 12y' + 9y = 169x\cos x$ given $y(0) = \frac{3}{13}$, $y'(0) = \frac{9}{26}$

(iv) $y'' - 8y' + 15y = 18xe^{2x}$ given $y(0) = -2$, $y'(0) = 0$

(v) $y'' + y' - 6y = 5e^{2x} - 6e^{3x} - 12x - 22$ given $y(0) = -1$, $y'(0) = 2$

5.4 OTHER LINEAR DIFFERENTIAL EQUATIONS WITH CONSTANT COEFFICIENTS

The methods described previously also apply to the nth order equation

$$ay^{(n)} + by^{(n-1)} + \ldots + cy' + dy = f(x)$$

The method of solution is as before, where the complementary function is added to the particular integral. The auxiliary equation is $am^n + bm^{n-1} + \ldots + cm + d = 0$, which has n roots $m_1, m_2, \ldots, m_n$. These roots may be real and different and/or real and equal and/or complex conjugate pairs.

(i) For k distinct roots $m_1, m_2, \ldots, m_k$ the corresponding terms in the complementary function are $c_1 e^{m_1 x} + c_2 e^{m_2 x} + \cdots + c_k e^{m_k x}$.

(ii) For a root m repeated k times the corresponding terms in the complementary function are $(c_1 + c_2 x + c_3 x^2 + \ldots + c_k x^{k-1})e^{mx}$.

(iii) For each pair of complex conjugate roots $m = \alpha \pm \beta i$ the corresponding terms in the complementary function are $e^{\alpha x}(A\cos \beta x + B\sin \beta x)$, as before.

Example 5W

State the form of the complementary function for each of the following auxiliary equations (i) $m - 5 = 0$, (ii) $(m-1)(m-2)(m-3) = 0$, (iii) $(m-1)(m^2 - 4m + 5) = 0$, (iv) $(m-1)(m+2)(m+1)^2 = 0$, (v) $m^3(m^2 + 9) = 0$.

Solution

	Auxiliary Equation	Roots	Complementary Function
(i)	$m - 5 = 0$	$m = 5$	Ae^{5x}
(ii)	$(m-1)(m-2)(m-3) = 0$	$m = 1, 2, 3$	$Ae^{x} + Be^{2x} + Ce^{3x}$
(iii)	$(m-1)(m^2 - 4m + 5) = 0$	$m = 1, 2 \pm i$	$Ae^{x} + e^{2x}(B\cos x + C\sin x)$
(iv)	$(m-1)(m+2)(m+1)^2 = 0$	$m = 1, -2, -1, -1$	$Ae^{x} + Be^{-2x} + (C + Dx)e^{-x}$
(v)	$m^3(m^2 + 9) = 0$	$m = 0, 0, 0, \pm 3i$	$A + Bx + Cx^2 + D\cos 3x + E\sin 3x$

□

Example 5X

Find the general solution of

(i) $y' - 2y = 6e^{4x}$ (ii) $y''' - 2y'' - 5y' + 6y = 4e^{-x}$

(iii) $y''' + 3y'' + 4y' + 2y = 2x + 2$

Solution

(i) The AE is $m - 2 = 0$ from which $m = 2$, giving the CF $y = Ae^{2x}$.
The PI is of the form $y = ae^{4x}$. Then $y' = 4ae^{4x}$ and, on substituting into the equation we find $a = 3$, giving the PI $y = 3e^{4x}$.
Hence the GS is $y = Ae^{2x} + 3e^{4x}$.
(Note that this equation can also be solved by the integrating factor method discussed in Chapter 2.5.)

(ii) The AE is $m^3 - 2m^2 - 5m + 6 = 0$ which factorizes to $(m + 2)(m - 1)(m - 3) = 0$ and so has roots $m = -2, 1, 3$. It follows that the CF is $y = Ae^{-2x} + Be^{x} + Ce^{3x}$.
The PI is of the form $y = ae^{-x}$, from which $y' = -ae^{-x}$, $y'' = ae^{-x}$ and $y''' = -ae^{-x}$.
Substitution into the equation shows that $8a = 4$, giving $a = \frac{1}{2}$ and the PI is $y = \frac{1}{2}e^{-x}$.
Hence the GS is $y = Ae^{-2x} + Be^{x} + Ce^{3x} + \frac{1}{2}e^{-x}$.

(iii) The AE is $m^3 + 3m^2 + 4m + 2 = 0$, or $(m + 1)(m^2 + 2m + 2) = 0$, with roots $m = -1, -1 \pm i$ and so the CF is $y = Ae^{x} + e^{-x}(B\cos x + C\sin x)$.
The form of the PI is $y = ax + b$ and, on substituting, we obtain the equations $2a = 2$ and $4a + 2b = 2$, with solution $a = 1$, $b = -1$, giving as PI $y = x - 1$.
The GS is $y = Ae^{-x} + e^{-x}(B\cos x + C\sin x) + x - 1$.

□

Example 5Y

Solve the equation $y^{iv} + 5y'' + 4y = 10e^{-2x}$ given $y(0) = \frac{1}{4}$, $y'(0) = -\frac{1}{2}$, $y''(0) = 4$, $y'''(0) = 0$.

Solution

The AE is $m^4 + 5m^2 + 4 = 0$, or $(m^2 + 4)(m^2 + 1) = 0$, with roots $m = \pm 2i, \pm i$. The CF is $y = A\cos 2x + B\sin 2x + C\cos x + D\sin x$. The PI is $y = ae^{-2x}$. After substitution we obtain $40a = 10$, so that $a = \frac{1}{4}$ and the PI is $y = \frac{1}{4}e^{-2x}$.

The GS is $y = A\cos 2x + B\sin 2x + C\cos x + D\sin x + \frac{1}{4}e^{-2x}$.

Applying $y(0) = \frac{1}{4}$ gives $A + C = 0$.

Applying $y'(0) = -\frac{1}{2}$ gives $2B + D = 0$.

Applying $y''(0) = 4$ gives $-4A - C = 3$.
Applying $y'''(0) = 0$ gives $-4B - D = 2$.
Solving the first and third of these simultaneously gives $A = -1$, $C = 1$ while the second and fourth give $B = -1$, $D = 2$.

The required solution is $y = -\cos 2x - \sin 2x + \cos x + 2\sin x + \frac{1}{4}e^{-2x}$.

□

5.5 LINEAR SECOND ORDER EQUATIONS IN GENERAL

A linear second order equation can be put in the form of equation (5.1), that is

$$a(x)y'' + b(x)y' + c(x)y = f(x) \tag{5.1}$$

There is no general method available for the solution of equations of this type, but solutions can be found in particular cases. This often includes the use of a substitution to transform the given equation into one of the standard types.

(i) Equations with *y* absent

In this case (5.1) becomes $a(x)y'' + b(x)y' = f(x)$ and the substitution $p = y'$ will reduce the equation to one of first order in p and x. The method can also be tried on non–linear equations.

Example 5Z

Find the general solution of $xy'' + y' = 4x + 1$.

Solution

Let $p = y'$, so that $p' = y''$ and the equation becomes $xp' + p = 4x + 1$, which is a first order linear equation which can be solved by the integrating factor method (Chapter 2.5). The equation is, in fact, already in the correct form and, using the product rule for differentiation, the equation becomes

$$\frac{d}{dx}(xp) = 4x + 1$$

Integrating: $xp = 2x^2 + x + A$ from which $p = 2x + 1 + \dfrac{A}{x}$

or $y' = 2x + 1 + \dfrac{A}{x}$

Integrating: $y = x^2 + x + A\ln x + B$

which is the general solution.

□

(ii) Cauchy–Euler Equations

These are equations of the form $ax^2y'' + bxy' + cy = f(x)$.
The substitution $x = \mathrm{e}^t$ transforms the equation into a constant coefficient equation.

Example 5AA

Solve $3x^2y'' + 11xy' - 3y = 7 - 6\ln x \quad (x > 0)$.

Solution

Let $x = \mathrm{e}^t$, so that $t = \ln x$. Then $\dfrac{dy}{dt} = \dfrac{dy}{dx}\dfrac{dx}{dt} = \dfrac{dy}{dx}\mathrm{e}^t = x\dfrac{dy}{dx}$. Hence $x\dfrac{dy}{dx} = \dfrac{dy}{dt}$.
Also

$$\frac{d^2y}{dt^2} = \frac{d}{dt}\left(\frac{dy}{dt}\right) = \frac{d}{dt}\left(x\frac{dy}{dx}\right) = \frac{dx}{dt}\frac{dy}{dx} + x\frac{d}{dt}\left(\frac{dy}{dx}\right) \qquad \text{(using the product rule)}$$

$$= \frac{dy}{dt} + x\frac{d^2y}{dx^2}\frac{dx}{dt} = \frac{dy}{dt} + x^2\frac{d^2y}{dx^2}$$

Hence $x^2\dfrac{d^2y}{dx^2} = \dfrac{d^2y}{dt^2} - \dfrac{dy}{dt}$

Substituting into the equation we find

$$3\left(\frac{d^2y}{dt^2} - \frac{dy}{dt}\right) + 11\frac{dy}{dt} - 3y = 7 - 6t$$

or $$3\frac{d^2y}{dt^2} + 8\frac{dy}{dt} - 3y = 7 - 6t$$

The AE is $3m^2 + 8m - 3 = 0$, or $(3m - 1)(m + 3) = 0$ with roots $m = \frac{1}{3}, -3$.

The CF is $y = A\mathrm{e}^{\frac{t}{3}} + B\mathrm{e}^{-3t}$.
To find a PI try $y = at + b$. Then $y' = a$, $y'' = 0$.

Substituting: $8a - 3(at + b) = 7 - 6t$.

For equality we require $-3a = -6$, so that $a = 2$ and $8a - 3b = 7$, giving $b = 3$.

Hence a PI is $y = 2t + 3$ and the GS is $y = Ae^{\frac{t}{3}} + Be^{-3t} + 2t + 3$.

Using $x = e^t$: $y = A(e^t)^{\frac{1}{3}} + B(e^t)^{-3} + 2t + 3$

The GS is $y = Ax^{\frac{1}{3}} + Bx^{-3} + 2\ln x + 3$

□

(iii) **Miscellaneous substitutions**

Example 5BB

Use the substitution $y = xv$ to solve the equation $x^2y'' - 2xy' + (x^2 + 2)y = x^3$.

Solution

If $y = xv$ then $\dfrac{dy}{dx} = x\dfrac{dv}{dx} + v$ and $\dfrac{d^2y}{dx^2} = x\dfrac{d^2v}{dx^2} + 2\dfrac{dv}{dx}$ and the equation becomes

$$x^2\left(x\frac{d^2v}{dx^2} + 2\frac{dv}{dx}\right) - 2x\left(x\frac{dv}{dx} + v\right) + (x^2 + 2)xv = x^3$$

$$x^3v'' + 2x^2v' - 2x^2v' - 2xv + x^3v + 2xv = x^3$$

which simplifies to $v'' + v = 1$ with solution $v = A\cos x + B\sin x + 1$ and so, using $y = xv$, the general solution of the original equation is

$$y = x(A\cos x + B\sin x + 1)$$

□

There are many other special cases where solutions can be found, but these will not be dealt with here. Solutions can also be found using infinite power series and the method of Frobenius, which allows solutions to well known equations such as the Legendre, Bessel and Chebychev equations to be found. Where no analytical methods exist, approximate numerical methods must be used.

DERIVE ACTIVITY 5C

(A) Use DERIVE to investigate the solution of

$$y'' + axy' + x^2y = 0$$

when a takes the values 2 or –2.

In each case find the particular solution given $y(0) = 0$ and $y'(1) = 2$. Find $y'(0)$ and $y(1)$, plot the graph of y and determine the range of values of x for which $|y| \geq 4$. What happens when a takes other values?

(B) Use DERIVE to attempt to solve the equations in problems 5, 8 and 10 of Exercise 5D.
Does DERIVE always provide a solution? If your answer is no, deduce for which differential equations DERIVE might work.

5.6 SYSTEMS OF LINEAR FIRST ORDER DIFFERENTIAL EQUATIONS

In Section 5 of Chapter 3 we looked briefly at the numerical solution of sets of first order equations. For a system with two dependent variables x and y which are dependent on one independent variable t the equations are of the form

$$\frac{dx}{dt} = f(t,x,y), \qquad \frac{dy}{dt} = g(t,x,y)$$

Here we will look at a special case when the equations can be solved analytically using the techniques of this chapter. This happens when the equations are linear with constant coefficients, in which case they will be of the form

$$a\frac{dx}{dt} + b\frac{dy}{dt} + cx + dy = p(t)$$
$$e\frac{dx}{dt} + f\frac{dy}{dt} + gx + hy = q(t),$$

where a, b, c, d, e, f, g and h are constants.

For convenience write $D \equiv \dfrac{d}{dt}$ and the equations become

$$aDx + bDy + cx + dy = p(t)$$

$$eDx + fDy + gx + hy = q(t)$$

or $\quad (aD + c)x + (bD + d)y = p(t)$

$$(eD + g)x + (fD + h)y = q(t)$$

We combine these two first order equations into one second order equation in x (or y), the other variable having been eliminated. This equation can be solved by the techniques seen earlier in the chapter. The second variable y (or x) is found in a similar way, or by substituting the known x (or y) into one of the equations, if this is easier.

The method is illustrated in the following example.

Example 5CC

Solve the equations $\dfrac{dx}{dt} - 2x - 3y = 1,\ \dfrac{dy}{dt} + x + 2y = 2t - 1$ given $x(0) = 0$, $y(0) = 2$.

Solution

Using $D \equiv \dfrac{d}{dt}$ the equations become

$$(D - 2)x - 3y = 1 \qquad (5.5)$$

$$x + (D + 2)y = 2t - 1 \qquad (5.6)$$

Eliminate x by forming (5.5) – $(D - 2)$ (5.6):

$$-3y - (D - 2)(D + 2)y = 1 - (D - 2)(2t - 1)$$

$$(D^2 - 1)y = 3 - 4t$$

The auxiliary equation is $m^2 - 1 = 0$, from which $m = \pm 1$ and the complementary function is

$$y = Ae^t + Be^{-t}$$

To find a particular integral try $y = at + b$. This gives

$$-at - b = 3 - 4t$$

giving $a = 4$, $b = -3$ and the particular integral is $y = 4t - 3$. Hence

$$y = Ae^t + Be^{-t} + 4t - 3$$

The easiest way to find x is from (5.6):

$$x = 2t - 1 - (D + 2)(Ae^t + Be^{-t} + 4t - 3)$$

$$x = -3Ae^t - Be^{-t} - 6t + 1$$

Applying $x(0) = 0$: $0 = -3A - B + 1$

Applying $y(0) = 2$: $2 = A + B - 3$

Solving these equations gives $A = -2$ and $B = 7$ and so the solutions are

$$x = 6e^t - 2e^{-t} - 6t + 1$$

$$y = -2e^t + 7e^{-t} + 4t - 3$$

□

5.7 NUMERICAL SOLUTION OF SECOND ORDER DIFFERENTIAL EQUATIONS

A second order equation is of the form

$$\frac{d^2y}{dx^2} = f\left(x, y, \frac{dy}{dx}\right) \quad \text{or} \quad y'' = f(x,y,y')$$

To find a numerical solution we also require two conditions. If the conditions are given at the **same** value of x we have an **initial value problem**, otherwise we have a **boundary value problem**. For example

(i) Solve $y'' = x^2 + y^2$ given $y(0) = 1$ and $y'(0) = 2$. This is an initial value problem as both conditions are given at $x = 0$.

(ii) Solve $y'' = x^2 + y^2$ given $y(0) = 1$ and $y(1) = 3$. This is a boundary value problem as the conditions are given at different values of x.

The methods of solution for initial and boundary value problems are different and here we will be concerned with initial value problems only. (All the applications considered in Chapter 6 are initial value problems.)

An initial value problem is solved by expressing the equation as a pair of first order equations and using the approach discussed in Section 3.5.

Consider the initial value problem

$$y'' = f(x,y,y') \quad \text{given} \quad y(x_0) = \alpha \quad \text{and} \quad y'(x_0) = \beta$$

Let $y' = z$, so that $y'' = z'$. The equation $y'' = f(x,y,y')$ becomes $z' = f(x,y,z)$. The condition $y'(x_0) = \beta$ becomes $z(x_0) = \beta$. Hence the problem is equivalent to

$$y' = z, \qquad y(x_0) = \alpha$$

$$z' = f(x,y,z), \qquad z(x_0) = \beta$$

The DERIVE Runge-Kutta command is of the form

$$\text{RK}([z,f(x,y,z)],\ [x,y,z],\ [x_0,\alpha,\beta],\ h,\ n)$$

Approximating gives a set of n $[x,y,z] = [x,y,y']$ triples and a plot of y against x can be made after extracting columns 1 and 2 from each.

Again h is reduced until the required accuracy is obtained.

Example 5DD

Solve $y'' + 6y' + 8y = 4x + 3e^{-x}$ given $y(0) = 0$, $y'(0) = 4$ to find $y(1)$ correct to six decimal places.

Solution

Putting $y' = z$ the equation is equivalent to the pair

$$y' = z \qquad y(0) = 0$$

$$z' = -6z - 8y + 4x + 3e^{-x} \qquad z(0) = 4$$

The required DERIVE command is

$$\text{RK}([z,-6z - 8y + 4x + 3e^{-x}],\ [x,y,z],\ [0,0,4],\ h,\ n)$$

where values of h and n have to be substituted. The following results were obtained.

Step size h	Number of steps n	Estimate of $y(1)$
0.1	10	0.598418
0.05	20	0.598450
0.02	50	0.598451
0.01	100	0.598451

Hence $y(1) = 0.598451$ to 6D.

In this case we can also solve the equation analytically as the original equation is linear, using the DSOLVE2_IV command. The equation was also solved in Example 5U (i). The analytical solution also gives $y(1) = 0.598451$ to 6D.

Figure 5.5 shows the approximate solution for $h = 0.01$ together with the analytical solution.

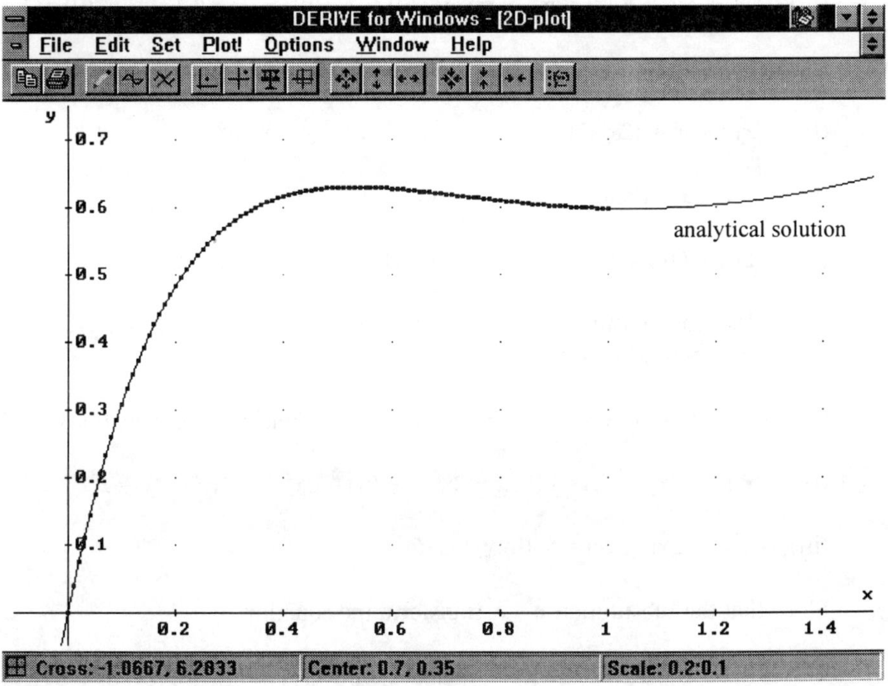

Figure 5.5 The solution of the equation in Example 5DD

Exercise 5D

1. Find the general solution of

(i)	$y' + 2y = 2x + 5$	(ii)	$2y' + 5y = 60e^{5x}$
(iii)	$y''' - y'' - 12y' = 0$	(iv)	$y''' - 3y'' + 3y' - y = 4\sin x$
(v)	$y''' - 2y'' + 5y' = 5 + 8e^{x}$	(vi)	$y^{iv} + 6y''' + 5y'' - 24y' - 36y = 0$
(vii)	$y^{iv} + 5y'' - 36y = 72x - 72$	(viii)	$y^{iv} - y = 5\sin 2x$

2. Solve

(i) $y' - y = 10\cos 2x$ given $y(0) = 0$

(ii) $y''' - 4y' = 4x$ given $y(0) = 3, y'(0) = 2, y''(0) = 5$

(iii) $y^{iv} + 5y'' + 4y = 40e^{2x}$ given $y(0) = 1, y'(0) = 2, y''(0) = 1, y'''(0) = 2.$

3. Use the substitution $p = y'$ to find the general solution of

(i) $xy'' + y' = 12x + 1$ (ii) $xy'' - y' = 8x^2e^{2x}$

(iii) $y''\cos x + y'\sin x = \cos^2 x$ (iv) $xy'' + 2y' = 2$

4. (i) Solve $(1 + x^2)y'' + 2xy' = 2$ given $y(0) = 1$ and $y'(0) = 2$.

(ii) Use the substitution $p = y''$ to solve $y''' + y'' = e^{-x}$ given $y(0) = 3$, $y'(0) = 0, y''(0) = 1$.

5. Use the substitution $x = e^t$ to solve the following Cauchy–Euler equations

(i) $x^2y'' + 7xy' - 7y = 14\ln x + 2$ (ii) $x^2y'' - 3xy' + 4y = x^3$

(iii) $x^2y'' + 5xy' + 5y = 50(\ln x)^2 + 16.$

6. Show that the substitution $x = e^t$ transform the equation

$$ax^2y'' + bxy' + cy = f(x) \quad \text{into} \quad a\frac{d^2y}{dt^2} + (b-a)\frac{dy}{dt} + cy = f(e^t)$$

and hence solve the equation $x^2y'' + 7xy' - 7y = 4x$ given $y(1) = 2$ and $y'(1) = \frac{1}{2}$.

7. Use the substitution $x = e^t$ to show that the third order Cauchy–Euler equation $ax^3y''' + bx^2y'' + cxy' + dy = f(x)$ transforms to

$$a\frac{d^3y}{dt^3} + (b-3a)\frac{d^2y}{dt^2} + (2a-b+c)\frac{dy}{dt} + dy = f(e^t)$$

and hence solve $x^3y''' - 2x^2y'' + 3xy' - 3y = 6\ln x + 1$ given $y(1) = 1, y'(1) = 0$, $y''(1) = 0$.

8. If $y = xv$ show that $\dfrac{dy}{dx} = x\dfrac{dv}{dx} + v$ and $\dfrac{d^2y}{dx^2} = x\dfrac{d^2v}{dx^2} + 2\dfrac{dv}{dx}$. Use the substitution $y = xv$ to solve the equations

(i) $x^2y'' - 2xy' + (4x^2 + 2)y = 4x^4$ (ii) $x^2y'' - 2xy' + (2 - 9x^2)y = 30x^3\cos x$

9. Show that the change of independent variable $t = \sqrt{x}$ transforms the equation $4xy'' + 2(1 - \sqrt{x})y' - 2y = 4\sqrt{x} + 1$ into

$$\frac{d^2y}{dt^2} - \frac{dy}{dt} - 2y = 4t + 1$$

and hence solve the equation.

10. Use the transformation $y = ve^{x^2}$ to solve the equation

$$y'' - 4xy' + 4x^2y = 6e^{x^2}$$

11. Solve the following systems of first order linear equations.

(i) $\dfrac{dx}{dt} - x - 5y = 18t$, $\dfrac{dy}{dt} + 2x + y = 9$ given $x(0) = 4$, $y(0) = -1$.

(ii) $\dfrac{dx}{dt} - 2x - 3y = 3\cos t$, $\dfrac{dy}{dt} + x + 2y = \sin t$ given $x(0) = -1$, $y(0) = 1$.

12. Solve the system of second order linear equations

$$\frac{d^2x}{dt^2} - x + 10y = 0, \qquad \frac{d^2y}{dt^2} - x + 6y = 0$$

given (i) $x(0) = 2$, $y(0) = 1$, (ii) $x(0) = 5$, $y(0) = 1$, (iii) $x(0) = 0$, $y(0) = 3$ together with $\dot{x}(0) = \dot{y}(0) = 0$ in each case.

13. Express the equation $x^2y'' + 7xy' - 7y = 14\ln x + 2$ as a pair of simultaneous first order equations. Given $y(1) = 3$ and $y'(1) = -1$ use the Runge-Kutta method to find $y(2)$ correct to six decimal places. Compare your answer with that obtained from the analytical solution found in question 5(i) of Exercise 5D.

6

Applications of Second Order and Simultaneous First Order Differential Equations

Application 4 of Chapter 1 showed how a second order differential equation can arise from the study of the vertical motion of a car travelling over a bumpy road. This is an example of a mechanical vibration.

Vibrations and oscillations occur in many physical systems and an understanding of their behaviour is important for the design engineer. In Section 1 we show how oscillations in mechanical systems and electrical circuits lead to the same type of differential equations. We also investigate the physical meaning of some solutions of these equations. In later sections we look at applications of simultaneous first order and second order equations to mixing problems and two dimensional projectile motion respectively.

6.1 OSCILLATIONS AND VIBRATIONS

Oscillations and vibrations occur in many everyday situations. For example

(i) a child on a swing oscillates backwards and forwards
(ii) a ship causes waves which die out as they move further away from the ship
(iii) a seismograph vibrates in response to an earthquake
(iv) a car passenger is affected by an undulating road or potholes
(v) a bridge vibrates due to traffic or wind
(vi) the wing of an aircraft flutters due to the vibration of the engines and rushing air
(vii) the electric current in a circuit.

Many mechanical oscillations and vibrations that occur can be modelled by a system consisting of a particle (the mass), a spring and a damper, or dashpot; this produces a resistive force and can model, for example, air resistance or friction. This system is called a spring-mass oscillator and we begin with such a system. We need to formulate sub-models for the force in an elastic spring and the force in a dashpot.

Sub-model for the force in a spring

Consider a weight such as a bag of coins suspended from the ceiling by a spring. The bag is pulled vertically downwards and released and then oscillates vertically. The diagram shows a "photograph" of the system during the oscillations, the photograph being taken at an instant when the bag is below its equilibrium position and moving downwards. We will use an x axis with origin at the ceiling and positive direction downwards.

The forces acting on the bag are

(i) its weight mg acting downwards

(ii) a restoring force of magnitude F_1 trying to restore the spring to its natural (unextended) length.

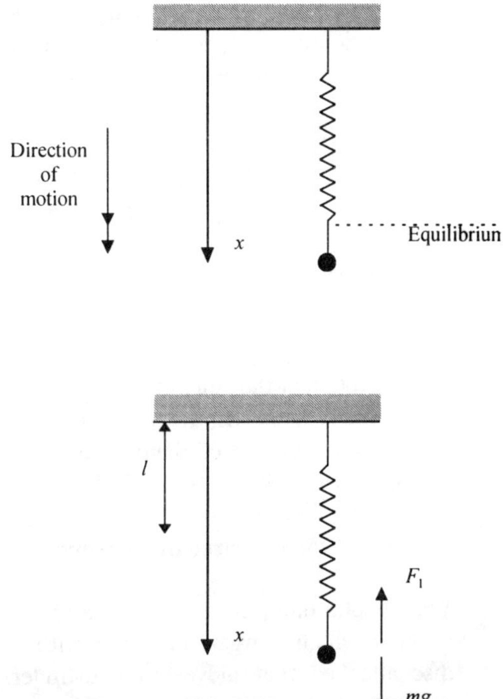

Figure 6.1

The restoring force produced by the spring is quantified by Hooke's law. This is an experimental law relating the magnitude of the force in an elastic string or spring to its extension. A perfect elastic string or spring is one for which the extension e is directly proportional to the force F applied, so that $F = ke$ where k is called the stiffness of the spring.

If the spring has natural length l then the magnitude of the resisting force is

$$F_1 = k(x - l)$$

Then, applying Newton's second law

$$m\ddot{x} = mg - F_1$$

$$m\ddot{x} = mg - k(x - l) \tag{6.1}$$

$$m\ddot{x} + kx = mg + kl$$

For a spring in **compression** Hooke's law states that the magnitude of the restoring force is $F = ke$, where e is the compression in the spring from its natural length, but this time the force is towards the particle.

In the above example the compression of the spring is $l - x$ and so $F_1 = k(l - x)$. From Newton's second law

$$m\ddot{x} = mg + F_1$$
$$m\ddot{x} = mg + k(l - x)$$
$$m\ddot{x} + kx = mg + kl$$

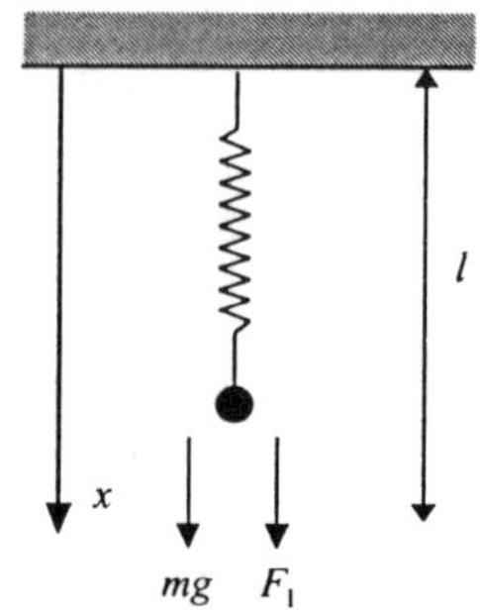

Note that the equation of motion is the same, whether the spring is extended or compressed. This will **always** be the case in spring-mass oscillators and so it is only necessary to do one of them. For consistency always assume that all springs are extended (whether this is physically possible or not).

Sub-model for the force in a dashpot

We model a damping force with a linear dashpot which consists of a rod with a disc attached that moves in a cylinder containing liquid. If the end A of the rod is pulled upwards the liquid will provide a resistance to this motion, this force acting downwards. Conversely, if A is pushed downwards the liquid will again provide a resistance but this time it will act upwards on A.

You can simulate this by holding a dinner plate horizontally in a bowl of water. Whether you lift it up or push it down you will feel a resistance from the water.

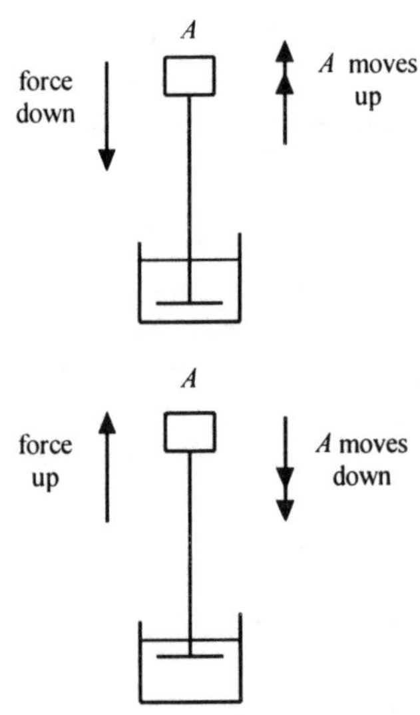

Figure 6.2

The magnitude F of the resistive force is assumed to depend on

(i) the type of liquid in the dashpot. You can easily imagine that resistance provided by oil would be more than that provided by water but would be less than for treacle. This is measured by the **dashpot constant** r. We assume that F is proportional to r; and

(ii) the speed with which the rod moves in the oil. Again the dinner plate analogy is useful. The dinner plate can be raised, keeping it horizontal, fairly easily if you do it slowly, but the faster you try to lift the plate the greater the force you will feel. Don't do it with all your force unless you are prepared to get wet! If the length of the dashpot is s then we assume that F is proportional to the magnitude of the rate of change of s; that is $F \propto |\dot{s}|$ (hence **linear**).

Thus our model for a damping force is that

(i) it opposes motion

(ii) its magnitude is $F = r|\dot{s}|$.

For a lengthening dashpot the force is towards the dashpot, $|\dot{s}| = \dot{s}$ and the magnitude of the force is $F = r\dot{s}$

For a shortening dashpot the force is away from the dashpot, $|\dot{s}| = -\dot{s}$ (because the length s is decreasing and so $\dot{s}$ is negative) and the magnitude of the force is $F = -r\dot{s}$.

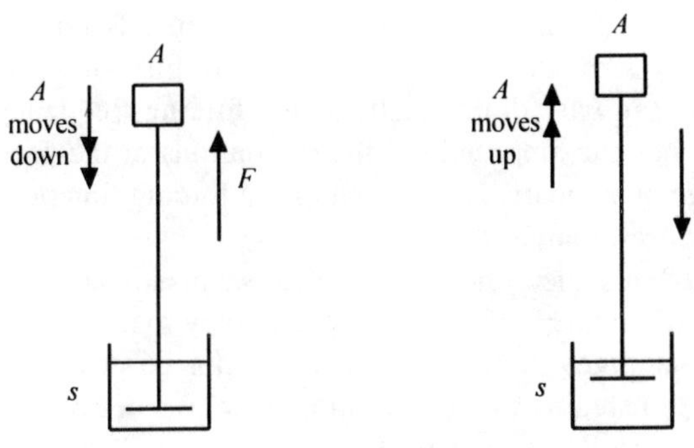

Figure 6.3

Returning to Figure 6.1, we will now include the effects of a resistive force. The forces in the spring and dashpot of the mass-spring-dashpot system are $F_1 = k(x - l)$ and $F_2 = r\dot{x}$. By Newton's second law

$$m\ddot{x} = mg - F_1 - F_2$$

$$= mg - k(x - l) - r\dot{x}$$

$$m\ddot{x} + r\dot{x} + kx = mg + kl \qquad (6.2)$$

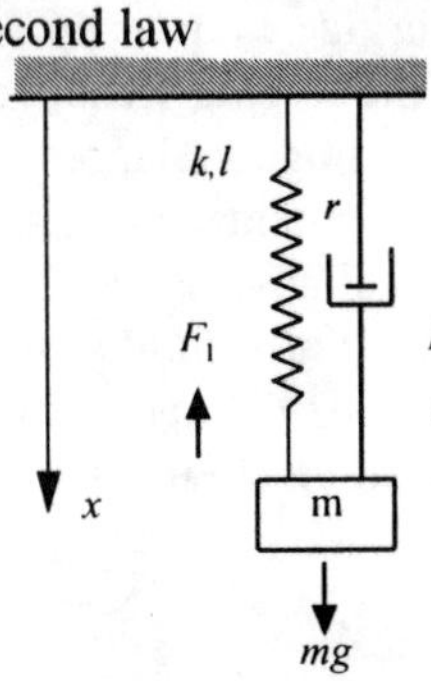

Finally we add a forcing term to the system. If the system experiences some external disturbance, for example by the ceiling being forced to vibrate, then the differential equation becomes

$$m\ddot{x} + r\dot{x} + kx = mg + kl + F(t) \tag{6.3}$$

where $F(t)$ is called the forcing term.

A couple of examples of forcing terms in real world situations are given below.

(i) The body of a car is forced to oscillate because of the vibration of the engine and also from being driven over a bumpy road. What effect does this have on the passengers?

(ii) A seismograph is forced to vibrate by earthquake waves in the earth's crust and hence measures the strength of the earthquake.

We shall assume that the forcing influence is periodic in nature so that it can be modelled by a sinusoid, and we wish to investigate the motion of a particle in a mass-spring-dashpot system caused by the forced oscillations.

We shall see that the motion of the particle depends on the amplitude and frequency of the forcing term and also, for systems with little or no damping, that the motion of the particle can vary dramatically as the forcing frequency changes. At a certain critical frequency the amplitude of the oscillations of the particle (the output) can become very large in comparison with that of the forcing function (the input); this behaviour is known as **resonance**.

Resonance is often something to be avoided, such as in car suspension systems, buildings or aircraft, when large vibrations due to bumpy roads or wind can do a great deal of damage. Sometimes, however, resonance can be desirable, such as when tuning a radio or when listening to an orchestra. Also, the best way to make a child's swing go higher is to push it in time with the oscillations, when small pushes quickly increase the height the swing reaches.

Two well known examples of resonance having disastrous effects were the collapse, for different reasons, of bridges at Broughton, near Manchester, in 1831 and Washington, USA in 1940. A column of soldiers marched, in step, over the wooden Broughton suspension bridge, setting up a periodic force which cause the bridge to vibrate. Unfortunately the forcing frequency was a critical value, causing resonance, with the result that the bridge collapsed. Since then columns of soldiers crossing bridges always break step; you may have seen this happening in newsreels and films.

The cause of the collapse of the Tacoma Narrows road bridge in Washington was wind. The bridge oscillated mildly from the day it opened, but four months later a critical 42 mph wind caused large vertical oscillations of the bridge. After three hours the bridge started to crack and an hour later it crashed down. You may have seen film of this collapse.

In Sections 6.1.2 to 6.1.4 we shall see how equations similar to (6.1), (6.2) and (6.3) can arise from the study of simple pendulums and of electric circuits. Sections 6.1.5 to 6.1.7 look at each equation in more detail.

6.1.2 Pendulums

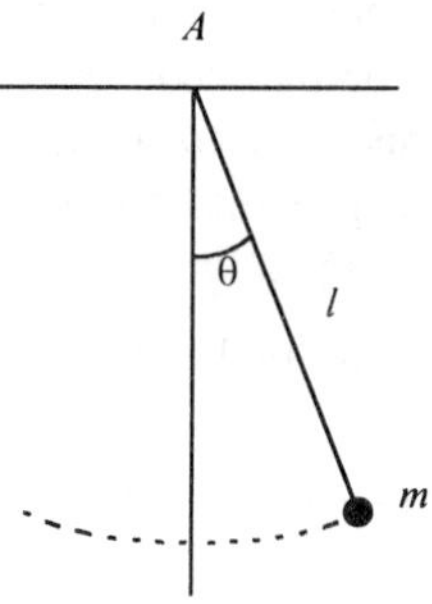

A pendulum consists of a small mass m at one end of a rigid light rod of length l, the other end A of which is fixed. If the mass is drawn to the side and released it will swing to and fro in a vertical plane, and we are interested in describing its motion. This can be done by finding the angle θ (measured anticlockwise positive from the downward vertical) at any time t.

We assume that the mass is a particle, that the rod has no mass and that there is no friction at A. To start the motion the particle is drawn aside through an angle α and released from rest, so that the initial conditions are $\theta(0) = \alpha$ and $\dot{\theta}(0) = 0$.

Undamped motion

We assume that there is no resistance on the particle from the air, water, or other fluid in which it moves. The forces acting on the particle are its weight mg and the tension T in the rod.

Applying Newton's second law tangentially

$$ma = -mg \sin \theta$$

As the arc length s is given by $s = l\theta$, the transverse acceleration is $a = \ddot{s} = l\ddot{\theta}$ and the equation of motion becomes

$$ml\ddot{\theta} = -mg \sin\theta$$

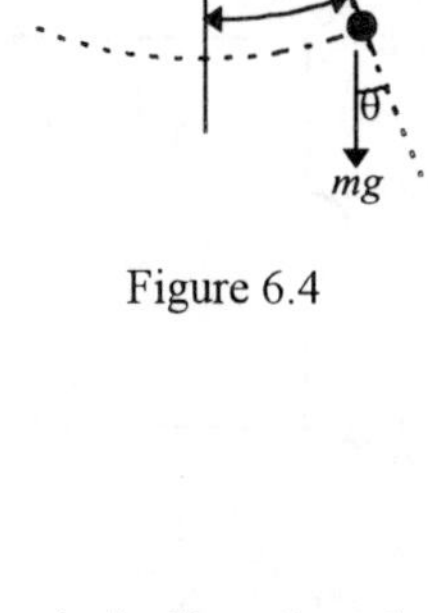

Figure 6.4

or $$\ddot{\theta} + \frac{g}{l}\sin\theta = 0 \qquad (6.4)$$

This equation is non-linear and cannot be solved analytically using the techniques of Chapter 5. It can, however, be simplified using the approximation $\sin\theta \approx \theta$, which is valid if θ is small. In this case the equation becomes

$$l\ddot{\theta} + g\theta = 0 \tag{6.5}$$

Damped motion

It is likely that the motion of the pendulum will be damped. Assume that there is a resistance to motion whose magnitude F_R is proportional to the speed of the particle. Then $F_R = c\dot{s} = cl\dot{\theta}$ and the equation of motion becomes

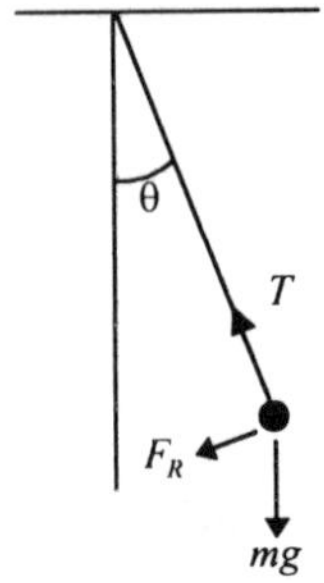

Figure 6.5

$$ml\ddot{\theta} + cl\dot{\theta} + g\sin\theta = 0 \tag{6.6}$$

which can be approximated by, for small θ,

$$ml\ddot{\theta} + cl\dot{\theta} + g\theta = 0 \tag{6.7}$$

If the pendulum is forced to vibrate, for example by a person pushing a swing, then the equation becomes

$$ml\ddot{\theta} + cl\dot{\theta} + g\theta = F(t)$$

6.1.3 Electrical circuits

Now we look briefly at electrical circuits containing a resistor, a capacitor, an inductor and an electromotive force (emf) in series (an LRC circuit). We are interested in how the electric charge and current vary with time.

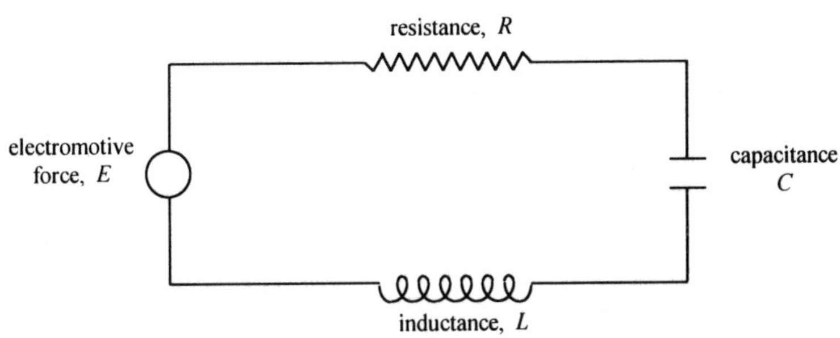

Figure 6.6

The applied emf E will cause a current to flow in the closed circuit. There will be voltage drops across the resistor, capacitor and inductor, which are quantified as follows:

(i) the voltage drop E_1 across a resistor of resistance R is given by $E_1 = Ri$, where i is the current flowing through the resistor

(ii) the voltage drop E_2 across a capacitor of capacitance C is given by $E_2 = \frac{q}{C}$, where q is the charge on the capacitor

(iii) the voltage drop E_3 across an inductor of inductance L is given by $E_3 = L\frac{di}{dt}$.

The units and abbreviations of the various quantities are given in the table below.

Quantity	Symbol	Unit
emf or voltage	E	volt (V)
current	i	ampere (A)
charge	q	coulomb
resistance	R	ohm (Ω)
inductance	L	henry (H)
capacitance	C	farad (F)

Kirchoff's laws describe LRC circuits. They are:

K1. The same current passes through each component (resistor, inductor or capacitor) of a circuit.

K2. The sum of the voltage drops across resistors, inductors and capacitors in a closed circuit is equal to the total emf E.

From K2 we have $E_1 + E_2 + E_3 = E$, giving

$$L\frac{di}{dt} + Ri + \frac{q}{C} = E \tag{6.8}$$

The current i is the rate of change of charge q, so that $i = \frac{dq}{dt}$. Eliminating i from (6.8) gives

$$L\frac{d^2q}{dt^2}+R\frac{dq}{dt}+\frac{1}{C}q=E \tag{6.9}$$

which is a linear second order differential equation that can be solved to find q in terms of t.

To find an equation for i in terms of t we first differentiate (6.9) to give

$$L\frac{d^3q}{dt^3}+R\frac{d^2q}{dt^2}+\frac{1}{C}\frac{dq}{dt}=\frac{dE}{dt}$$

Using $i=\dfrac{dq}{dt}$ this becomes

$$L\frac{d^2i}{dt^2}+R\frac{di}{dt}+\frac{1}{C}i=\frac{dE}{dt} \tag{6.10}$$

The forcing term $\dfrac{dE}{dt}$ can lead to resonance as described for mechanical systems.

6.1.4 Summary

The following table shows that the above three systems lead to essentially the same differential equation models and type of solution.

spring mass oscillator	pendulum	electric circuit
$m\ddot{x}+r\ddot{x}+kx=F(t)$	$ml\ddot{\theta}+cl\dot{\theta}+g\theta=F(t)$	$L\dfrac{d^2i}{dt^2}+R\dfrac{di}{dt}+\dfrac{1}{C}i=F(t)$
x: displacement	θ: angular displacement	i: current
m: mass	m: mass	L: inductance
r: dashpot constant	l: length	R: resistance
k: spring stiffness	c: damping constant	C: capacitance
F: forcing term	F: forcing term	F: forcing term $\left(=\dfrac{dE}{dt}\right)$

6.1.5 Simple harmonic motion

Simple harmonic motion is the most basic type of oscillation and is modelled by the differential equation

$$\ddot{x} + \Omega^2 x = h \tag{6.11}$$

Each of the models in the summary table is of this form when there is no damping and when $F(t) = h$ = constant.

To solve the equation, load the utility file ode2.mth and declare Ω to be a positive variable. The command DSOLVE2(0, Ω^2,h,t,B,C) gives the general solution

$$x = B\cos\Omega t + C\sin\Omega t + \frac{h}{\Omega^2}$$

which can also be put in the form

$$x = D\cos(\Omega t + \phi) + \frac{h}{\Omega^2}$$

The graph of x against t is a sinusoid, and is shown qualitatively in Figure 6.1.

$D = \sqrt{B^2 + C^2}$ is the amplitude
Ω is the angular frequency
ϕ is the phase, which measures the lag of $\cos(\Omega t + \phi)$ behind $\cos \Omega t$
$T = \dfrac{2\pi}{\Omega}$ is the period of the oscillations
$f = \dfrac{1}{T} = \dfrac{\Omega}{2\pi}$ is the frequency of the oscillations.

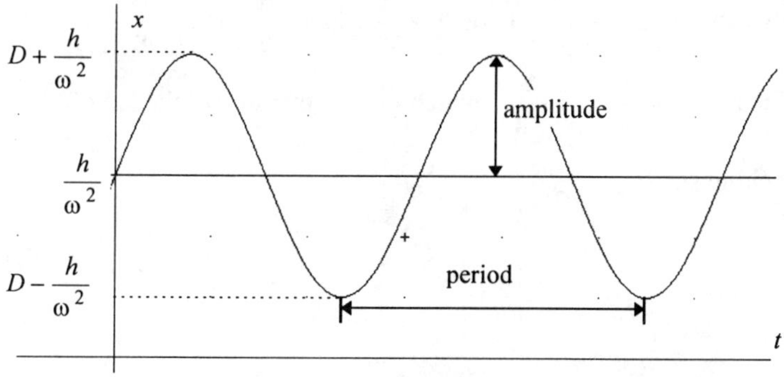

Figure 6.7 Graph of displacement against time

The interpretation of this solution is that the system e.g. the mass, pendulum or current in the circuit oscillates back and forth 'forever' with constant amplitude and period.

Example 6A Properties of the solution

Solve $\ddot{x}+4x=12$ given $x(0) = 4$ and $\dot{x}(0)=2$.

Solution

Comparing with (6.11) the angular frequency is 2 radians per second, the period $T = \pi$ seconds and the frequency $f = \dfrac{1}{\pi}$ per second. The equation can be solved with the command

DSOLVE2_IV(0,4,12,*t*,0,4,2)

Declare Algebra State choose **Simplification** set **Trigonometry** to **Collect** and **Mode** to **Exact**. Simplifying the solution gives

$$\sqrt{2}\,\sin\left(2t+\frac{\pi}{4}\right)3$$

The oscillations have amplitude $\sqrt{2}$ and phase $\dfrac{\pi}{4}$ about the centre $x = 3$.

x has a maximum value of $3+\sqrt{2}\approx 4.41$ when $\sin\left(2t+\dfrac{\pi}{4}\right)=1$, or

$2t+\dfrac{\pi}{4}=2n\pi+\dfrac{\pi}{2}$, or $t=\dfrac{\pi}{8}+n\pi$ (n integer). x has a minimum value of

$3-\sqrt{2}\approx 1.59$ when $\sin\left(2t+\dfrac{\pi}{4}\right)=-1$, or $2t+\dfrac{\pi}{4}=2k\pi-\dfrac{\pi}{2}$, or $t=\dfrac{5\pi}{8}+k\pi$

(k integer). The phase is $\dfrac{\pi}{4}$, or $-\dfrac{7\pi}{4}$, which gives a measure of the lag of

$\sin\left(2t+\dfrac{\pi}{4}\right)$ behind $\sin 2t$.

□

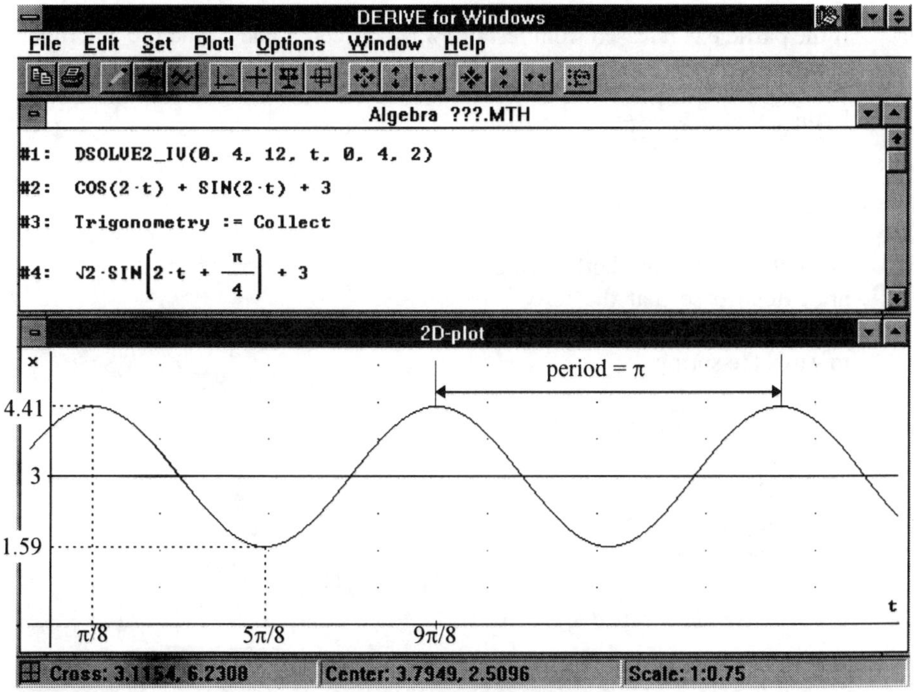

Figure 6.8 DERIVE solution of Example 6A

Example 6B A spring-mass system

A particle of mass m is attached to two fixed points A and B as shown.
The upper spring has stiffness $2k$ and natural length l.
The lower spring has stiffness k and natural length $2l$.
The distance from A to B is $3l$.
$\left(\text{Also } k < \dfrac{2mg}{3l}\right)$.

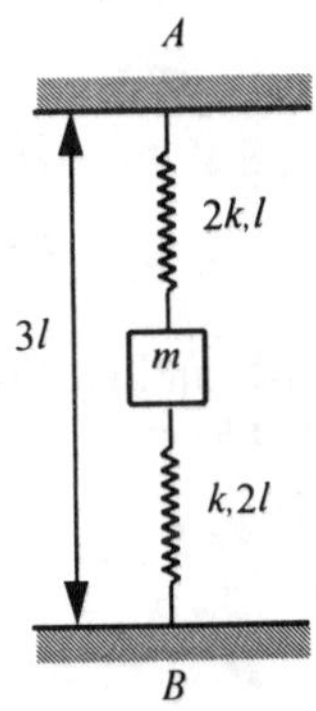

Figure 6.9

(i) Find the equation of motion of the particle when it oscillates along AB and deduce its equilibrium position with (a) x measured from the floor; (b) x measured from the ceiling.

(ii) If the particle is released from rest midway between A and B find the amplitude and period of the oscillations. How close does the particle get to A, and to B? Deduce that $k > \dfrac{4mg}{15l}$.

Solution

(i) We will assume that both springs are extended so that the force on the particle from each spring is towards the spring.

(a) With x measured from B.

By Hooke's law the magnitudes of the spring forces are

$$T_1 = 2k[(3l - x) - l]$$

$$= 2k(2l - x)$$

and $T_2 = k(x - 2l)$

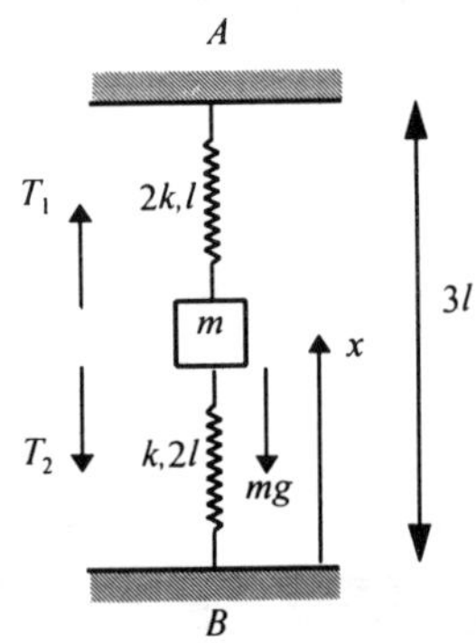

Figure 6.10

From Newton's second law

$$m\ddot{x} = T_1 - T_2 - mg$$
$$m\ddot{x} = 2k(2l - x) - k(x - 2l) - mg \qquad (6.12)$$
$$m\ddot{x} + 3kx = 6kl - mg$$

The equilibrium position x_e is found either by equating the forces, so that $T_1 = T_2 + mg$ which gives

$$x_e = 2l - \frac{mg}{3k}$$

or by putting $\dot{x} = 0$, $\ddot{x} = 0$ in the equation of motion, which also gives

$$x_e = 2l - \frac{mg}{3k} \qquad \text{(above the floor).}$$

(b) With x measured from A the magnitudes of the forces are

$$T_1 = 2k(x - l)$$

and

$$T_2 = k[(3l - x) - 2l] = k(l - x)$$

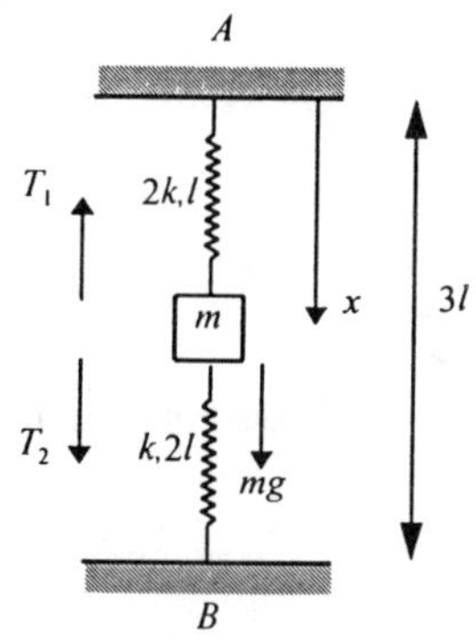

Figure 6.11

Then

$$m\ddot{x} = T_2 + mg - T_1$$
$$m\ddot{x} = k(l - x) + mg - 2k(x - l)$$
$$m\ddot{x} + 3kx = 3kl + mg \qquad (6.13)$$

The equilibrium position is $x_e = l + \dfrac{mg}{3k}$ (below the ceiling).

Note that the left hand sides of equations (6.12) and (6.13) are the same, that their right hand sides reflect the change of origin and that the equilibrium positions are compatible.

(ii) Using (a), the initial conditions are $x(0) = \dfrac{3}{2}l$ and $\dot{x}(0) = 0$. Dividing (6.12), by m we obtain

$$\ddot{x} + \frac{3k}{m}x = \frac{6kl}{m} - g$$

The command DSOLVE2_IV$\left(0, \dfrac{3k}{m}, \dfrac{6kl}{m} - g, t, 0, 1.5l, 0\right)$ gives, after tidying up,

$$x = \left(\frac{mg}{3k} - \frac{l}{2}\right)\cos\sqrt{\frac{3k}{m}}t - \frac{mg}{3k} + 2l$$

Hence the particle oscillates with amplitude $\left|\dfrac{mg}{3k} - \dfrac{l}{2}\right| = \dfrac{mg}{3k} - \dfrac{l}{2}$, as $k < \dfrac{2mg}{3l}$.

Closest approach to A = largest x value (when $\cos\sqrt{\frac{3k}{m}}t = 1$)

$$= \frac{3}{2}l$$

Closest approach to B = smallest x value (when $\cos\sqrt{\frac{3k}{m}}t = -1$)

$$= \frac{5}{2}l - \frac{2mg}{3k}$$

For complete oscillations the particle must not strike the floor. Hence we need $\frac{5}{2}l - \frac{2mg}{3k} > 0$, which gives $k > \frac{4mg}{15l}$.

□

Example 6C A simple pendulum

A pendulum consists of a particle attached to a light rod of length 1 metre. The particle is drawn aside and released from rest when the rod makes an angle α with the downward vertical.

By comparing the solutions of the exact equation of motion and the approximate equation for small angle θ, investigate what values of α gives the period from the approximate model to within 5% of its exact value. (Take g = 10ms^{-2}.)

Solution

With the given values the equations of motion (6.4) and (6.5) are

$$\ddot{\theta} + 10\sin\theta = 0 \qquad (6.14)$$

and $$\ddot{\theta} + 10\theta = 0 \qquad (6.15)$$

and the initial conditions are $\theta(0) = \alpha,\ \dot{\theta}(0) = 0$.

(6.15) can be solved with the DERIVE command DSOLVE2_IV(0,10,0,t,0,α,0) to give $x = \alpha\cos\sqrt{10}t$. The particle is predicted to oscillate simple harmonically with period $\frac{2\pi}{\sqrt{10}} \approx$ 1.99 seconds.

Equation (6.14) must be expressed as a pair of first order equations and solved numerically using the classical fourth order Runge-Kutta method, as described in Section 7 of Chapter 5. Putting $\dot{\theta} = y$, (6.14) is equivalent to

$$\dot{\theta} = y \qquad \theta(0) = \alpha$$

$$\dot{y} = -10\sin\theta \qquad y(0) = 0$$

The utility file ODE_APPR must be loaded, and the required DERIVE command is RK([y,−10sinθ], [t,θ,y], [0,α,0], h, n), where values for α, h and n must be substituted before the command is **Approximated**.

Investigation shows that for very small α the period is about 1.99 seconds, but that it increases with α. A step length $h = 0.01$ finds the period correct to two decimal places for all α in the range $0 < \alpha < 180°$ and n needs to have a value of at least 200. The following results were obtained with DERIVE.

α	10°	30°	50°	70°	90°	120°	150°	170°
Period T	1.99	2.02	2.09	2.19	2.35	2.73	3.50	4.84
% error $= \dfrac{\left(T - \dfrac{2\pi}{\sqrt{10}}\right)}{\dfrac{2\pi}{\sqrt{10}}} \times 100$	0	1.7	5.2	10	18	37	76	144

Hence the solution to the simpler equation (6.28) gives the period correct to within 5% for values of α less than about 50°. As α increases the error becomes unacceptably large and the approximation should not be used.

□

Exercise 6A

1. A perfect spring has a natural length of 1 metre. A mass of 2kg is attached to one end of the spring and the other end is suspended from the ceiling, and with the mass in equilibrium the length of the spring is 1.25 metres.

   (i) Find the stiffness of the spring.

   (ii) Find the equilibrium length of the spring when the 2kg mass is replaced by one of 3kg.

2. A particle of mass m is attached to a wall by a spring of stiffness k and natural length l and slides on a smooth horizontal table.

   (i) Find the equation of motion of the particle while it is in motion and at an instant when (a) the spring is extended, (b) the spring is compressed.

(ii) Find the distance x of the particle from the wall at time t if it is released from rest when it is at a distance $\frac{4}{3}l$ from the wall.

(iii) Find the amplitude, period and centre of the oscillations.

3. A mass stretches an elastic string of natural length l by 15cm when at rest. It is set in vertical motion at a point 25cm below the equilibrium position with a downward velocity of 10cms^{-1}.

(i) Show that the displacement x of the particle below the support at time t satisfies the equation

$$\ddot{x} + 65.33x = 9.8 + 65.33l$$

(ii) Solve the differential equation and draw a graph of the solution.

(iii) When does the mass return to its starting position?

(iv) When does it reach its highest point?

4. A circuit contains an inductance of 0.2 henry and a capacitance of 0.8 farad. The circuit has a battery of constant emf.

(i) Show that the equation for the current in the circuit is $\dfrac{d^2i}{dt^2} + 6.25i = 0$.

(ii) Solve the differential equation and describe the behaviour of the current.

5. Two mass-spring systems are hanging vertically side by side. For one system a mass stretches the spring 8cm and for the other the same mass stretches the spring 20cm. The springs have the same natural lengths but different stiffnesses. From the equilibrium positions, each mass is pulled down 2cm. The first particle is released from rest. The second particle is given an initial speed of 1ms^{-1} downwards.

(i) Solve the differential equation and describe the properties of the motion of each system.

(ii) Draw the graph of each solution on the same axes.

(iii) Find the times of the first two occasions that the masses are on the same horizontal level.

6. A pendulum consists of a particle attached to a light rod of length 0.7 metres. Taking $g = 9.8\text{ms}^{-2}$ show that the angle θ the rod makes with the downward vertical at time t satisfies the equation $\ddot{\theta} + 14\sin\theta = 0$ (1), which can be approximated by $\ddot{\theta} + 14\theta = 0$ (2). The particle is initially projected horizontally from its lowest point with angular speed v.

(i) Solve equation (2) and predict, for various values of v, the maximum value of θ and the time to get there.

(ii) Multiply equation (1) by $2\dot{\theta}$, integrate with respect to time and use the initial conditions to obtain $\dot{\theta}^2 = v^2 + 28\cos\theta - 28$. Solve this equation for various values of v to find the maximum value of θ and the time to get there.

(iii) Compare the results in (i) and (ii). For what values of v does the maximum value of θ from the approximate equation (2) lie within 5% of the exact value from (1).

6.1.6 Damped vibrations

In simple harmonic motion the particle or current is predicted to oscillate indefinitely, but in practice the oscillations will inevitably die away. That is, the vibrations are damped, the damping being provided by some resistance force such as friction, air resistance or resistance in a circuit.

The equation of motion of a damped system takes the form

$$m\ddot{x} + c\dot{x} + ax = d \qquad \text{(see equation (6.2))}$$

where *m, c, a* and *d* are constants, with *m, c* and *a* positive. We shall illustrate the possible forms of the solution of this equation with an example which, for fixed values of *m, a* and *d*, shows what can happen as the magnitude of the resistive force increase.

Example 6D Properties of a damped system

Solve the equation $\ddot{x} + c\dot{x} + x = 0$ given $x(0) = 1$ and $\dot{x}(0) = 0$ for (i) $c = 0$, (ii) $c = 0.5$, (iii) $c = 2$, (iv) $c = 4$.

Solution

The required DERIVE command is DSOLVE2_IV(*c*,1,0,*t*,0,1,0). Replacing *c* by each of the above values in turn, solving or approximating, and plotting each graph results in the DERIVE screen below.

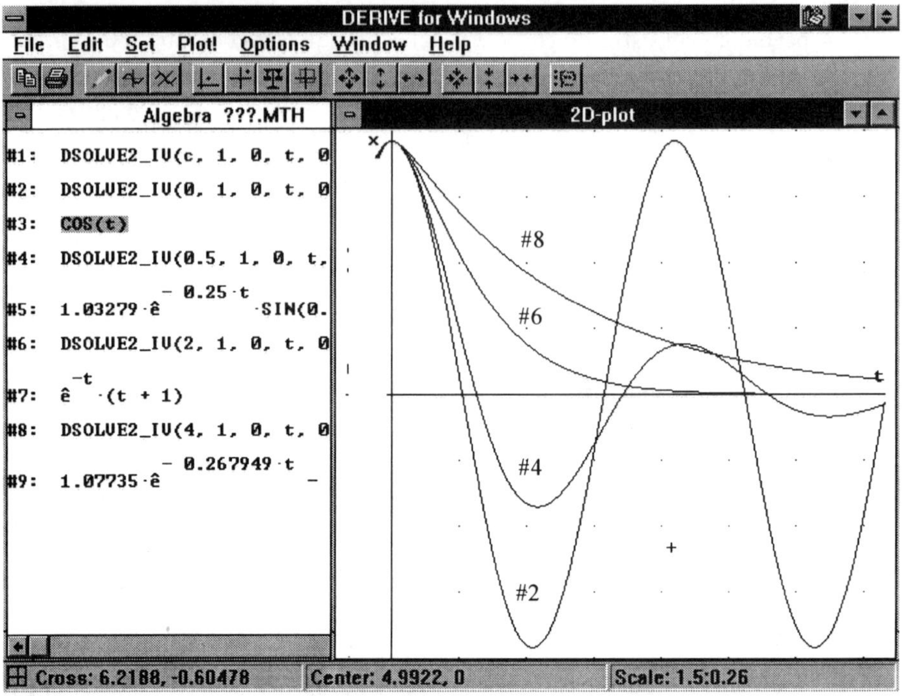

Figure 6.12 DERIVE solutions of Example 6D

(i) With $c = 0$, $x = \cos t$ and the graph is sinusoidal.

(ii) With $c = 0.5$, $x = 1.033e^{-0.25t}\sin(0.968t + 1.318)$ (3D).
The negative exponential term $e^{-0.25t}$ ensures that $x \to 0$ as $t \to \infty$.
The sine term indicates oscillations.
This solution represents oscillations with a decreasing amplitude and this is known as weak damping or underdamping.
Note that the roots of the auxiliary equation are complex.

(iii) With $c = 2$, $x = e^{-t}(t + 1)$.
Again $x \to 0$ as $t \to \infty$, but this time there are no oscillations.
This represents critically damped motion. c has increased to the point where the roots of the auxiliary equation are no longer complex, but are now equal.

(iv) With $c = 4$, $x = 1.077e^{-0.268t} - 0.077e^{-3.373t}$.
Again $x \to 0$ as $t \to \infty$.
This represents strongly damped or over damped motion.

The value of c can be thought of as representing the amount of resistance present in the system. With no resistance ($c = 0$) the particle moves with simple harmonic motion. With some resistance ($0 < c < 2$) the resistance gradually kills off the oscillations, but the particle **does** oscillate.

$c = 2$ is the point where the resistance is just sufficient to prevent oscillations. More resistance ($c > 2$) results in a strongly damped system. When $c = 2$ and when $c > 2$ there are no oscillations and $x \to 0$ as $t \to \infty$. The mathematical solution suggests that it takes an infinite time for the oscillations to completely die away, but we know from experience that they stop much sooner. Suppose we define a damped oscillation as being negligible when its amplitude is 5% of the initial value. Show that x decreases to this value in a shorter time when $c = 2$ than it does for any value of $c > 2$, which is probably opposite to what intuition might suggest.

The general case

For the equation $m\ddot{x} + r\dot{x} + kx = 0$, where m, r and k are positive, the auxiliary equation is $m\lambda^2 + r\lambda + k = 0$ with solution

$$\lambda = \frac{-r \pm \sqrt{r^2 - 4mk}}{2m}$$

The type of solution depends on the value of $r^2 - 4mk$.

(i) **Weak damping** occurs when $r < 2\sqrt{mk}$.
If $r < 2\sqrt{mk}$ then $r^2 - 4mk$ is negative and the auxiliary equation has a pair of complex conjugate roots $\alpha \pm i\beta$, where

$$\alpha = -\frac{r}{2m} \quad \text{and} \quad \beta = \frac{\sqrt{4mk - r^2}}{2m}$$

Then $x = e^{\alpha t}(B \cos \beta t + C \sin \beta t)$

or $x = Ae^{\alpha t} \sin(\beta t + \phi)$

Since $\alpha < 0$, $e^{\alpha t} \to 0$ as $t \to \infty$ and so $x \to 0$ as $t \to \infty$.
The sine term gives oscillations and so we have damped oscillations (as illustrated by graph #4 in Figure 6.3).

(ii) **Critical damping** occurs when $r = 2\sqrt{mk}$.
Then $r^2 - 4mk = 0$ and the auxiliary equation has equal roots $-\dfrac{r}{2m}$ and

$$x = e^{-\frac{r}{2m}t}(A + Bt) .$$

As t increases, x decreases towards zero exponentially, as illustrated by graph #6 in Figure 6.3.

(iii) **Strong damping** occurs when $r > 2\sqrt{mk}$.
Then $r^2 - 4mk$ is positive and the auxiliary equation has two negative roots

$$\lambda_1 = \frac{-r + \sqrt{r^2 - 4mk}}{2m} \quad \text{and} \quad \lambda_2 = \frac{-r - \sqrt{r^2 - 4mk}}{2m}$$

x decreases towards zero exponentially, but not as quickly as when critically damped. This is illustrated by graph #8 in Figure 6.3.

Example 6E A damped pendulum

A pendulum consists of a particle of mass 0.25kg at the end of a light rod of length 1 metre. The particle is released from rest when the rod makes an angle of 60° with the downward vertical. Show that equations (6.6) and (6.7) become $\ddot{\theta} + 4c\dot{\theta} + 10\sin\theta = 0$ and $\ddot{\theta} + 4c\dot{\theta} + 10\theta = 0$ and compare their solutions in the cases (i) c = 0.125 and (ii) c = 2. Take g = 10ms^{-2}. Find the time t_1, for the particle to effectively come to rest, in this case defined as the time for the amplitude of the oscillations to be less than 0.1 radian for all $t > t_1$.

Solution

With the given values equations (6.6) and (6.7) become, respectively $0.25\ddot{\theta} + c\dot{\theta} + 2.5\sin\theta = 0$ and $0.25\ddot{\theta} + c\dot{\theta} + 2.5\theta = 0$, which rearrange to those given in the question.

(i) With c = 0.125 these become

$$\ddot{\theta} + 0.5\dot{\theta} + 10\sin\theta = 0 \qquad (6.16)$$

and $$\ddot{\theta} + 0.5\dot{\theta} + 10\theta = 0 \qquad (6.17)$$

The initial conditions are $\theta(0) = \frac{\pi}{3}$ and $\dot{\theta}(0) = 0$ and so equation (6.17) can be solved using DSOLVE2_IV$\left(0.5,10,0,t,0,\frac{\pi}{3},0\right)$. The solution predicts damped oscillations with period $\dfrac{2\pi}{3.15237} \approx 1.99$ seconds. The graph of the solution is the solid curve shown in Plot 1:1 of Figure 6.13. Plot 1:2 shows that $|\theta(t)| < 0.1$ for all $t > 9.1$ (1D), so that $t_1 = 9.1$ for equation (6.17).

To solve (6.16) we again use the Runge-Kutta method. As a pair, the equations become

$$\dot{\theta} = y \quad \text{and} \quad \dot{y} = -0.5y - 10\sin\theta$$

and the required DERIVE command is

$$\text{RK}([y,-0.5y - 10\sin\theta],\ [t,\theta,y],\ \left[0, \frac{\pi}{3}, 0\right],\ h,\ n)$$

where values of h and n need to be substituted and the command **Approximated**. This results in n sets of $[t,\theta,y]$ triples, with t being incremented by h each time. To plot values of θ against t the values of $y = \dot{\theta}$ need to be removed. This was done in Figure 6.13 using the command EXTRACT_2_COLUMNS(#4,1,2), which removes columns 1 and 2 from the triples in #4, and **Approximating**.

Figure 6.13 shows a plot of the solution for $h = 0.1$. Plot 1 indicates that the period is slightly more than for the solution to (6.17) and Plot 2 is an enlargement after about 5 cycles, indicating that the approximate solution has a period about 0.02 seconds less than the solution to (6.16). Plot 3 shows that $|\theta(t)| < 0.1$ for all $t > 9.2$ ($= t_1$).

(ii) With $c = 2$ equations the models are

$$\ddot{\theta} + 8\dot{\theta} + 10\sin\theta = 0 \tag{6.18}$$

and

$$\ddot{\theta} + 8\dot{\theta} + 10\theta = 0 \tag{6.19}$$

The resistance is now large enough to prevent any oscillations and θ tends to 0 exponentially. The solution to (6.19) is $\theta = 1.38e^{-1.55t} - 0.33e^{-6.45t}$ (2D) and $\theta < 0.1$ when $t \geq 1.7$ seconds.

We can solve (6.18) numerically, and extract and plot the θ, t values, as in (i). The plot shows that $|\theta| < 0.1$ when $t \geq 1.8$ seconds. Agreement between the two solutions is good, the largest deviation being about 0.06 when $t \approx 0.8$. Figure 6.14 shows these results.

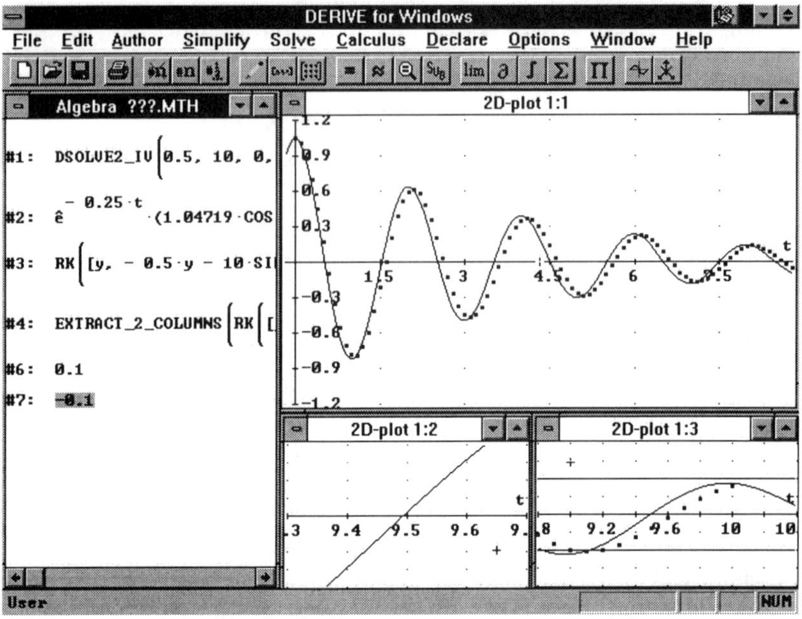

Figure 6.13 Solution to Example 6E (i)

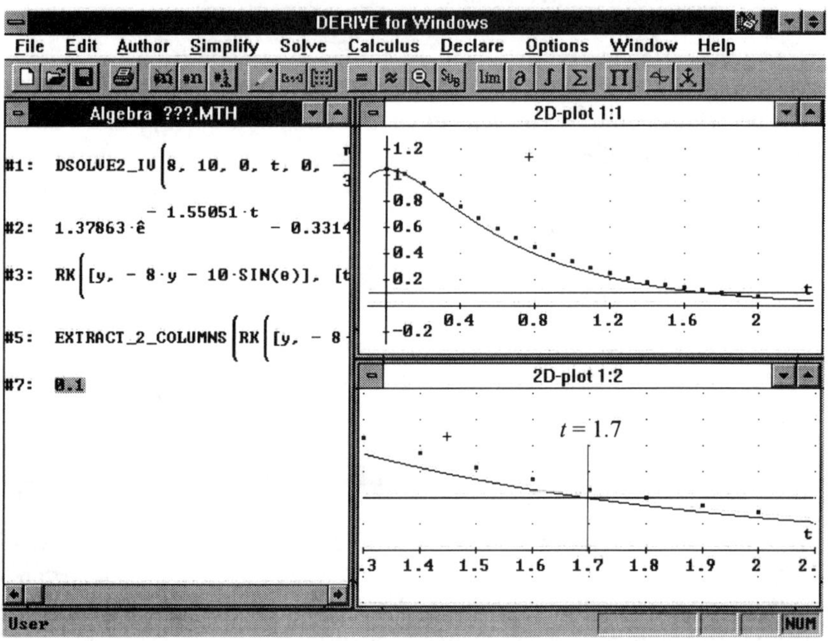

Figure 6.14 Solution to Example 6E (ii)

Example 6F A damped circuit

An LRC circuit has a constant emf of 240V, a resistance of 10Ω, an inductance of 0.5H and a capacitance of 0.002F in series. If the initial current and initial charge on the capacitor are both zero find the current i and charge q in terms of the time t.

Solution

With the given values equation (6.9) for an electric circuit becomes

$$0.5\frac{d^2q}{dt^2}+10\frac{dq}{dt}+\frac{1}{0.002}q=240$$

or

$$\frac{d^2q}{dt^2}+20\frac{dq}{dt}+1000q=480$$

and the initial conditions are $q(0) = 0$ and $\frac{dq}{dt}(0)=0$.

Solving using the DERIVE command DSOLVE2_IV(20,1000,480,t,0,0,0) gives

$$q = 0.48 - e^{-10t}(0.48 \cos 30t + 0.16 \sin 30t)$$

As we know q, i can be found most easily by differentiating this expression. Then

$$i=\frac{dq}{dt}=16e^{-10t}\sin 30t$$

The solutions and graphs show that the current soon decays to zero whereas the charge approaches its steady state value of 0.48 coulombs.

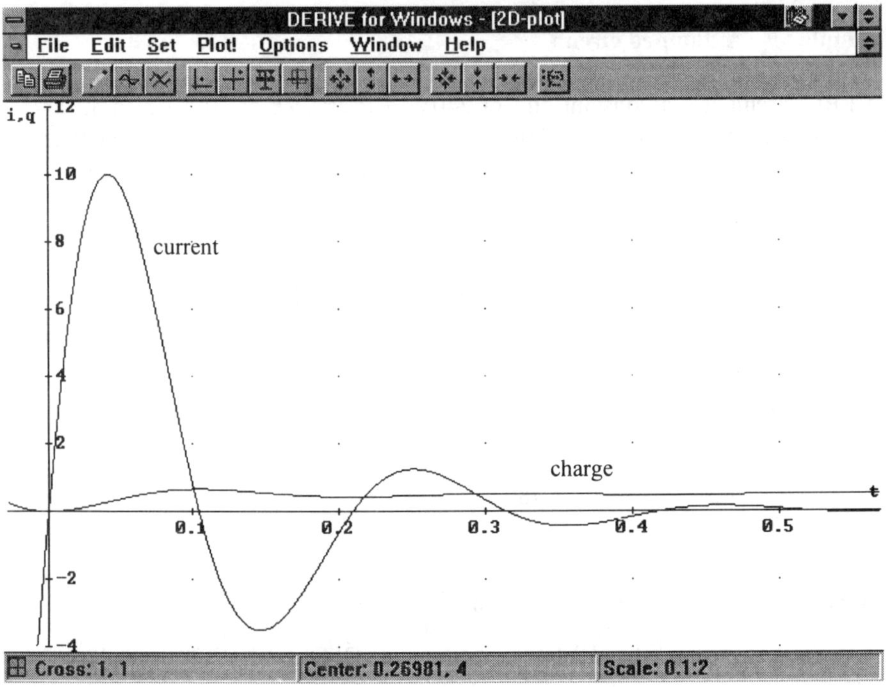

Figure 6.15 Current and charge in Example 6F

6.1.7 Forced vibrations

Now we investigate the effects of forcing a system to vibrate. The first example is an undamped system and then we include the effects of damping.

Example 6G Forced, free vibrations

A particle of mass m is suspended by a spring of stiffness k and natural length l from the ceiling. Investigate the motion of the particle when the ceiling is forced to vibrate sinusoidally with amplitude a and angular frequency α, so that $y = a \cos \alpha t$.

Consider the cases (i) $\alpha \neq \sqrt{\dfrac{k}{m}}$ and (ii) $\alpha = \sqrt{\dfrac{k}{m}}$.

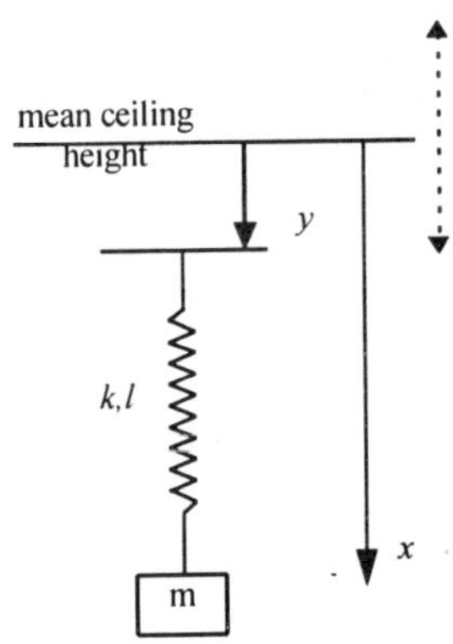

Figure 6.16

Solution

Assuming that the spring is extended the force on the particle is upwards with magnitude

$$T = k(x - y - l)$$

(The length of the spring is $x - y$, so its extension is $x - y - l$.)

By Newton's second law

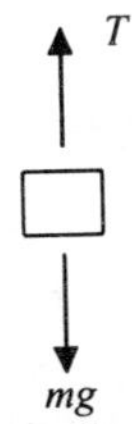

$$m\ddot{x} = mg - T$$
$$m\ddot{x} = mg - k(x - y - l)$$
$$m\ddot{x} + kx = mg + kl + ky$$
$$\ddot{x} + \frac{k}{m}x = g + \frac{kl}{m} + \frac{ky}{m}$$

or $\quad \ddot{x} + \Omega^2 x = g + \Omega^2 l + \Omega^2 y \qquad$ where $\Omega^2 = \dfrac{k}{m}$.

Note that if the particle was set in motion with the ceiling **not** vibrating ($y = 0$) then it would execute simple harmonic motion with angular frequency $\Omega = \sqrt{\dfrac{k}{m}}$. This is called the **natural frequency** of the system.

Using $y = a \cos \alpha t$ the equation becomes

$$\ddot{x} + \Omega^2 x = g + \Omega^2 l + \Omega^2 a \cos \beta t \tag{6.20}$$

(i) $\quad \beta \neq \sqrt{\dfrac{k}{m}} = \Omega$

When using DERIVE to solve this equation, and other similar ones, you need to use the DECLARE command to define Ω to be positive and also **Declare Algebra** state **Simplification Trigonometry** to **Expand**. Figure 6.5 shows how DERIVE can be used to solve the equation. The approach used here is one of many possible.

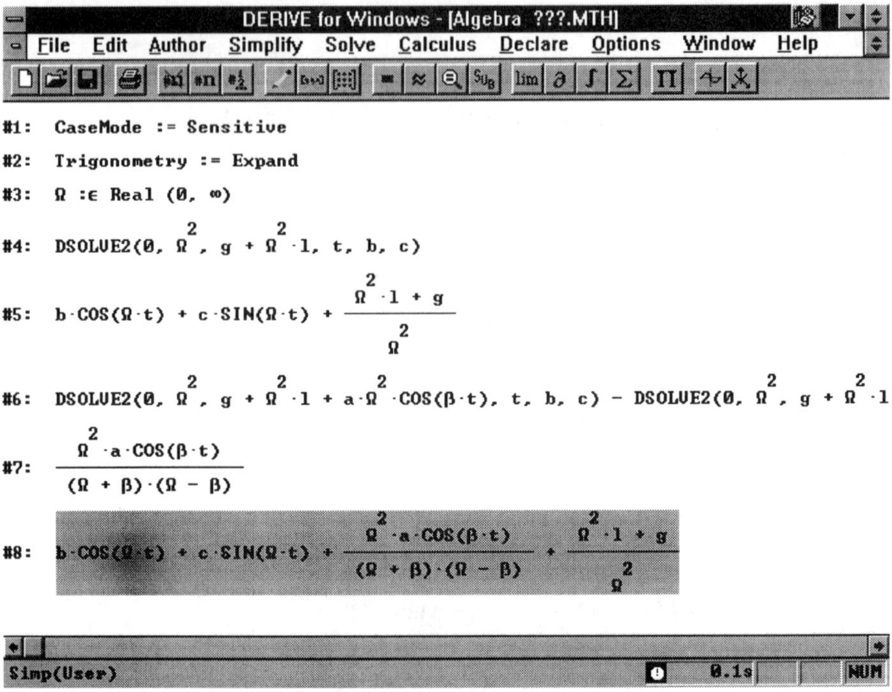

Figure 6.17 Using DERIVE to solve equation (6.20)

Lines #1-#3 set up DERIVE with the appropriate settings. Line #4 is the command to find the general solution of $\ddot{x}+\Omega^2 x = g+\Omega l$. **Simplify Basic** to obtain Line #5.

Line #6 is the command to find the particular integral corresponding to the term $\Omega^2 a\cos(\beta t)$; (the second DSOLVE2 command is line #4). **Simplify Basic** to give line #7. Line #8 is the most convenient form of the general solution of (6.20), found by adding #5 and #7.

(i) $B\cos\Omega t + C\sin\Omega t$ describe the free oscillations when there is no forcing term,

(ii) $\dfrac{g}{\Omega^2}+l$ gives the equilibrium position of the particle,

(iii) $\dfrac{a\Omega^2}{\Omega^2-\alpha^2}\cos\beta t$ represent oscillations caused by the vibrating ceiling.

The amplitude of the forcing term y is a, whereas that of the resulting oscillations is $\dfrac{a\Omega^2}{\Omega^2-\beta^2}$. The ratio of these terms is called the **relative response** of the system. That is

$$\text{relative response } R = \frac{\text{output amplitude}}{\text{input amplitude}} = \frac{\Omega^2}{\Omega^2-\beta^2}$$

and the magnitude of R is $|R| = \left|\dfrac{\Omega^2}{\Omega^2-\beta^2}\right|$.

Note that when $\beta = 0$, $|R| = 1$
when $\beta = \Omega$, $|R|$ is not defined
as $\beta \to \Omega$, $|R| \to \infty$
as $\beta \to \infty$, $|R| \to 0$

Figure 6.18 shows a graph of $|R|$ against β for the case $\Omega = 2$. (The shape of the curve will be similar for other values of Ω.) The particle resonates as $\beta \to \Omega$.

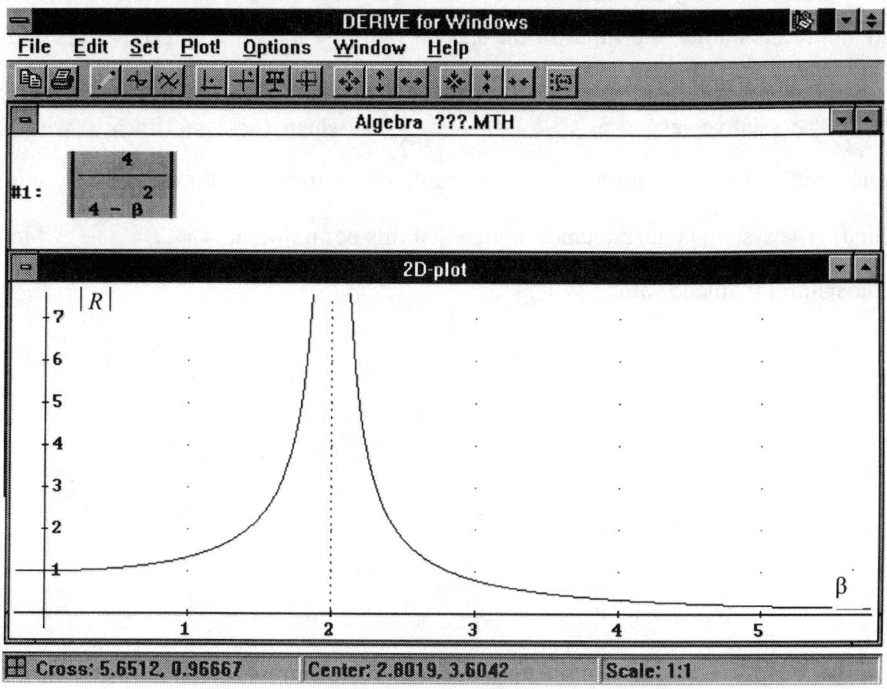

Figure 6.18 Graph of $|R|$ against β with $\Omega = 2$

(ii) When $\beta = \sqrt{\dfrac{k}{m}} = \Omega$ equation (6.10) becomes

$$\ddot{x} + \Omega^2 x = g + \Omega^2 l + \Omega^2 a \cos\Omega t \tag{6.21}$$

We are now in a case of failure situation as the sinusoidal part of the right hand side, $\Omega^2 a \cos \Omega t$, is one of the terms in the complementary function.

The equation can be solved using DERIVE as in Figure 6.17, but this time we will use a different approach. As we are only interested in the solution corresponding to the forcing term we wish to eliminate the constant terms and the complementary function. In Figure 6.19:

#1 is the command to solve (6.21), ignoring the constant
#2 is the general solution
#3 is #1 after editing to change the right hand size to zero
#4 is the solution of #3, or the complementary function of #1
#5, found from #2-#4, is the required solution

Hence the output corresponding to an input $a \cos \Omega t$ is $\frac{1}{2} a\Omega t \sin\Omega t$ (#5). #6 and #7 are the output and input in the special case $a = 1$, $\Omega = 2$ and these functions are plotted in the graphing window.

The relative response is $R = \dfrac{\frac{1}{2} a\Omega t}{a} = \dfrac{1}{2}\Omega t$ which increases linearly with t. Hence while the input amplitude is constant, the output amplitude grows without bound. The system is in resonance; notice that this occurs when $\alpha = \Omega = \sqrt{\dfrac{k}{m}}$, which is the natural frequency of the system.

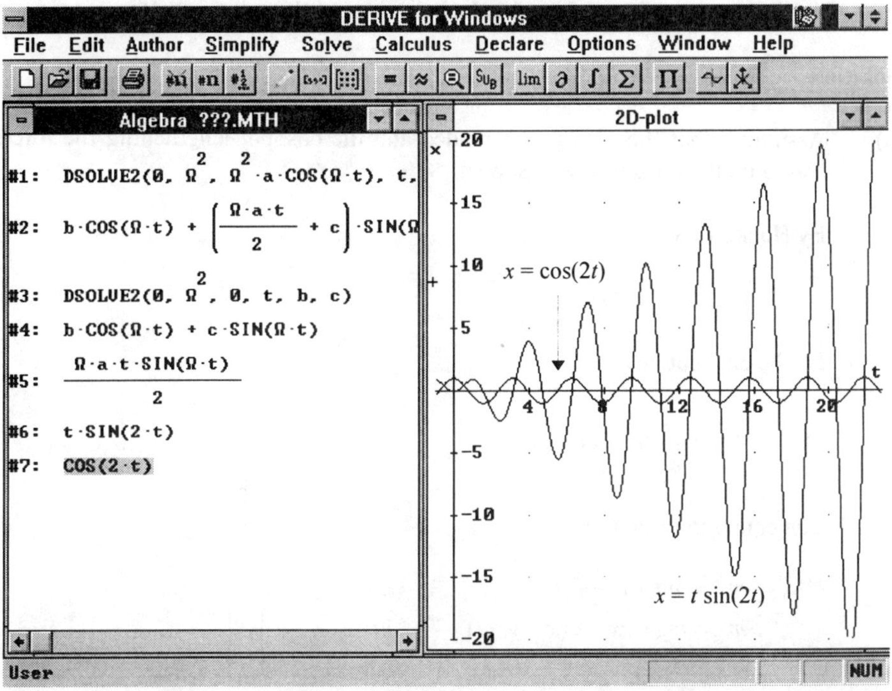

Figure 6.19 Comparing the sinusoidal input and output in (6.21)

□

Example 6H Damped, forced vibrations

The motion of the particle in Example 6G is damped by means of a dashpot of constant r, attached to the floor, as shown in the diagram.

(i) Obtain the equation of motion of the particle.

The parameters are now given values, in SI units, $m = 0.25$, $k = 25$, $l = 0.5$, $r = 0.25$ and $g = 10$, and $y = 0.04 \cos 8t$. The particle is set in motion with $x(0) = 0.8$ and $\dot{x}(0) = 1$.

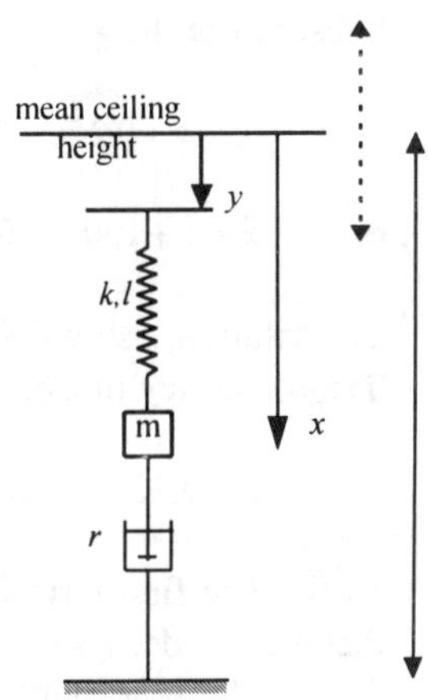

Figure 6.20

(ii) Solve the equation and describe the behaviour of the solution as t increases.

Solution

(i) Assuming that the spring is extended and the dashpot lengthening the forces acting on the particle are as shown.

By Hooke's law

$$T = k(x - y - l)$$

By the dashpot law

$$F = r\frac{d}{dt}(d - x) = -r\dot{x}$$

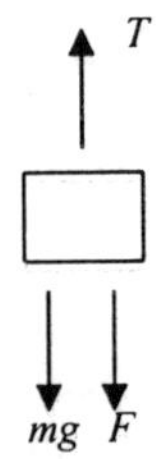

The equation of motion is

$$m\ddot{x} = mg + F - T$$
$$m\ddot{x} = mg - r\dot{x} - k(x - y - l)$$
$$m\ddot{x} + r\dot{x} + kx = mg + kl + ky$$

or, using $y = a\cos\alpha t$

$$m\ddot{x} + r\dot{x} + kx = mg + kl + ka\cos\alpha t \tag{6.22}$$

(ii) Substituting the given values into equation (6.22) gives

$$0.25\ddot{x} + 0.25\dot{x} + 25x = 15 + \cos 8t$$

or $$\ddot{x} + \dot{x} + 100x = 60 + 4\cos 8t \tag{6.23}$$

The solution, using DSOLVE2_IV(1,100,60 + 4cos 8t,t,0,0.8,1) and **Trigonometry** in **Collect** mode is

$$x = 0.127e^{-0.5t}\sin(9.99t + 0.830) + 0.108\sin(8t + 1.352) + 0.6 \quad \text{(3D)}$$

The first term is the complementary function x_c which, because of the factor $e^{-0.5t}$, decays to zero. For this reason it is called a **transient**.

The other term x_p comprises the particular integral. As the transient term dies away the motion of the particle becomes essentially that of the particular integral, which is called the **steady state solution**. This predicts that the particle oscillates simple harmonically about a centre $x = 0.6$ (the equilibrium position) with amplitude 0.108 and phase 1.352.

Note that the output corresponding to the input $y = 0.4 \cos 8t$, which is equivalent to $y = 0.04\sin\left(8t + \frac{\pi}{2}\right) = 0.04\sin(8t + 1.571)$ is $0.108\sin(8t + 1.352)$.

The relative response is $\frac{0.108}{0.04} = 2.7$ and the phase shift is $1.352 - 1.571 \approx -0.219$.

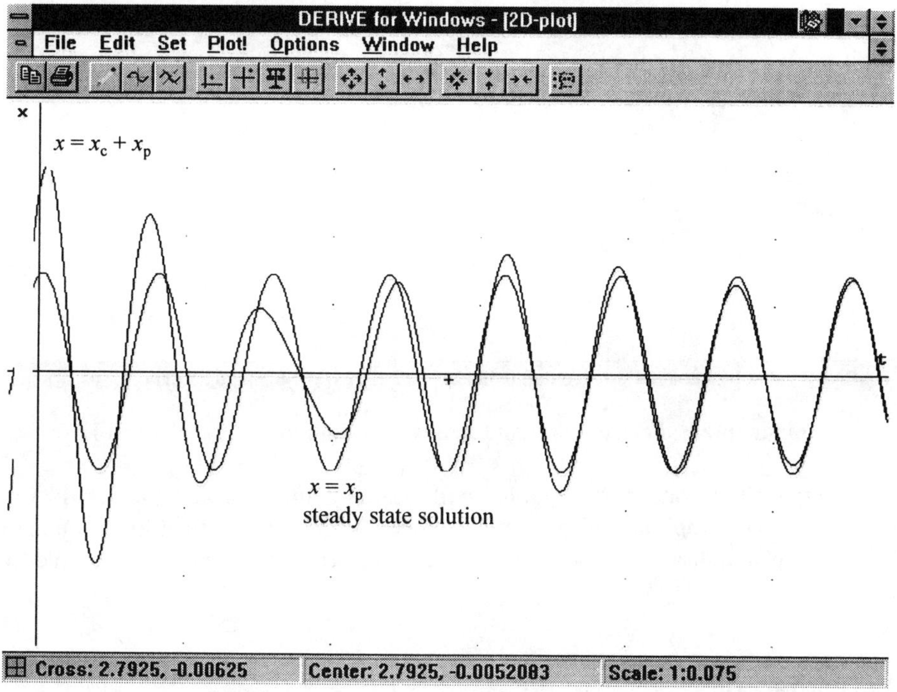

Figure 6.21 Solution and steady state solution for Example 6H

Figure 6.21 shows the graphs of the solution x and the steady state solution x_p (with the constant 0.6 removed from each). This clearly shows how $x \to x_p$ as $t \to \infty$.

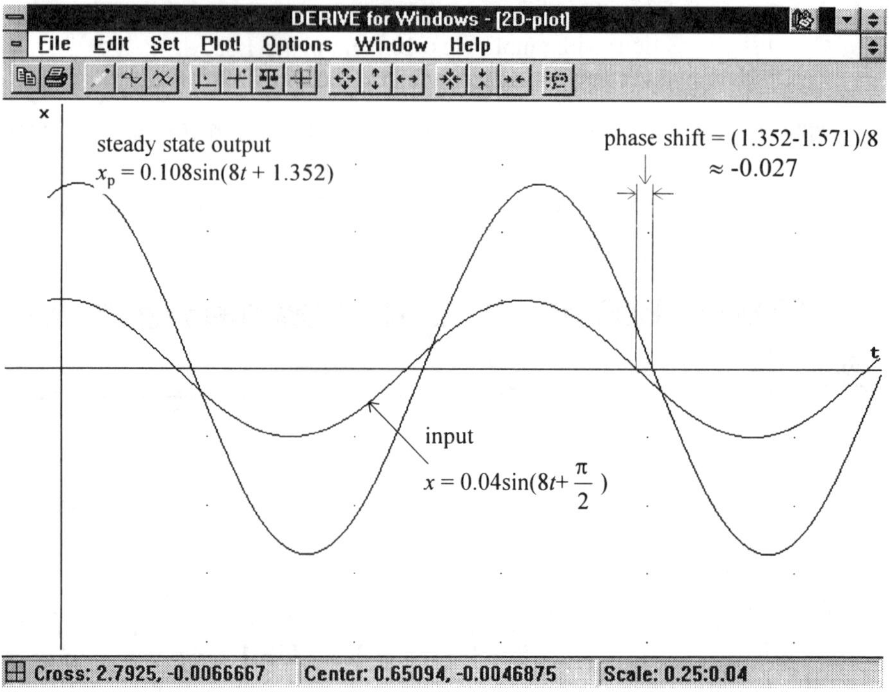

Figure 6.22 Forcing term and steady state solutions for Example 6H

Figure 6.22 shows the graphs of the forcing function and the steady state solution x_p. The amplitude of the latter is about 2.7 times that of the former while the horizontal phase shift is a measure of the delay between equivalent points on the two curves.

□

The general case

The equation of motion for a damped and forced system will be of the form

$$m\ddot{x} + r\dot{x} + kx = c + d\cos(\alpha t + \beta) \tag{6.24}$$

The general solution is the sum of the complementary function x_c and particular integral x_p. That is, $x = x_c + x_p$. The complementary function x_c, the solution of $m\ddot{x} + r\dot{x} + kx = 0$, was discussed in Section 6.1.2 where it was shown that the solution tends to zero as t increases, for all positive values of m, r and k. That is $x_c \to 0$ as $t \to \infty$, x_c is therefore called **transient**. The solution x then effectively becomes $x = x_p$, where the particular integral x_p will be of the form $x_p = e + f\cos(\alpha t + \gamma)$, which is a sinusoid of amplitude f oscillating about the centre $x = e$. x_p is the **steady state**

solution. Hence after a while the solution of equation (6.24) is the steady state $x = x_p$, which is independent of the initial conditions (the values of $x(0)$ and $\dot{x}(0)$).

Example 6I Resonance

Investigate the motion of the particle in Example 6H as

(i) the forcing angular frequency α changes
(ii) the damping constant r changes.

Solution

(i) If the forcing term is $y = 0.04 \cos \alpha t$ the differential equation (6.23) in Example 6H becomes

$$\ddot{x} + \dot{x} + 100x = 60 + 4 \cos \alpha t$$

As we are only interested in the response of the system to the forcing term we can ignore the constant and solve the equation with

DSOLVE2(1,100,4 cos αt, t,b,c)

Subtracting the complementary function gives the steady state solution

$$x = \frac{9721924(\alpha \sin(\alpha t) - (\alpha^2 - 100)\cos(\alpha t))}{(1559\alpha^2 - 31141\alpha + 155900)(1559\alpha^2 + 31141\alpha + 155900)}$$

Figure 6.23 shows the graphs of the steady state solution for α = 4, 8 and 12. Figure 6.24 shows what happens as α approaches its value for resonance, which is approximately 10.

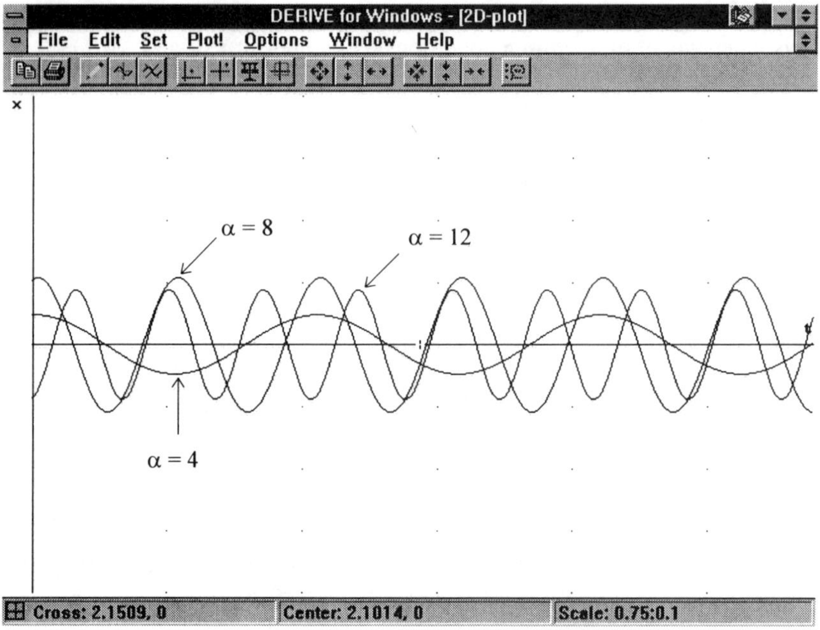

Figure 6.23 Steady state solutions in Example 6I (i) for α = 4, 8 and 12

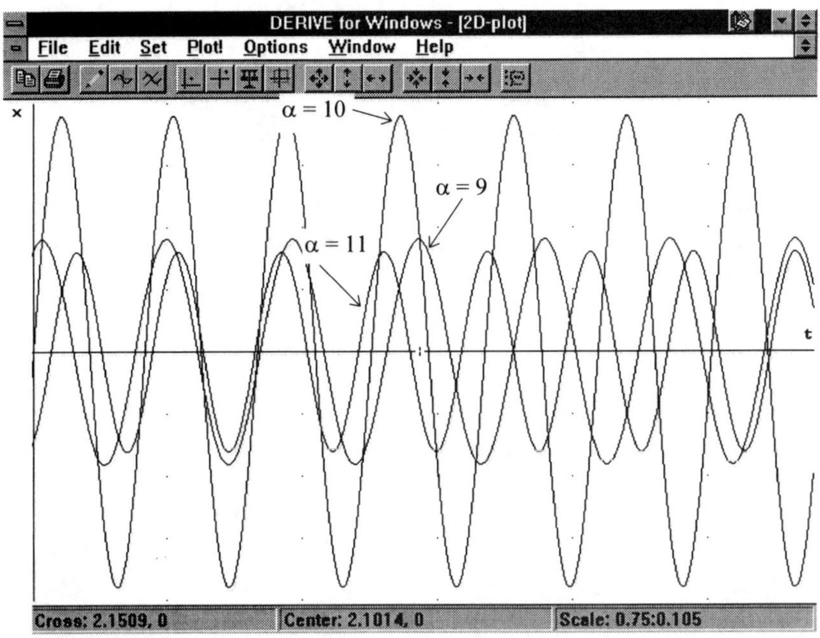

Figure 6.24 Steady state solutions in Example 6I (i) for α = 9, 10 and 11

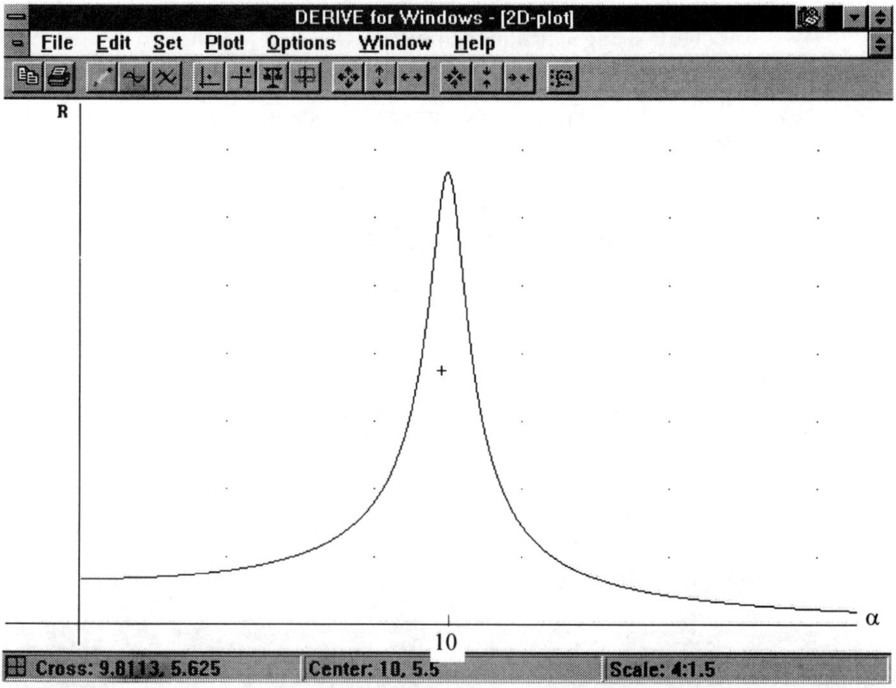

Figure 6.25 Plot of R against α in Example 6I (i)

DERIVE shows that the relative response of the system is

$$R = \frac{243048100\sqrt{\alpha^4 - 199\alpha^2 + 10000}}{24304881\alpha^4 - 4836656881\alpha^2 + 24304810000}$$

which is equivalent to $R = \dfrac{100}{\sqrt{\alpha^4 - 199\alpha^2 + 10000}}$.

A graph of R against α is shown in Figure 6.25. R has a maximum value near $\alpha = 10$, which is the natural frequency of the system and a search with the cross shows that $R_{MAX} \approx 10$ when $\alpha \approx 9.97$. This result is true in general; for lightly damped systems resonance occurs near the natural frequency of the system, but the presence of some damping prevents the relative response becoming infinite, as it did in Example 6G.

A check using calculus shows that $\frac{dR}{d\alpha} = 0$ when $\alpha = \frac{\sqrt{398}}{2} \approx 9.97$, that $\frac{d^2R}{d\alpha^2} < 0$ at this point, showing that R has a maximum value at $\alpha = 9.97$ and $R_{MAX} \approx 10.01$.

(ii) With parameters other than r taking their previous values, equation (6.13) becomes

$$\ddot{x} + 4r\dot{x} + 100x = 4\cos\alpha t$$

We can use DERIVE to show that the relative response is

$$R = \frac{100}{\sqrt{\alpha^4 + 8\alpha^2(2r^2 - 25) + 10000}}$$

and Figure 6.26 shows how R varies with α for different values of r.

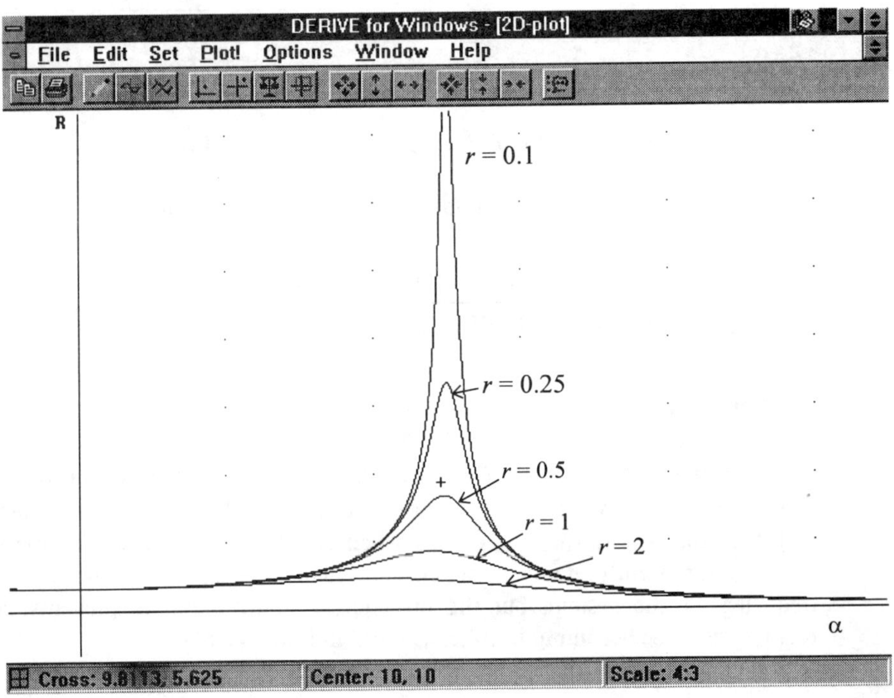

Figure 6.26 Graph of R against α for different values of r

The graphs show that the smaller r is (i.e. the lighter the damping), the larger the relative response at resonance, but as damping (r) increases the maximum relative response decreases, so that eventually there is no resonance. In all cases

$R = 1$ when $\alpha = 0$

$R \to 0$ as $\alpha \to \infty$

resonance, when it occurs, takes place near the natural frequency of the system ($\alpha = 10$)

Systems which are subject to an external forcing influence should be sufficiently heavily damped if resonance is to be avoided. The bridge collapses at Broughton and Washington were due to the periodic forcing term, caused by footsteps in the first case and wind in the second, being close to one of the natural frequencies of the bridge.

□

Example 6J

We shall look at the model of car suspensions discussed in Application 4 in Chapter 1. A car is driven at different speeds over an undulating road and we wish to predict the vertical motion of the car (the particle) caused by the undulations. The equation of motion of the car is, from Application 4, $m\ddot{x} + r\dot{x} + kx = ky + r\dot{y}$, where $y = a\cos\dfrac{2\pi ut}{d}$. (The constant terms $kl - mg$ can be ignored as we are interested in the amplitude of the forced oscillations, not their centre.)

Assume that the values of the parameters are m = 1000 kg, k = 10000 Nm^{-1}, r = 10000 Nms^{-1}, a = 0.2m, d = 4m. Investigate the vertical motion of the car when it is driven at speeds of 5, 10, 20 and 30ms^{-1} (approximately 11, 34, 45 and 68 mph).

Solution

With the given values, $y = 0.2\cos\dfrac{\pi ut}{2}$, $\dot{y} = -0.1\pi u \sin\dfrac{\pi ut}{2}$ and the differential equation simplifies to $\ddot{x} + 10\dot{x} + 10x = 2\cos\dfrac{\pi ut}{2} - \pi u \sin\dfrac{\pi ut}{2}$. Assuming that the car is leaving a flat road initially, reasonable initial conditions for the vertical displacement are $x(0) = 0$ and $\dot{x}(0) = 0$, although, as we have seen, these are largely irrelevant as it is the steady state solution in which we are interested. The equation can be solved using

$$\text{DSOLVE2\_IV}\left(10{,}10{,}2\cos\frac{\pi ut}{2} - \pi u\sin\frac{\pi ut}{2}, t{,}0{,}0{,}0\right)$$

Inserting the given values of u and finding the relative response gives the following results:

speed u ms^{-1}	relative response (2D)
5	0.84
10	0.55
20	0.31
30	0.21

The effect of the suspension system is to reduce the amplitude by 16% at 5ms^{-1}, increasing to 79% at 30ms^{-1}, with the reduction increasing with the speed. It must be remembered, however, that as the speed increases the period $\dfrac{2\pi}{\frac{1}{2}\pi u} = \dfrac{4}{u}$ decreases, so whether the car driver feels more comfortable needs further scrutiny. In a modelling situation we would obtain real car data and investigate how combinations of spring stiffness and dashpot constant can improve the ride of passengers over various types of road surface. It may also be necessary to improve the model by, for example, including the effects of the tyres.

□

Exercise 6B

1. A particle of mass m is suspended from a fixed point A by a spring and a dashpot, as shown in the diagram. Find the equation of motion of the particle assuming that the spring is extended and (i) the dashpot is lengthening, (ii) the dashpot is shortening. Given that m = 1 kg, l = 0.4 m, k = 20 Nm^{-1} predict, qualitatively, the motion of the particle when (i) r = 8 Nms^{-1}, (ii) r = 10 Nms^{-1}.

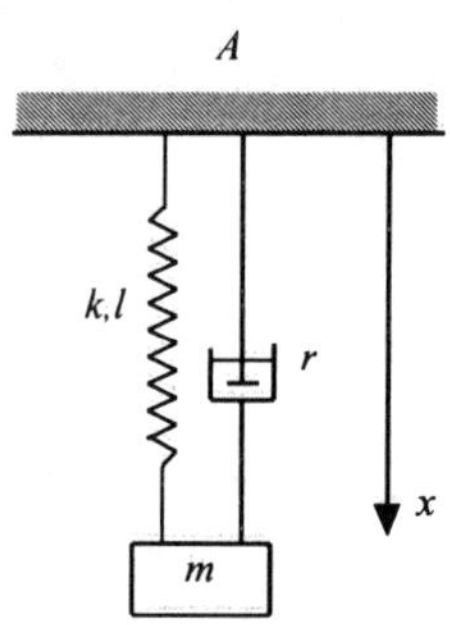

2. A particle of mass 3 kg is attached to the ceiling by a spring and a dashpot and to the floor by another dashpot, as shown in the diagram. The spring has stiffness 120 Nm^{-1} and natural length 0.6m. The upper and lower dashpots have dashpot constants 2 Nms^{-1} and 1 Nms^{-1} respectively. The distance from the floor to the ceiling is 1.8m. Take g = 10 ms^{-2}.

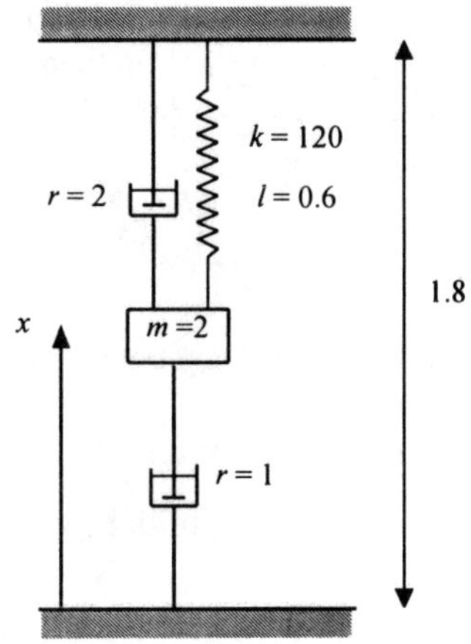

(i) Find the equation of motion of the particle and describe its motion qualitatively.

(ii) Solve the equation given that the particle is projected upwards with speed 1 ms^{-1} when it is 0.9m above the floor.

(iii) Sketch the graph of x against t.

3. Consider again the system in Example 6B. The parameters have the values m = 0.3 kg, k = 10 Nm^{-1}, g = 10 ms^{-2} and l = 0.25m. When the particle is at rest in its equilibrium position the ceiling is forced to vibrate so that $y = 0.15\sin\alpha t$. Show that the equation of motion is $\ddot{x}+100x = 35+10\sin\alpha t$. Find the relative response and interpret the solution graphically when (i) $\alpha = 3$, (ii) $\alpha = 9$, (iii) $\alpha = 10$.

4. On one definition, resonance occurs when the relative response has a maximum value which is at least double its value at $\alpha = 0$. Find the largest value of r for the system in Example 6I (ii) to exhibit resonance.

5. An LRC circuit has an alternating emf $E = 50\sin 50t$ V, a resistance of 10Ω, an inductance of 0.5H and a capacitance of 0.002F in series. If the initial current and initial charge on the capacitor are both zero determine the current and charge at any time t. Identify the steady state and transient charge and current.

6. A swinging door has a damping device so that the mathematical model for the angular displacement $\theta(t)$ is

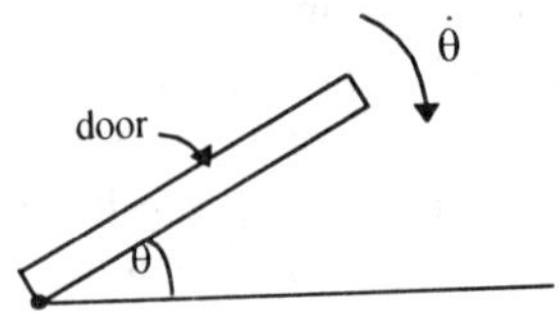

$$I\ddot{\theta} = -a\theta - b\dot{\theta}$$

where I, a and b are constants.

Consider a system for which $I = 1.5$ kg s^{2}m^{-1}, $a = 6$ kg m^{-1}, $b = 7.5$ kg s m^{-1} and for which the initial conditions are $\theta(0) = \dfrac{\pi}{3}$ and $\dot{\theta}(0) = 0$.

(i) Formulate and solve an initial value problem that models this system.

(ii) Draw a graph of the displacement θ with time.

(iii) What does your model predict as t becomes large?

7. An LRC circuit consists of an inductance $L = 0.2$H, $R = 1\Omega$ and $C = 0.8$F. The initial charge in the circuit is 2 coulombs and the initial current is 4 amps. The applied voltage is given by $E(t) = 1 + e^{-t}$ volts.

(i) Formulate an initial value problem which models the charge $q(t)$ on the capacitor at any time, t.

(ii) Solve the problem to find the charge $q(t)$.

(iii) Draw a graph of the solution.

(iv) What is the charge on the capacitor over after a long period of time?

8. The shock absorber on a car is modelled as a spring-dashpot system with spring stiffness 120000 kgs^{-2} and dashpot constant 3000 kgs^{-1}. The shock absorber supports a mass of 300 kg.

A car travels over a bumpy road so that the forcing term on the shock absorber is $15e^{-2t}\sin 3t$ kg ms^{-2}.

(i) Formulate an initial value problem to model the displacement of the car.

(ii) Solve the problem and draw a graph of the solution for $0 \leq t \leq 4$s.

(iii) What is the maximum displacement from the equilibrium of the system?

(iv) Describe the motion of a person in the car as it travels over the bumpy road.

6.2 MIXING PROBLEMS

In some problems there will be two, or more, variables dependent on a single independent variable, usually time, and the mathematical analysis leads to a set of interconnected (coupled) simultaneous differential equations. Two examples are given below.

(i) A drug injected into the bloodstream will spread through the body and, hopefully, have the desired effect. The blood will pass through the tissues of the body, but also through the kidneys which will gradually filter out the drug. It is important to know the period for which the drug remains effective so that repeat doses can be administered at the correct time. A model of the system is formulated by treating the body as a number of compartments and tracking the concentrations x and y of the drug into and out of the blood and tissues.

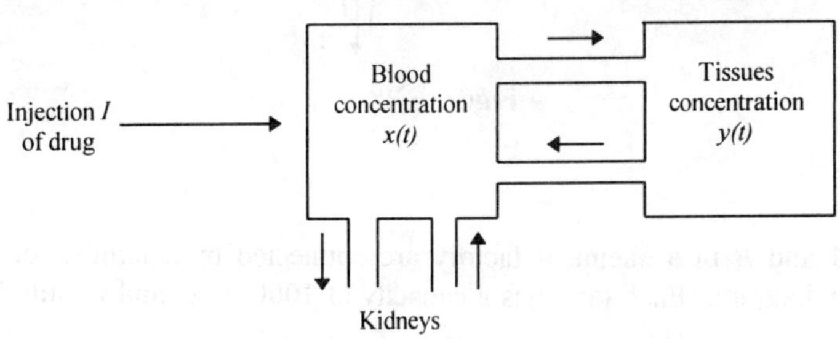

Figure 6.27

On the assumption that the rate of transfer of the drug is proportional to the concentrations x and y, and by considering the input and output to each compartment, we will obtain a pair of equations of the form

$$\frac{dx}{dt} = I - ax + by, \quad \frac{dy}{dt} = cx - dy$$

where a, b, c and d are positive constants.

The long term concentrations x and y at time t can be found by solving these equations together with the initial conditions. This is done by first combining the two equations into a single second order linear equation in x (or y).

(ii) The outflow from a power station on the shore of a lake causes pollution in that and neighbouring lakes, together with the river flowing through. By considering the inputs to and outputs from each lake (compartment) we can obtain a set of three equations from which the concentrations of pollutant (c_1,c_2,c_3) can be predicted.

The following example illustrates the technique.

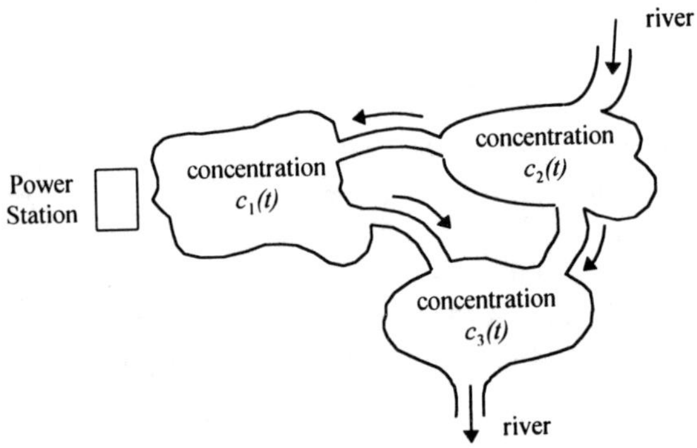

Figure 6.28

Example 6K

Two tanks A and B in a chemical factory are connected by a number of pipes, as shown in the diagram. Each tank has a capacity of 1000 litres and is initially full of pure water.

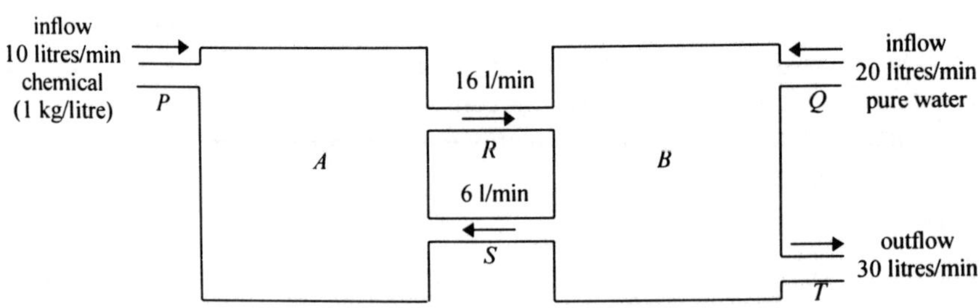

Figure 6.29

When the pumps are started water containing 1 kg/litre of chemical is pumped into tank A at P at a rate of 10 litres per minute and pure water is pumped into tank B at Q at a rate of 20 litres per minute. There is a flow of 16 litres per minute from A to B along pipe R, 6 litres per minute from B to A along pipe S and an outflow of 30 litres per minute from tank B at T. The mixture in each tank is kept uniform by stirring. Investigate the build up of chemical in each tank as time passes.

Solution

Suppose that the amount of chemical in tanks A and B at time t minutes is x kg and y kg respectively. As each tank always contains 1000 litres the concentrations of the chemical at time t are $\dfrac{x}{1000}$ and $\dfrac{y}{1000}$ kilograms per litre respectively.

The concentration of chemical in tank A increases at a rate of 10 kg per minute due to the inflow at P and $\dfrac{6y}{1000}$ kg/min due to the flow from tank B along pipe S. It decreases at a rate of $\dfrac{16x}{1000}$ kg/min due to the flow to tank B along pipe R. Hence we obtain the differential equation

$$\frac{dx}{dt} = 10 + \frac{6y}{1000} - \frac{16x}{1000} \tag{6.25}$$

Treating tank B in a similar way we obtain

$$\frac{dy}{dt} = \frac{16x}{1000} - \frac{36y}{1000} \tag{6.26}$$

This is a pair of first order linear equations which can be solved analytically by first combining them to form a single second order linear equation. From (6.26),

$$x = 62.5\frac{dy}{dt} + 2.25y \tag{6.27}$$

and substitution into (6.25) gives

$$\frac{d^2y}{dt^2} + 0.052\frac{dy}{dt} + 0.00048y = 0.16$$

Initially $x(0) = y(0) = 0$ and so, from (6.26), $\dfrac{dy}{dt}(0) = 0$.

Using DERIVE to solve this equation gives, approximately,

$$y = -476e^{-0.012t} + 143e^{-0.04t} + 333\tfrac{1}{3}$$

x can be found most easily by using DERIVE to differentiate y and to substitute for y and $\dfrac{dy}{dt}$ in (6.27). This gives

$$x = -714e^{-0.12t} - 35.7e^{-0.04t} + 750$$

It is clear from the solutions and their graphs in Figure 6.30 that as t increases x and y approach their equilibrium values of 750 and $333\frac{1}{3}$ kg respectively. You can enlarge the graph or use **Substitute** to show that both concentrations are within 1% of these values after 7 hours.

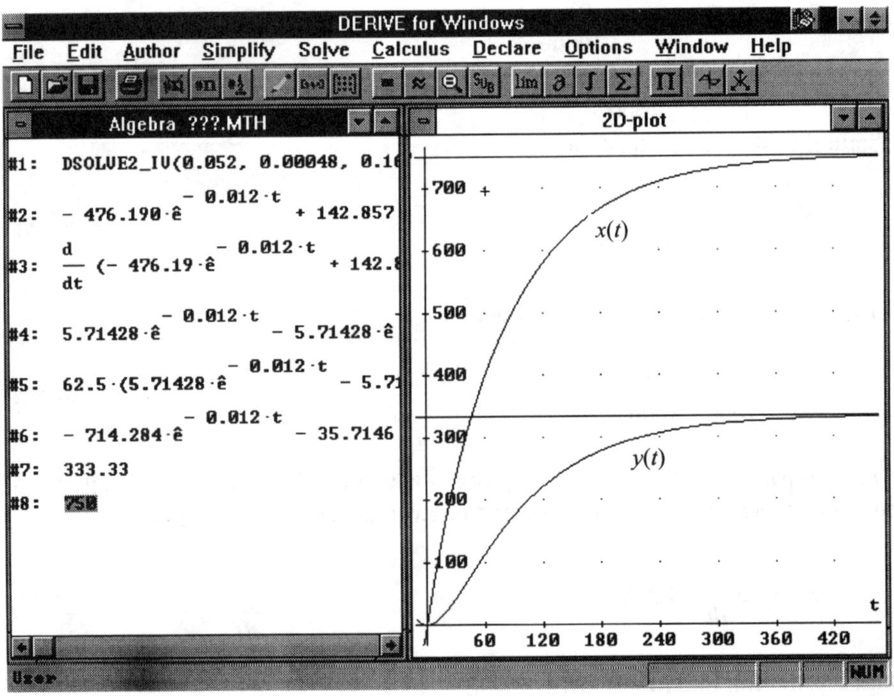

Figure 6.30 Solution to Example 6K

□

6.3 PROJECTILE MOTION IN TWO DIMENSIONS

In Chapter 4.4 we looked at the motion of a particle in one dimension, with and without the effects of air resistance (drag). Here we will extend this to motion in two dimensions. For motion ignoring drag the problems can be solved analytically, as illustrated in Examples 6L. Including the effects of drag substantially complicates the equations of motion and solutions usually have to be found numerically. Analytical solutions can, however, be found when the drag is assumed to be proportional to the velocity of the particle, and the technique is illustrated in Example 6M.

It was noted in Chapter 4 that for many everyday objects the quadratic approximation for drag is more appropriate, and in this case we obtain a set of coupled differential equations which have to be solved numerically. The approach is illustrated in Example 6N and the preceding analysis.

The effect of spin on the motion of a particle is clearly also very important. Applying top spin or back spin are standard shots in tennis and table tennis while any golfer dreads inadvertently applying side spin and hence hooking or slicing his shot. When including the effects of back or top spin the problem remains two dimensional and this is illustrated in problem 4 of Exercise 6C. Side spin turns the problem into a three dimensional one and that will not be considered here.

6.3.1 Projectile motion without drag

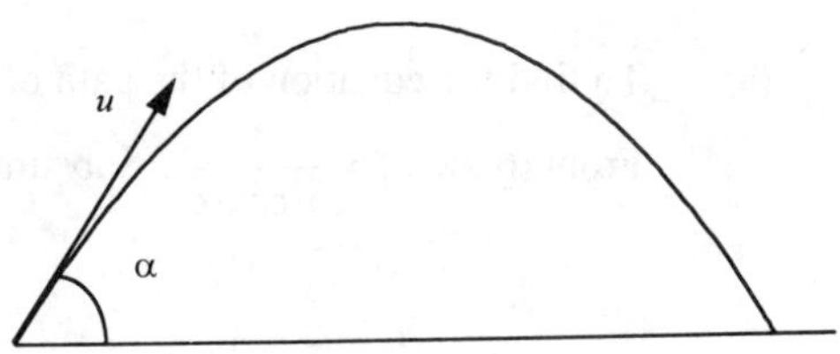

A particle is projected with speed u at an angle α to the horizontal. We are interested in determining such things as its greatest height, the time of flight and the range.

We assume that the only force acting on the particle is its weight.

Choose the origin at the point of projection. Applying Newton's second law horizontally gives

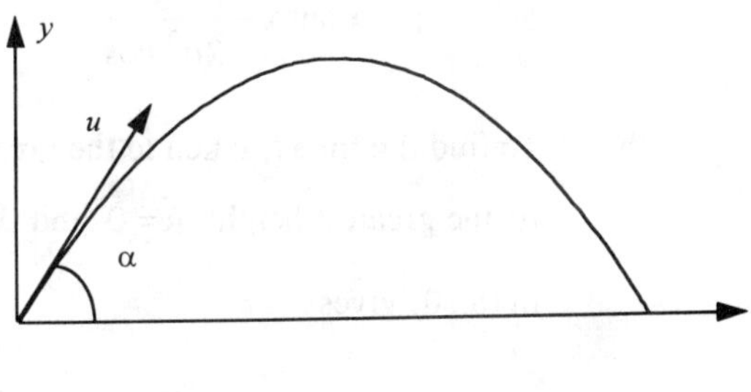

$$m\ddot{x} = 0 \quad \text{or} \quad \ddot{x} = 0$$

Integrating:

$$\dot{x} = A \quad \text{and} \quad x = At + B$$

The initial conditions $x(0) = 0$ and $\dot{x}(0) = u\cos\alpha$ give $A = u\cos\alpha$ and $B = 0$.

Hence

$$x = ut\cos\alpha \tag{6.28}$$

and $$\dot{x} = u\cos\alpha \tag{6.29}$$

Applying N2 vertically:

$$m\ddot{y} = -mg \quad \text{or} \quad \ddot{y} = -g$$

Integrating:

$$\dot{y} = -gt + C \quad \text{and} \quad y = -\frac{1}{2}gt^2 + Ct + D$$

The initial conditions $y(0) = 0$ and $\dot{y}(0) = u\sin\alpha$ give $C = u\sin\alpha$ and $D = 0$. Then

$$y = ut\sin\alpha - \frac{1}{2}gt^2 \tag{6.30}$$

and $$\dot{y} = u\sin\alpha - gt \tag{6.31}$$

(a) To find the equation of the path of the particle.
From (6.28), $t = \dfrac{x}{u\cos\alpha}$. Substituting in (6.30),

$$y = u\left(\frac{x}{u\cos\alpha}\right)\sin\alpha - \frac{1}{2}\left(\frac{x}{u\cos\alpha}\right)^2$$

or $$y = x\tan\alpha - \frac{gx^2}{2u^2\cos^2\alpha} \quad \text{(a parabola)} \tag{6.32}$$

(b) To find the time t_1 taken to the greatest height, and this height y_1.
At the greatest height $\dot{y} = 0$ and then, from (6.31), $t_1 = \dfrac{u\sin\alpha}{g}$. Substituting in (6.30) gives

$$y_1 = u\sin\alpha\left(\frac{u\sin\alpha}{g}\right) - \frac{1}{2}g\left(\frac{u\sin\alpha}{g}\right)^2 = \frac{u^2\sin^2\alpha}{2g}$$

(c) To find the time of flight t_2 and the maximum range R.
The particle hits the ground when $y = 0$. From (6.30)

$$0 = ut_2 \sin\alpha - \frac{1}{2}gt_2^2$$

$$\frac{1}{2}t_2(2u\sin\alpha - gt_2) = 0$$

giving $t_2 = 0$ or $t_2 = \dfrac{2u\sin\alpha}{g}$. To find the range substitute $t_2 = \dfrac{2u\sin\alpha}{g}$ in (6.28). Then range

$$R = u \times \frac{2u\sin\alpha}{g}\cos\alpha = \frac{u^2 \times 2\sin\alpha\cos\alpha}{g} = \frac{u^2 \sin 2\alpha}{g}$$

□

Example 6L

A basketball player throws a basketball at a basket which is 5 metres away from him horizontally and 1.5 metres above him vertically.

(i) If he throws the ball at an angle $\alpha = 60°$ find the speed of projection u so that the ball enters the basket.

(ii) If he throws the ball at 10ms^{-1} find the angle α for the ball to enter the basket.

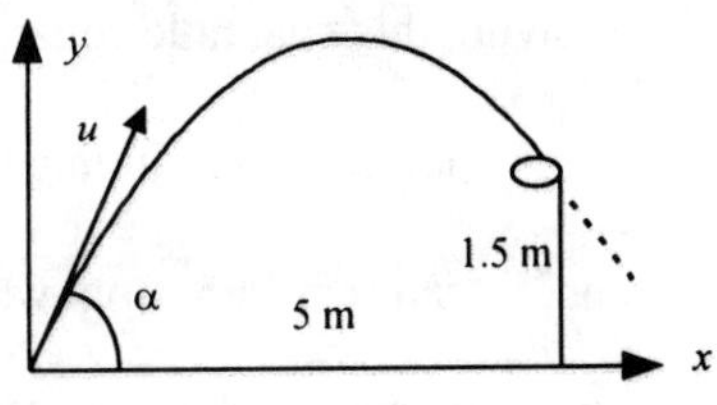

Solution

With the origin at the point of projection equations (6.28) to (6.32) apply.

(i) With $\alpha = 60°$, so that $\cos\alpha = \dfrac{1}{2}$ and $\tan\alpha = \sqrt{3}$, and $g = 9.81$, (6.32) becomes

$$y = \sqrt{3}x - \frac{9.81}{2u^2 \times \frac{1}{4}}x^2 \quad \text{or} \quad y = \sqrt{3}x - \frac{19.62x^2}{u^2}$$

To reach the basket we require $y = 1.5$ when $x = 5$. Then

$$1.5 = 5\sqrt{3} - \frac{19.62 \times 25}{u^2}$$

from which $u = \sqrt{\dfrac{19.62 \times 25}{5\sqrt{3} - 1.5}} \approx 8.28$, so that ball must be thrown at 8.28ms^{-1}.

(ii) With $u = 10$, $g = 9.81$, (6.32) becomes

$$y = x\tan\alpha - \frac{9.81}{2 \times 10^2 \cos^2\alpha} x^2$$

Putting $y = 1.5$, $x = 5$ and using $\dfrac{1}{\cos^2\alpha} = \sec^2\alpha = 1 + \tan^2\alpha$:

$$1.5 = 5\tan\alpha - \frac{9.81 \times 25}{2 \times 100}(1 + \tan^2\alpha)$$

$$\frac{9.81}{8}\tan^2\alpha - 5\tan\alpha + \left(1.5 + \frac{9.81}{8}\right) = 0$$

Solving this quadratic equation for $\tan\alpha$ gives

$\tan\alpha = 3.43$, from which $\alpha \approx 74°$

or $\tan\alpha = 0.65$, from which $\alpha \approx 33°$

Since the ball must be coming down to enter the basket, the angle of projection must be 74°.

□

6.3.2 **Projectile motion with drag** (linear approximation)

The effects of drag (but not spin) will now be included, and here we will assume that the magnitude of the drag force is proportional to the speed of the particle. As we shall see in Example 6M this complication still results in equations that can be solved analytically; DERIVE will be used to solve these equations.

Example 6M

A particle of mass 0.5kg is projected from ground level with speed $50ms^{-1}$ at an angle of 60° to the horizontal. The forces acting on the particle are its weight (take $g = 10ms^{-2}$) and air resistance, which is equal to $-0.2 \times$ velocity.

(i) Apply N2 vertically and horizontally and hence find the position of the particle at time t.

(ii) Find the time the particle takes to reach its highest point and the coordinates of this point.

(iii) Find the range of the particle and the speed with which it hits the ground.

(iv) If the particle was projected off the edge of a tall cliff find its terminal speed and maximum range.

(v) Plot the graph of the trajectory of the particle.

Solution

The direction of the force due to air resistance is directly opposite to that of the velocity. When the velocity makes an angle θ with the horizontal the horizontal and vertical components of R are

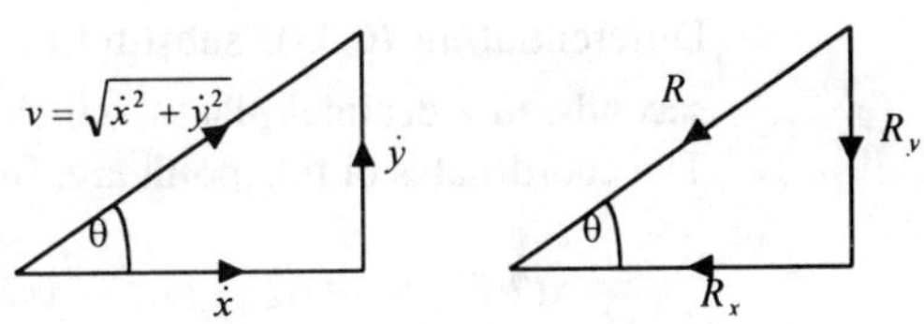

$$R_x = 0.2v\cos\theta = 0.2\sqrt{\dot{x}^2 + \dot{y}^2} \times \frac{\dot{x}}{\sqrt{\dot{x}^2 + \dot{y}^2}}$$

$$R_y = 0.2v\sin\theta = 0.2\sqrt{\dot{x}^2 + \dot{y}^2} \times \frac{\dot{y}}{\sqrt{\dot{x}^2 + \dot{y}^2}}$$

where v is the speed of the particle.

The weight of the particle is

$$mg = 0.5 \times 10 = 5 \text{ newtons.}$$

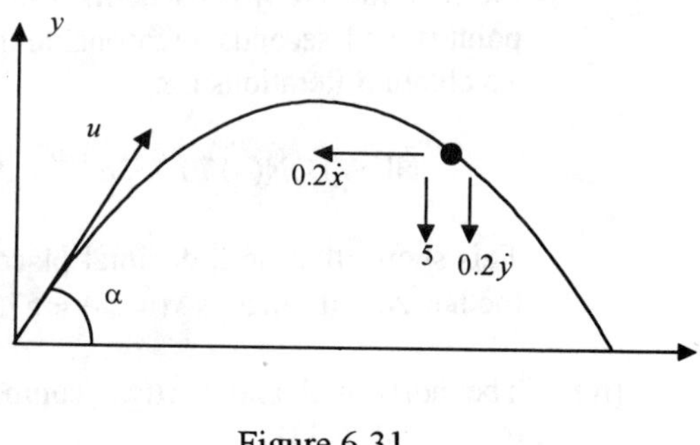

Figure 6.31

(i) Applying Newton's second law horizontally

$$0.5\ddot{x} = -0.2\dot{x} \quad \text{or} \quad \ddot{x} + 0.4\dot{x} = 0$$

The initial conditions are $x(0) = 0$ and $\dot{x}(0) = u\cos\alpha = 50\cos 60° = 25$.
To solve this equation use DSOLVE2_IV(0.4,0,0,t,0,0,25) from which

$$x = 62.5 - 62.5e^{-0.4t} \tag{6.32}$$

Applying Newton's second law vertically

$$0.5\ddot{y} = -5 - 0.2\dot{y} \quad \text{or} \quad \ddot{y} + 0.4\dot{y} = -10$$

Using the initial conditons $y(0) = 0$ and $\dot{y}(0) = 50\sin 60° = 25\sqrt{3}$ the required command is DSOLVE2_IV$(0.4,0,-10,t,0,0,25\sqrt{3})$ from which

$$y = -170.753e^{-0.4t} - 25t + 170.753 \tag{6.33}$$

(ii) At the highest point $\dot{y} = 0$.
Differentiating (6.33), substituting 0 for $\dot{y}$ and solving for t gives t = 2.51 seconds, to 2 decimal places, which is the time to the highest point.
The coordinates of this point are, from (6.32) and (6.33),

$$x(2.51) = 39.62 \quad \text{and} \quad y(2.51) = 45.44 \qquad \text{(2D)}$$

(iii) The particle hits the plane when y = 0. Equation (6.33) will be solved numerically in this case using the Newton-Raphson method with the help of the function NEWTON defined in Example 4M. As the time to the highest point is 2.51 seconds a reasonable initial guess for the total flight time is t_0 = 5. To obtain 5 iterations use

$$\text{NEWTON}(-170.753e^{-0.4t} - 25t + 170.753, t, 5, 5)$$

This shows that, to 2 decimal places, the total flight time is 6.28 seconds and the horizontal range is $x(6.28) \approx 57.43$ metres.

(iv) The horizontal and vertical components of the velocity are, differentiating (6.32) and (6.33)

$$\dot{x} = 25e^{-0.4t} \quad \text{and} \quad \dot{y} = 68.30e^{-0.4t} - 25$$

As t increases $e^{-0.4t}$ decreases towards zero and so, as $t \to \infty$, $\dot{x} \to 0$ and $\dot{y} \to -25$. Hence the terminal velocity of the particle is 25ms^{-1} vertically downwards. From (6.32), $x = 62.5(1 - e^{-0.4t})$ and, as $t \to \infty$, $x \to 62.5$ and the particle can never travel more than 62.5 metres horizontally.

(v) To plot the trajectory we must obtain the equation between x and y by eliminating t between (6.32) and (6.33). **Solve** (6.32) for t to give $t = 12.0707 - 2.5 \ln (125 - 2x)$. **Substitute** for t in (6.33) and obtain

$$y = 62.5 \ln(125 - 2x) + 2.73214x - 301.775$$

A graph of y against x is shown in Figure 6.32.

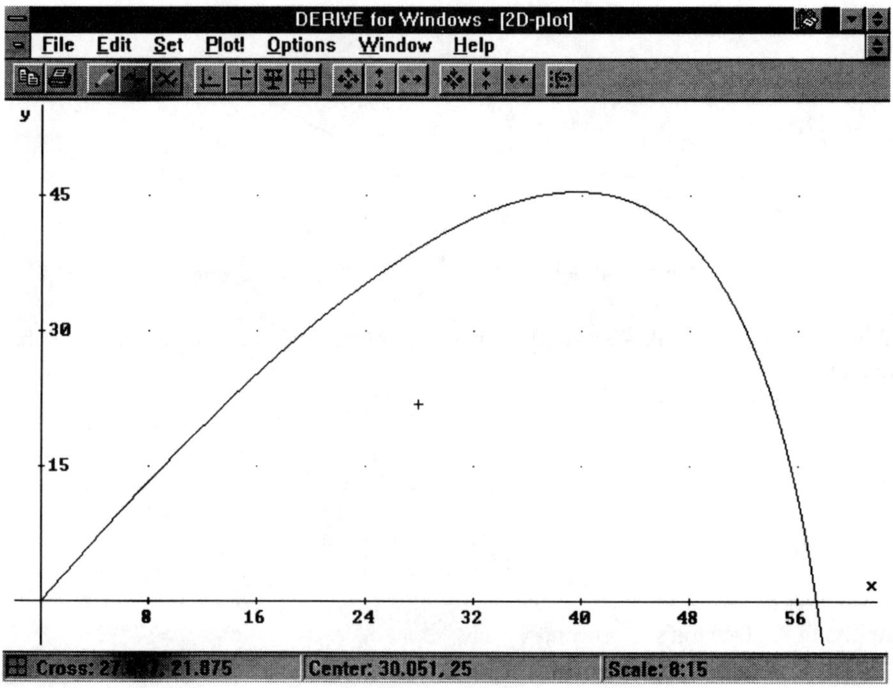

Figure 6.32 The flight of the particle in Example 6M

The presence of drag destroys the symmetry of the trajectory; note the steepness of the particle's descent. It is instructive to compare this trajectory with those obtained without drag, and when the amount of drag is varied.

□

Projectile motion with drag (quadratic approximation)

It was noted previously that for many everyday situations the quadratic approximation for drag was more appropriate, and we shall now assume that the magnitude of the drag force is proportional to the square of the speed, with its direction opposite to velocity. Note that the effects of spin, although sometimes important, are not considered. Example 6N investigates the motion of a shuttlecock for which, because of the feathers, drag is of major importance but spin is negligible.

Suppose that at time t the particle is at the point (x,y) and has a velocity with components (u,v), so that $\dfrac{dx}{dt} = u$ and $\dfrac{dy}{dt} = v$.

The direction of the resistive force is directly opposite to that of the velocity.

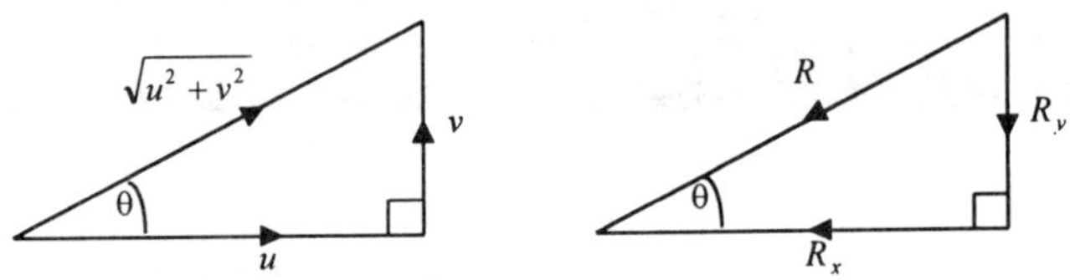

velocity components force components

When the velocity makes an angle θ with the horizontal the components of R have magnitudes

$$R_x = k(u^2+v^2)\cos\theta = k(u^2+v^2)\times\frac{u}{\sqrt{u^2+v^2}} = ku\sqrt{u^2+v^2}$$

$$R_y = k(u^2+v^2)\sin\theta = k(u^2+v^2)\times\frac{v}{\sqrt{u^2+v^2}} = kv\sqrt{u^2+v^2}$$

Applying Newton's second law horizontally and vertically:

$$m\frac{du}{dt} = -ku\sqrt{u^2+v^2}$$

$$m\frac{dv}{dt} = -mg - kv\sqrt{u^2+v^2}$$

To find the trajectory of the particle we must solve, together with the initial conditions, the four coupled equations

$$\frac{dx}{dt}=u,\quad \frac{dy}{dt}=v,\quad \frac{du}{dt}=-cu\sqrt{u^2+v^2},\quad \frac{dv}{dt}=-g-cv\sqrt{u^2+v^2} \tag{6.34}$$

where $c=\dfrac{k}{m}$.

If the particle is projected from the point $A(a,b)$ with speed V at an angle α to the horizontal the initial conditions are

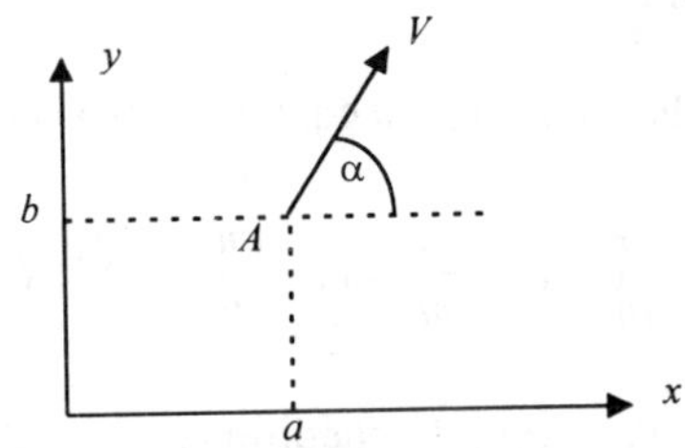

$$x(0)=a,\quad y(0)=b,$$

$$u(0)=V\cos\alpha,\quad v(0)=V\sin\alpha$$

To solve the equations numerically we will use the classical fourth order Runge-Kutta method in the DERIVE utility file ODE_APPR.

Example 6N

For a shuttlecock the value of c is approximately 0.2. Taking g = 10ms^{-2} investigate the trajectory of the shuttlecock for three different strokes. The relevant dimensions of a badminton court are shown in the diagram.

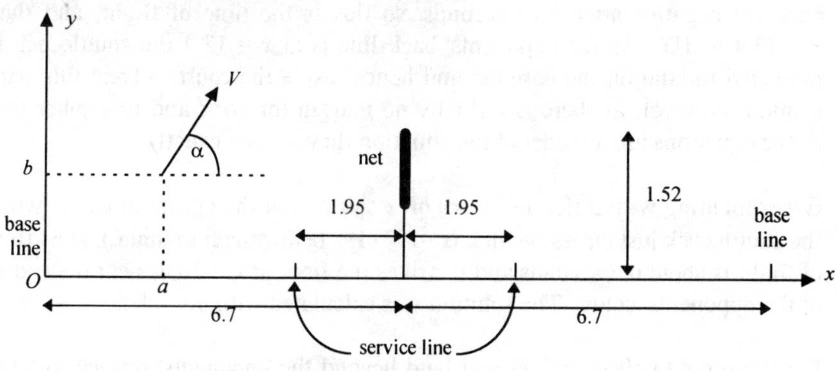

Figure 6.33

The origin O is taken at the base line.

(i) Investigate the path of an overhead clearance from the back of the court, for which $(a,b)=(1,2.5)$, V = 45ms^{-1}, $\alpha=35°$.

(ii) For a mid court smash (a,b) = (4,2.5) and V = 50ms^{-1}. Find α so that the shuttlecock just clears the net. Where does it land?

(iii) For an underarm serve with $(a,b) = (5,0.6)$ and $\alpha = 45°$ find V so that the shuttlecock clears the net and lands just inside the service court.

Solution

With the values given equations (6.34) become

$$\frac{dx}{dt} = u, \quad \frac{dy}{dt} = v, \quad \frac{du}{dt} = -0.2u\sqrt{u^2+v^2}, \quad \frac{dv}{dt} = -10-0.2v\sqrt{u^2+v^2}$$

(i) The initial conditions are $x(0) = 1$, $y(0) = 2.5$, $u(0) = 45\cos 35°$, $v(0) = 45\sin 35°$. To use 44 steps of the classical Runge-Kutta method with $h = 0.05$ (appropriate values found by experimentation) the DERIVE command is

$$\text{RK}([u,v,-0.2u\sqrt{u^2+v^2},-10-0.2v\sqrt{u^2+v^2}],[t,x,y,u,v],$$
$$[0,1,2.5,45\cos 35°,45\sin 35°],0.01,220)$$

Approximating this gives 220 sets of $[t,x,y,u,v]$ values.

By scanning down the third column of the results you can see that y becomes negative after 2.17 seconds, so this is the time of flight, and then $x = 13.4$ to 1D. As the opponents' back line is at $x = 13.4$ the shuttlecock is predicted to land on the baseline and hence inside the court. (Treat this with caution, however, as there is virtually no margin for error and remember that as the equations are a model of the situation they are not exact!)

(ii) Experimenting with different values of α shows that the optimum value, when the shuttlecock just clears the net, is $-19°$ ($19°$ below the horizontal). The time of flight is about 0.3 seconds and it strikes the floor at $x = 10.4$, near the centre of the opponents' court. The solution was calculated with $h = 0.01$.

(iii) For the serve to clear the net and land beyond the opponents' service line the optimum value of v is 8ms^{-1}. The shuttlecock passes about 16cm over the net and lands just inside the service area after about 1.7 seconds. Here the step length used was $h = 0.025$.

The paths of the three trajectories in (i), (ii) and (iii) are shown in Figure 6.34.

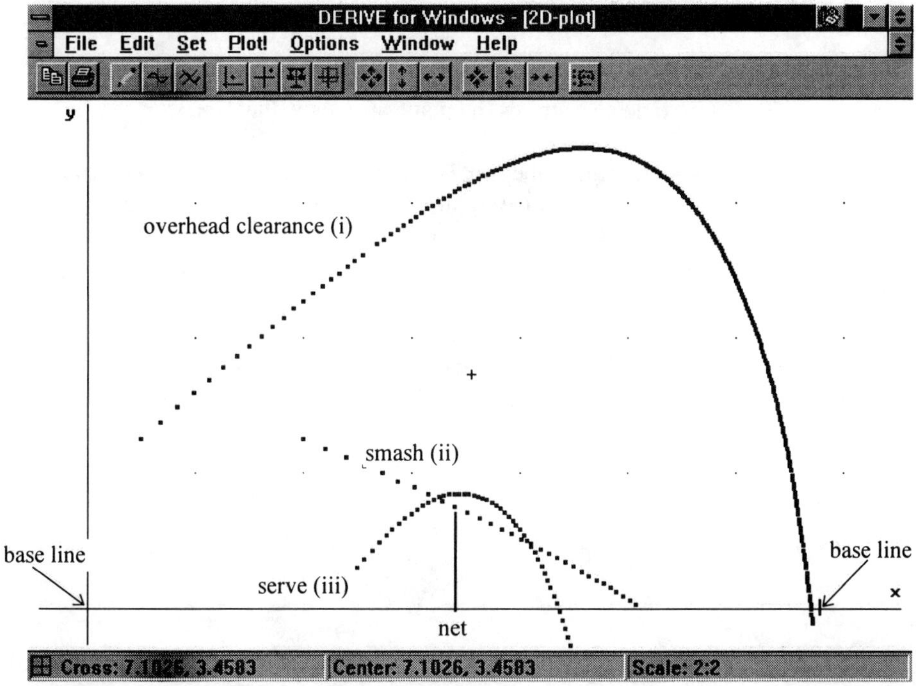

Figure 6.34 Paths of the shuttlecock in Example 6N

□

Exercise 6C

1. A and B are two points 850 metres apart on horizontal ground and C is 150 metres vertically above A. A particle is projected from B with speed u ms^{-1} at an angle of 45° above the horizontal. Take $g = 10\text{ms}^{-2}$.

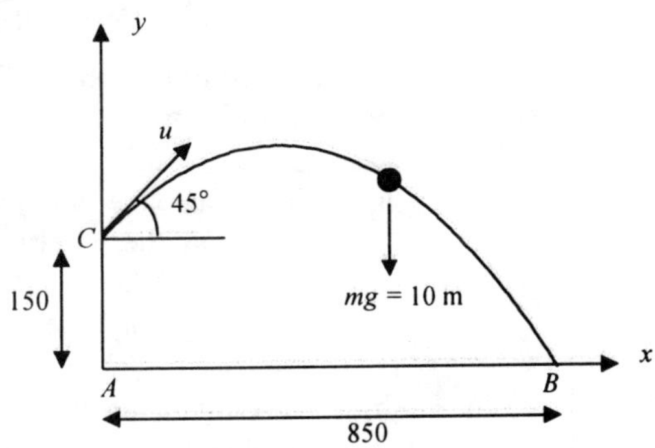

   (i) Show that $x = \dfrac{ut}{\sqrt{2}}$,

   $y = 150 + \dfrac{ut}{\sqrt{2}} - 5t^2$ and

   find the equation of the path of the particle.

(ii) If the particle strikes the ground at B show that $u = 85\text{ms}^{-1}$.

(iii) A second particle is projected from C at 85ms^{-1} at an angle θ ($\neq 45°$). If this particle also strikes the plane at B show that $\theta \approx 35°$.

2. Consider the situation described in Example 6M with air resistance equal to $-\alpha \times$ velocity. Solve the problem with values of α equal to 0.1, 0.3, 0.5 and 1. Draw the trajectories of the particle in each case and also when air resistance is ignored ($\alpha = 0$). Describe the effects of increasing the air resistance constant α.

3. Applying spin to a ball is a useful tactic in some sports. In particular we know that putting top spin on a tennis ball reduces its range while backspin increases it.

It can be shown that the equations of motion of a ball with top or back spin are

$$\frac{dx}{dt} = u, \quad \frac{dy}{dt} = v, \quad \frac{du}{dt} = -\alpha v, \quad \frac{dv}{dt} = \alpha u - 10$$

where (x,y) and (u,v) are the components of the position and velocity of the ball at time t and $g = 10\text{ms}^{-2}$. α measures the magnitude of the spin and is positive for backspin and negative for top spin.

(i) A ball is projected from the origin with speed 25ms^{-1} at an angle $20°$ to the horizontal. Sketch the trajectory of the ball for $\alpha = -0.3, -0.15, 0, 0.15$ and 0.3. Note the effect of α on the range and time of flight.

(ii) For a tennis ball $|\alpha| \approx 0.2$. Take the origin to be at one baseline; the net is about 1 metre high and is 12 metres from the baseline. Consider two shots hit from a point 1 metre above the baseline.

(a) Investigate the initial speed and angle of projection of a top spin lob from the baseline if it is to be out of reach of a player standing 2 metres inside his court but still land inside the baseline. Compare with a similar shot without spin.

(b) Investigate the initial speed and angle of projection of a backspin defensive shot to an opponent standing near his baseline. The ball must clear the net and land near the baseline. Compare with a similar shot without spin.

4. Example 6N and question 3 above consider respectively, the effects of air resistance and spin on a particle. Obtain the equations of motion for a particle subject to both air resistance and spin and investigate how different tennis shots should be made to achieve the desired effect.

Appendix 1

Glossary of DERIVE FOR WINDOWS Commands

Approximate

This command approximates any expression within the algebra window. Click on the expression you wish to approximate to highlight it, then click on the  icon. The approximated expression will appear within the algebra window.

Author

This command allows expressions to be entered into the algebra window. Ensure that you have the Algebra toolbar

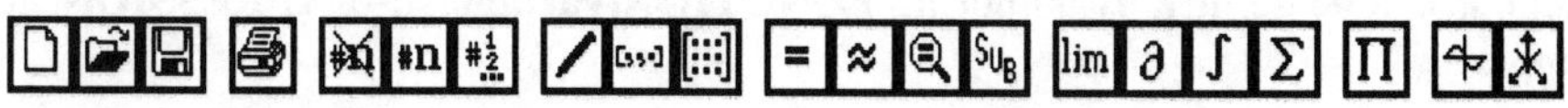

on screen.

If you should have a 2 or 3D plot screen in view click on the  icon. If there should be more than one algebra window open select the desired algebra window from the **Window** menu.

Expressions are entered by clicking on the  icon, after which the Author dialogue box

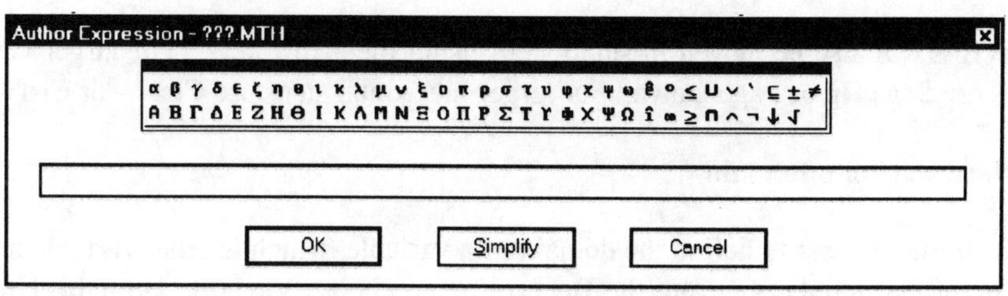

will appear. Type in your expression and then click on **OK** or press **enter.**

If your syntax is correct DERIVE will make mathematical sense of what you type. For example **sinxlncosx** will be interpreted as $\sin x \ln(\cos x)$.

Previous expressions can be authored by typing # followed by the number of the expression in the algebra window. For example, if expression 1 is $\sin x$ and expression 2 is $\cos x$ then $\dfrac{\sin x}{\cos x}$ can be authored by: clicking on ; typing **#1/#2**; and clicking on OK.

The F3 and F4 keys are very useful for editing previous expressions. After clicking on , F3 will copy the highlighted expressions which may be a line or part of a line, to the **Author** box so that it can be modified. F4 is similar but puts the whole expression in parentheses.

Special symbols can be entered onto the author line by clicking on the desired symbol in the Author box. Note for example that the number e (= 2.718...) is ê and that square root is $\sqrt{\ }$.

Functions can be defined by authoring **FUNCTION(variables):=EXPRESSION**, for example

$$f(x) := x\sin x - \cos x \quad \text{or} \quad z(a,b,c) := \sqrt{a^2+b^2+c^2} \quad \text{or} \quad \text{DOUBLE}(x) := 2x\,.$$

Centre

This redraws a plot (or a series of plots) with the cross in the centre of the plot. It can only be used in a 2D plot window. If you wish the plot to be centred at (2,4) use the **Set Cross** command to transfer the cross to the point (2,4) and then click on 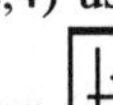 to centre on the cross.

The cross can also be moved in small steps using the arrow keys. For larger vertical steps use Page Up or Page Down. For larger horizontal steps use Ctrl ← or Ctrl →.

Declare Variable Domain

This allows the user to define the domain of a variable, which is otherwise taken to be the set of real numbers. From the **Declare** menu choose **Variable Domain**, type the variable name and click on the **Declare** button. Choose between a list of options including positive, non-negative and different types of interval. For an interval type the lower and upper bounds *l* and *u*.

Delete

To delete the last plot click on the 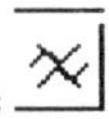 icon.

To delete some or all the plots in a plot window, within a plot window select **Edit** then **Delete Plot** and choose from the options given.

Differentiate

This command differentiates an authored expression. In the algebra window highlight the desired expression and then click on the  icon. The Differentiate dialogue box

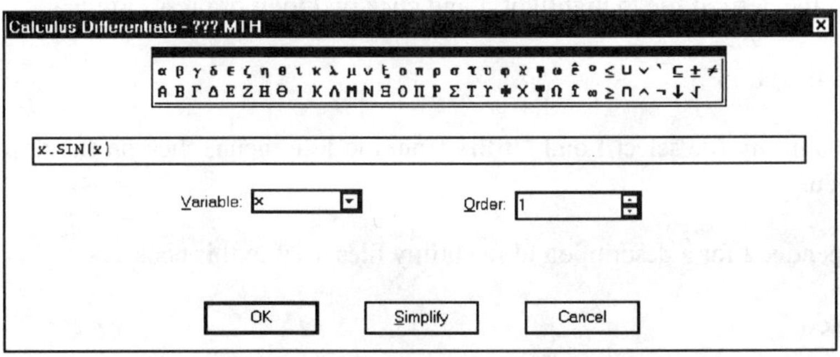

will appear. The highlighted expression will automatically appear within the dialogue box (here this is $x\sin x$). Select the **Variable** by clicking on the arrow within the variable field and selecting the required variable. The **Order** of differentiation defaults to 1, but this can be changed by clicking on the arrows in the **Order** box.

Clicking on **OK** will enter the unsimplified expression into the algebra window while clicking on **Simplify** will enter the calculated derivative into the algebra window.

Load

This loads an existing DERIVE file into an algebra window and displays the contents. Click on the  icon and the following dialogue box will appear:

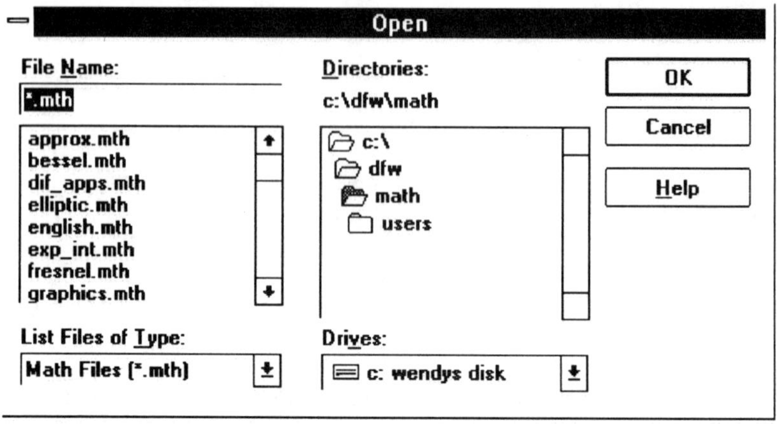

Click on the desired file to highlight it and click on **Open** or press **enter**.

Load (Utility)

To load a utility file select **Load Utility** from the **File** menu, click on the desired file and **Open**.

See Appendix 2 for a description of the utility files used in this book.

Open New

To open a new algebra window click on the  icon.

Plot

To produce a 2D graph of a function (1 variable) select the function in the algebra window and click on the 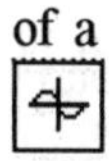 icon. A plot window will appear with a new menu bar:

To plot the selected function click on the 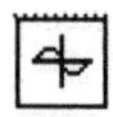 icon and the function will be drawn. With some functions the scale may have to be adjusted to view the graph (see Scale).

Points

This determine whether or not plotted points are joined by straight lines. The default is that they are, but to check or change this select **Points** from the **Options** menu in the 2-D plot window and the dialogue box

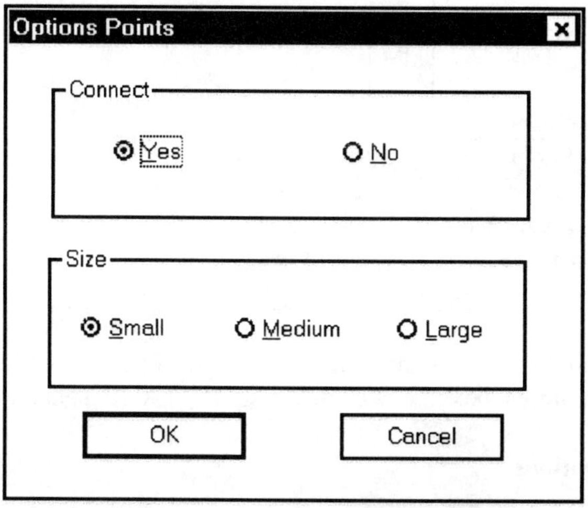

appears. Select **Yes** or **No** and also the size of the points.

Remove

This command allows you to remove unwanted expressions from an algebra window. Firstly, select the expression or expressions you wish to remove. To select one expression just click on it. To select a number of *consecutive* expressions click and drag down to the last expression; release the mouse button. Now click on the  icon and the expression(s) will be removed from the algebra window.

Save

To save a DERIVE activity as an MTH file, from the algebra window click on the  icon.

The **Save As** dialogue box

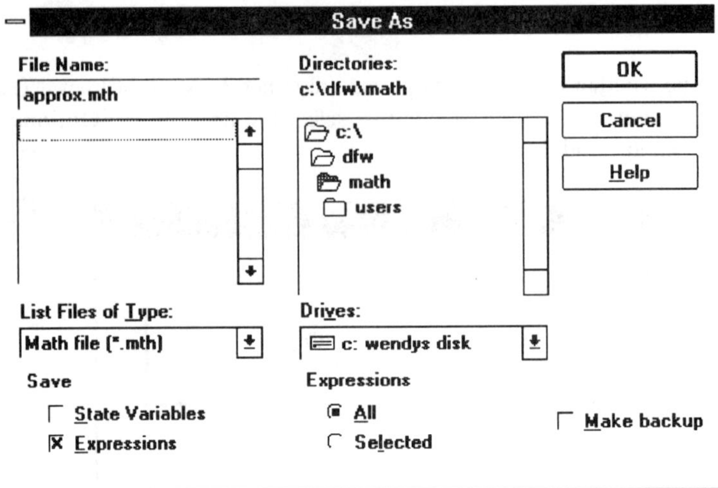

will appear. Choose the folder into which you wish to place your file, enter the file name into the **File name**: box and click on **Save**. Type A: filename to save the floppy.

Simplification Options

From the **Declare** menu select **Algebra State** then **Simplification...**

The following dialogue box will appear

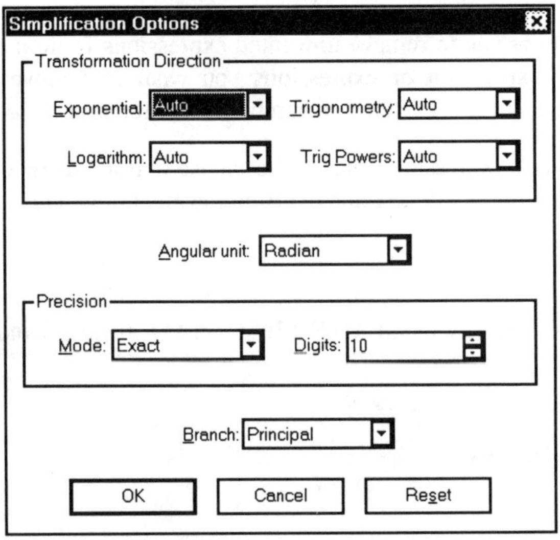

(i) **Precision**

Choose between **Mixed**, **Approximate** or **Exact Mode** and the number of digits of precision to which DERIVE will calculate.

(ii) **Angles**

Choose between degrees and radians.

(iii) **Transformation Direction**

For **Trigonometry** and **Logarithms** choose between **Auto**, **Collect** and **Expand**. For example choosing **Collect** for both (and x, y to the positive)

$\sin(x)\cos(y) + \cos(x)\sin(y)$	simplifies to $\sin(x + y)$
$\ln(x) + \ln(y)$	simplifies to $\ln(xy)$
$2\ln(x)$	simplifies to $\ln(x^2)$

Choosing **Expand** for both

$\sin(x + y)$	simplifies to $\sin(x)\cos(y) + \cos(x)\sin(y)$
$\ln(xy)$	simplifies to $\ln(x) + \ln(y)$
$\ln(x^2)$	simplifies to $2\ln(x)$

Choosing **Auto** for both

Simplifying $\sin(x + y)$ and $\sin(x)\cos(y) + \cos(x)\sin(y)$ leaves them both unchanged.
Simplifying $\ln(x) + \ln(y)$ gives $\ln(xy)$.
Simplifying $\ln(x^2)$ gives $2\ln x$.

Scale

This command allows you to select a particular scale for a 2D graph. From a 2D-plot window click on the icon and the dialogue box

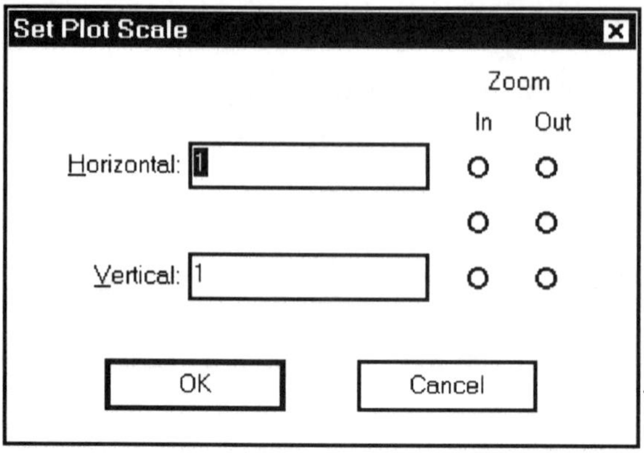

will appear. Change the horizontal and vertical scale as required and click on OK.

Simplify

This command allows the algebraic simplification of an authored expression. Highlight the desired expression and click on the  icon.

Solve

This command solves an equation in an algebra window (if the expression has no equals sign then DERIVE assumes that the expression = 0). Select the equation you wish to solve. Click on the  icon and the **Solve** dialogue box will appear:

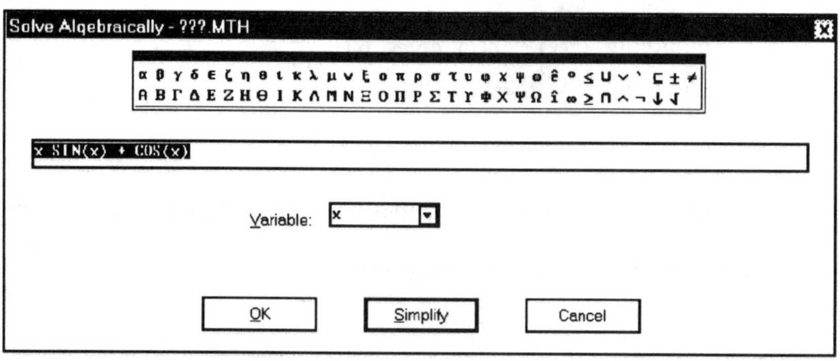

Select the required variable and click on **Simplify**. If the equation has an analytical solution DERIVE will return it. However, if there is no analytical solution DERIVE will return the equation unsolved.

To solve an equation numerically highlight it and select **Numerically** from the **Solve** menu. The Solve Numerically dialogue box

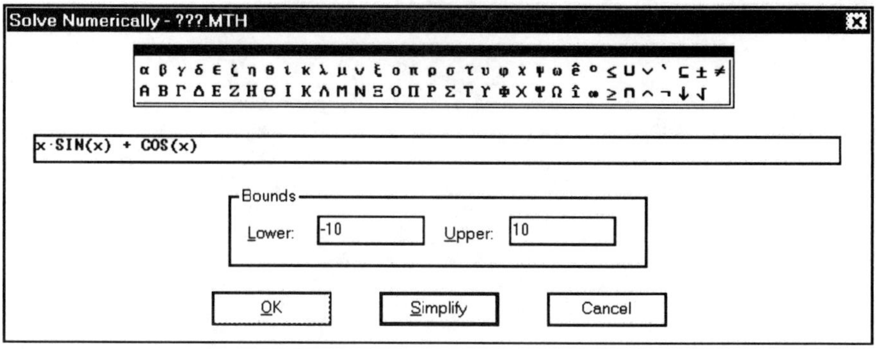

will appear.

Enter the lower and upper bounds for the solution search and select **Simplify**. A solution (if one exists) will be returned in the algebra window.

Substitute

To substitute for variables in an authored expression highlight it and click on the 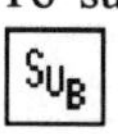 icon. The dialogue box

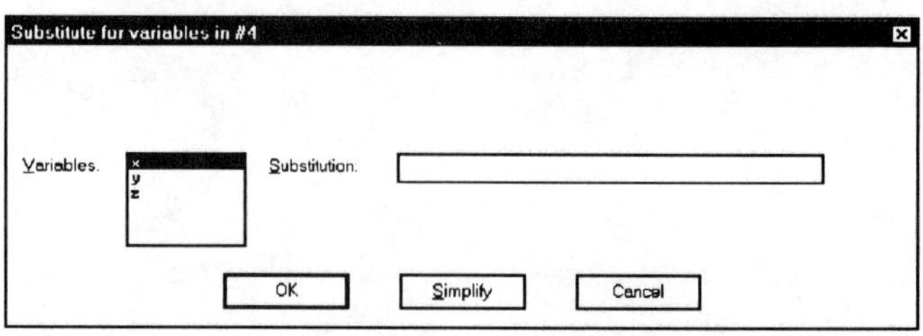

will appear. All the variables within the expression are listed in the **Variables** box. To substitute for a particular variable within the expression: click on it in the **Variable** box and then enter the substitution in the **Substitution** box. When all variables have been substituted for, select **OK**.

Zoom (in/out)

Within a plot window click on the  icons to zoom in or out respectively.

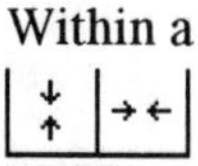 zoom in vertically and horizontally, respectively.

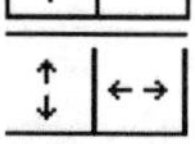 zoom out vertically and horizontally, respectively.

Appendix 2

FUNCTIONS AND UTILITY FILES

1. VECTOR

The function VECTOR(u,k,m,n,s) simplifies to a vector (or list) of elements found by evaluating u for values of k stepping from m to n in steps of s. If $s = 1$ it may be omitted.

Examples:

Simplifying VECTOR(x^2,x,1,5) gives [1,4,9,16,25].
Simplifying VECTOR([n,sin(nx)],n,1,5,2) gives the (3 × 2) matrix

$$\begin{bmatrix} 1 & \sin(x) \\ 3 & \sin(3x) \\ 5 & \sin(5x) \end{bmatrix}$$

2. ODE1

The utility file ODE1.MTH contains the following functions for finding general and particular solutions to first order differential equations.

(i) SEPARABLE_GEN(p,q,x,y,c) finds the general solution of $y' = p(x)q(y)$.
For example, to find the general solution of $y' = \dfrac{2x}{3y^2}$
Author SEPARABLE_GEN($2x$,1/($3y^2$),x,y,c).
Simplify to obtain $y^3 = x^2 + c$.

(ii) SEPARABLE(p,q,x,y,a,b) solves $y' = p(x)q(y)$ given $y(a) = b$.
For example, to solve $e^y y' = \cos x$ given $y(0) = 0$
Author SEPARABLE(cos$x$,e^{-y},x,y,0,0)
Simplify and **Solve** for y to get $y = \ln(\sin x + 1)$.

(iii) LINEAR1_GEN(p,q,x,y,c) finds the general solution of the linear equation $y' + p(x)y = q(x)$. For example, to solve $y' + y = e^{-x}$
Author LINEAR1_GEN(1,e^{-x},x,y,c).
Simplify to $y = e^{-x}(x + c)$.

(iv) LINEAR1(p,q,x,y,a,b) solves $y' + p(x)y = q(x)$ given $y(a) = b$.
For example, to solve $xy' - 3y = x^5$ given $y(1) = 0$
Author LINEAR1($-3/x,x^4,x,y,1,0$).
Simplify to $y = x^3(x^2 - 1)/2$.

(v) BERNOULLI_GEN(p,q,n,x,y,c) solves $y' + p(x)y = q(x)y^n$.

For example, to find the GS of $y' - y = \dfrac{xe^{2x}}{y}$

Author BERNOULLI_GEN($-1,xe^{2x},-1,x,y,c$).
Simplify to $y^2 = e^{2x}(x^2 + c)$.

(vi) BERNOULLI(p,q,n,x,y,a,b) solves $y' + p(x)y = q(x)y^n$ given $y(a) = b$.
For example, to solve $y' + y = xy^2$ given $y(0) = 0.5$
Author BERNOULLI($1,x,2,x,y,0,0.5$). **Simplify**, **Solve** for y and obtain

$$y = \frac{1}{e^x + x + 1}.$$

(vii) HOMOGENEOUS_GEN(r,x,y,c) solves $y' = r(x,y)$, where $r(x,y)$ is the ratio of two homogeneous functions of the same degree.

For example, to solve $y' = \dfrac{x^3 + y^3}{xy^2}$

Author HOMOGENEOUS_GEN($(x^3 + y^3)/(xy^2),x,y,c$).

Simplify to obtain the real solution $y = 3^{1/3}x(\ln x + c)^{1/3}$. (Exact mode for Simplification required.)

(viii) HOMOGENEOUS(r,x,y,a,b) solves $y' = r(x,y)$ given $y(a) = b$.

For example, to solve $y' = \dfrac{x + y}{x}$, $y(1) = 2$

Author HOMOGENEOUS($(x + y)/x,x,y,1,2$).
Simplify and **Solve** for y to give $y = x\ln x + 2x$.

(ix) EXACT_GEN(p,q,x,y,c) solves the equation $p(x,y) + q(x,y)y' = 0$.
If the equation is not exact DERIVE returns "inapplicable".

For example, to solve $y' = -\dfrac{2xy}{x^2 + 4y^3}$

Author EXACT_GEN($2xy,x^2 + 4y^3,x,y,c$).
Simplify to $x^2y + y^4 = c$.

(x) EXACT(p,q,x,y,a,b) solves $p(x,y) + q(x,y)y' = 0$ given $y(a) = b$.
For example, to solve $x\cos y\dfrac{dy}{dx} + (e^x + \sin y) = 0$ given $y(0) = 1$
Author EXACT($e^x + \sin y, x\cos y, x, y, 0, 1$).
Simplify to $e^x + x\sin y - 1 = 0$.

(xi) DSOLVE1_GEN(p,q,x,y,c) is a general purpose function for finding general solutions of first order differential equations of the form $p(x,y) + q(x,y)y' = 0$.
For example, to solve $y' = -\dfrac{y}{x}$
Author DSOLVE1_GEN(y,x,x,y,c). **Simplify** to $xy = c$.

(xii) DSOLVE1(p,q,x,y,a,b) finds the particular solution of $p(x,y) + q(x,y)y' = 0$ given $y(a) = b$.
For example, to solve $2xy + x^2y' = 0$ given $y(1) = 2$
Author DSOLVE1($2xy,x^2,x,y,1,2$).
Simplify to $x^2y - 2 = 0$. **Solve** for y to $y = \dfrac{2}{x^2}$.

3. ODE2

The utility file ODE2.MTH contains functions for solving second order differential equations.

(i) DSOLVE2(p,q,r,x,a,b) finds the general solution of the equation $y'' + p(x)y' + q(x)y = r(x)$.
For example, to solve $2y'' - y' - y = 4x$ (or $y'' - 0.5y' - 0.5y = 2x$)
Author DSOLVE2($-0.5,-0.5,2x,x,a,b$).
Simplify to ($y =$) $ae^x + be^{-x/2} - 4(x - 1)$.

(ii) DSOLVE2_IV(p,q,r,x,x_0,y_0,v_0) solves the equation $y'' + p(x)y' + q(x)y = r(x)$ given the initial conditions $y(x_0) = y_0$ and $y'(x_0) = v_0$.
For example, to solve $y'' + 4y = 10e^x$ given $y(0) = 1$ and $y'(0) = 0$
Author DSOLVE2_IV($0,4,10e^x,x,0,1,0$).
Simplify to ($y =$) $2e^x - \cos 2x - \sin 2x$.

4. ODE_APPR

The utility file ODE_APPR.MTH contains functions for finding numerical solutions to first order, and systems of first order, differential equations.

(i) DIRECTION_FIELD($r,x,x_0,x_m,m,y,y_0,y_n,n$) produces data for a direction field for the equation $y' = r(x,y)$. x varies from x_0 to x_m in m steps and y varies from y_0 to y_n in n steps.

For example, to plot a direction field for $y' = 5 - 2y$ for

$$0 \le x \le 5 \quad \text{and} \quad 0 \le y \le 6$$

Author DIRECTION_FIELD(5 – 2*y*,*x*,0,5,10,*y*,0,6,12). **Open** a 2-D window with  and from the **Options** menu choose **Points** and select **Yes** for connect and **Small** for size. **Plot!** to show the direction field. **Approximate** this expression. Move the cursor and change the scales to improve the plot.

(ii) EULER(r,x,y,a,b,h,n) finds approximations to the solution of $y' = r(x,y)$ given $y(a) = b$ using n steps of Euler's method with step size h.

For example, to estimate $y(1)$ from $y' = \dfrac{3x}{y}$, $y(0) = 1$ with 10 steps of Euler's method **Author** EULER(3*x*/*y*,*x*,*y*,0,1,0.1,10) and **Approximate** to give an (11×2) matrix showing that $y(1) \approx 1.95311$.

(iii) TAY_ODE1(r,x,y,a,b,n) finds a Taylor polynomial of degree n approximating to the solution of $y' = r(x,y)$, $y(a) = b$.

For example, to estimate $y(1)$ from $y' = \dfrac{x}{y}$, $y(0) = 1$ using a Taylor polynomial of degree 6 **Author** TAY_ODE1(*x*/*y*,*x*,*y*,0,1,6). **Simplify**, **Substitute** 1 for x and **Approximate** to give $y(1) \approx 1.4375$.

(iv) The RK function finds approximations to the solution of a first order differential equation, or to a system of such equations, using the classical fourth order Runge-Kutta method.

(a) RK([r],[x,y],[a,b],h,n) uses n steps of size h to approximate to the solution of $y' = r(x,y)$, $y(a) = b$.
For example, to estimate $y(1)$ from $y' = e^x - y$, $y(0) = 1$ using 10 steps of the Runge-Kutta method
Author RK([$e^x - y$],[x,y],[0,1],0.1,10). **Approximate** to give an (11×2) matrix, from which $y(1) \approx 1.54308$.

(b) RK($[r_1,r_2],[t,x,y],[t_0,a,b],h,n$) uses n steps of size h to approximate to the solution of the system of two equations $x' = r_1(t,x,y)$, $y' = r_2(t,x,y)$ given $x(t_0) = a$, $y(t_0) = b$.
For example, to estimate $x(1)$ and $y(1)$ with $h = 0.1$ from $x' = y$, $y' = 1 - 2x - 3t$, $x(0) = 2$, $y(0) = 1$
Author RK($[y,1 - 2x - 3t],[t,x,y],[0,2,1],0.1,10$).
Approximate to give an (11×3) matrix which shows that $x(1) \approx 0.980060$ and $y(1) \approx -3.20549$.
A system of three or more equations is treated in a similar way.

(v) EXTRACT_2_COLUMNS can be used to extract two of the columns from the Runge-Kutta solution in (iv) (b) so that, for example, a plot of x against t can be made.
EXTRACT_2_COLUMNS(#k,m,n) extracts columns m and n from the matrix in line k to form a matrix with 2 columns.
For example to plot the $[t,x]$ pairs from the solution of the example in (iv) (b) **Author** EXTRACT_2_COLUMNS(#k,1,2), where k is the number of the RK line. **Approximate** to give a matrix with 2 columns. This matrix can now be plotted if required.

Answers to the Exercises

Exercise 1A

1. Ball moves along a straight line. No side wind.

2.

Variables/ Parameters	Represents	Units
k	stiffness of spring	Nm^{-1}
l	natural length of spring	m
r	dashpot constant	$Nm^{-1}s$
m	mass of car	kg
F_w	weight of car	N
F_s	force in spring	N
F_D	force in dashpot	N
g	acceleration due to gravity	ms^{-2}
x	displacement of mass above mean road level	m
y	displacement of road surface above mean road level	m
t	time	s
a	amplitude of sine curve modelling road shape	m
ω	angular frequency of since curve modelling road shape	s^{-1}
d	wavelength of sine curve modelling road shape	m
u	car speed	ms^{-1}
T	time for car to cover one wavelength	s

3. $$\frac{dm}{dt} = -\alpha m$$

   m is the mass of the radioactive substance at time t and α is a constant of proportionality.

4. assumptions
   drug instantly absorbed into the bloodstream
   drug clearance rate is proportional to amount of drug in the blood

model

$$\frac{dq}{dt} = -kq$$

t is time (in hours)
q is quantity of drug (in mg) in the blood at time t
k is constant of proportionality (in hours^{-1})

5.
$$\frac{dI}{dt} = \alpha I(P - I)$$

t is the time (in days)
P is the population of students
I is the number of infected students
α is a constant of proportionality (in days^{-1})

Exercise 1B

	differential equation	order	linearity	homogeneity
1.	$\frac{dx}{dt} - 4x = 0$	first	linear	honogeneous
2.	$\frac{dy}{dx} + 3y^2 = x$	first	non-linear	non-homogeneous
3.	$y\frac{dy}{dx} = 4$	first	non-linear	non-homogeneous
4.	$\frac{d^2y}{dx^2} + 7\frac{dy}{dx} + 2y = 0$	second	linear	homogeneous
5.	$\frac{d^2x}{dt^2} + 8\frac{dx}{dt} + 6x = t\sin(4t)$	second	linear	non-homogeneous
6.	$\frac{d^4y}{dx^4} + 3\frac{d^2y}{dx^2} + 2y = e^{4x}$	fourth	linear	non-homogeneous
7.	$\ddot{x} - 3\dot{x} = \cos(2t)$	second	linear	non-homogeneous
8.	$\frac{dx}{dt} + x^2 = 4t - 3$	first	non-linear	non-homogeneous

Exercise 2A

1. (i), (iii), (iv).

2. (i) $y = x^2 + c$
(ii) $y = \ln x + \cos x + c$
(iii) $y = 2x^4 + x^3 - x^2 + cx + d$
(iv) $y = 2x - \ln x + c$
(v) $y = \arcsin x + c$
(vi) $y = -\ln|x + 1| + \ln|x - 2| + c$

3. (i) $y = x^3 - \ln x + 1$
(ii) $y = 1 - x^2 + 2\sin 3x$
(iii) $y = 1 - xe^x - e^{-x}$
(iv) $x = t^3 + t + 1$

4. (i) 2 seconds
(ii) 20 metres
(iii) 4 seconds
(iv) 20ms^{-1}

5. $y = -\dfrac{wx^3}{12EI} + \dfrac{wl^2x}{16EI}$

$\dfrac{wl^3}{48EI}$

6. 5.73 metres

7. $v = u\ln 2 - \dfrac{gM}{2c}$

distance $= \dfrac{uM}{2c}(1 - \ln 2) - \dfrac{gM^2}{8c^2}$

8. (i) $y = \dfrac{1}{8}(1 + x^2)^4 + c$

(ii) $y = \cos(2x) + 2x\sin(2x) + c$

(iii) $y = \ln x + e^{x^2} + c$

(iv) $y = 2\ln(1 + x^2) - 2\arctan x + 3x + c$

9. (i) $y = x + 2x\tan^{-1}x - \ln(1 + x^2)$

(ii) $y = x - \frac{\sqrt{4 - x^2}}{4} + \frac{1}{2}$

Exercise 2B

1. (i), (iii), (iv), (vi)

2. (i) $y^2 = x^2 + c$
(ii) $y = Ae^x$
(iii) $y = Ae^{2x^2} - 1$
(iv) $y = \ln(c - 2\cos x)$
(v) $y = \frac{1}{2}\cos^{-1}(Ae^{-2x})$
(vi) $y = \tan(c + 2x^4)$
(vii) $y = A\exp\left(\arcsin\frac{x}{2}\right)$
(viii) $y^2 = \frac{x^2}{1 - cx^2}$
(ix) $y = Ae^{6\sqrt{x}} - \frac{1}{3}$

3. (i) $y = 2e^x$
$y = 3e^{2x^2} - 1$
$y = 2\exp\left(\arcsin\frac{x}{2}\right)$
$y = \frac{1}{3}(7e^{6\sqrt{x}} - 1)$

(ii) $y = \ln(3 - 2\cos x)$
$y = \frac{1}{2}\cos^{-1}(e^{-2x})$
$y = \tan(2x^4)$

4. (i) 4484
(ii) 7.67 years

5. (i) $\theta = 20 + 80e^{-0.04t}$
$\theta(20) = 55.9°C$

(ii) $t = 25\ln\left(\dfrac{80}{\theta - 20}\right)$

34.7 minutes

6. (i) $v = 1000e^{-1.2t}$
549ms^{-1}

(ii) 376 metres

(iii) 1.06 seconds
280 ms^{-1}

7. 16.6 minutes
$h = 4(1 - e^{-0.125t})$

8. 143.2 seconds

9. (i) $v = 40e^{-\frac{1}{2}t} - 20$

(ii) $h = 40 - 40e^{-\frac{1}{2}t} - 20t$
$t = 3.2$ seconds, $v = 11.9$ms^{-1}

10. (i) $m = Me^{-\lambda t}$

(ii) 23105 years

(iii) $\lambda = 0.000124$
Table made about 1220.
Not King Arthur's!

11. $u \geq 1.118 \times 10^4$ms^{-1}

12. (i) $N = \dfrac{25000}{1 + 1.5e^{-0.25t}}$, 17485

(ii) 7.17 years

(iii) Population approaches 25000 trout

13. (i) $x = \dfrac{\alpha\beta(e^{\alpha ct} - e^{\beta ct})}{\alpha e^{\alpha ct} - \beta e^{\beta ct}}$

(ii) (a) $x \to \beta$ if $\alpha > \beta$
(b) $x \to \alpha$ if $\alpha < \beta$

(iii) $x = \dfrac{\alpha^2 ct}{1+\alpha ct} \to \alpha$ as $t \to \infty$

(iv) (a) $x \to 1$ $t = 0.405$
(b) $x \to 2$ $t = 0.157$
(c) $x \to 4$ $t = 0.25$

14. (i) $c = c_0 e^{-\lambda t}$

$$t = \frac{1}{\lambda}\ln\left(\frac{c_0}{c_1}\right)$$

(ii) 8.15 hours
dose 10mgl^{-1} every 4.1 hours (example)

(iii) (a) $c = \dfrac{I}{\lambda} + \left(c_0 - \dfrac{I}{\lambda}\right)e^{-\lambda t}$

$$c \to \frac{I}{\lambda}$$

(b) $I = 2$ and $c_0 = 15\text{mgl}^{-1}$ (example)

Exercise 2C

1. (i), (iv), (vi)

2. (i) $y = (c + 2x)e^{-x}$
(ii) $y = \dfrac{1}{5}x^2 + cx^{-3}$
(iii) $y = 2x^4 + cx$
(iv) $y = \dfrac{1}{3}(x+1) + \dfrac{c}{(x+1)^2}$
(v) $y = (x + c)\cos x$
(vi) $y = ce^{-2e^{-x^2}} - 2$

3. (i) $y = (1 + 2x)e^{-x}$

$$y = \frac{1}{3}(x+1) + \frac{2}{3(x+1)^2}$$

$y = (x + 1)\cos x$

$y = 3e^{2(1-e^{-x^2})} - 2$

(ii) $y = \frac{1}{5}x^2 + \frac{9}{5}x^{-3}$

$y = 2x^4$

4. (i) $x = 200(1 - e^{-0.1t})$
(ii) $y = 1346.4(1 - e^{-0.1t}) - 100t$
(iii) (52,49)
(iv) 93.5 metres

6. $v = \frac{mg}{k\eta a}(1 - e^{-k\eta at/m})$

$t = \frac{m\ln(2)}{k\eta a}$

7. $\theta = 85.25e^{-0.1t} - 0.25\cos(2t) + 0.0125\sin(2t) + 15$
or $\theta = 85.25e^{-0.1t} - 0.25\cos(2t + 0.05) + 15$

9. (a) $\frac{dp}{dt} = k\left(I - \frac{P}{R}\right)$

(b) $p = p_0 e^{-\frac{K}{R}(t-t_0)}$ $\quad t_0 \le t \le T$

(c) $p = (p_0 - RI_0)e^{\frac{K}{R}(t_0-t)} + RI_0$ $\quad 0 \le t \le t_0$

(d) $R = \frac{p_0\left(e^{-\frac{K}{R}(T-t_0)} - e^{\frac{K}{R}t_0}\right)}{I_0\left(1 - e^{\frac{K}{R}t_0}\right)}$

10. (i) (a) $s = Pe^{rt/100}$

(b) 2158.92 compounded annually
2191.12 compounded every six months
2219.64 compounded monthly
2225.35 compounded daily
2225.54 compounded continuously

(ii) (a) $s = Pe^{\frac{rt}{100}} + \frac{100k}{r}\left(e^{\frac{rt}{100}} - 1\right)$

(b) Total value = £612846, Interest = £531846

(c) 15.27 years

It will last 7.83 years longer

Exercise 2D

1. (i), (ii)

2. (i) $y = \frac{e^x}{c - e^x}$

(ii) $y = \left(2\sqrt{x} + \frac{c}{x^2}\right)^2$

3. (i) $y = (3x^2 + 24x^3)^{-\frac{1}{3}}$

(ii) $y = \sqrt{\frac{\sin x}{5 - 4\sin x}}$

4. $N = \frac{25000}{1 + 1.5e^{-0.25t}}$

7.17 years

population of trout decreases towards a new equilibrium value of 16000

5. (i), (iii)

6. (i) $y = x\sin^{-1}(Ax^2)$

(ii) $y = \frac{A^2x^8 - 1}{2Ax^3}$

7. (i) $y = \sqrt{x^2 + 3x}$

(ii) $y = 2x\tan\left(2\ln x + \frac{\pi}{4}\right)$

8. (i), (iv)

9. (i) $x^4y^2 = c$
(ii) $e^y + xe^x - x^2y = c$

10. (i) $y^2e^x - y^3x^2 = 1$
(ii) $y + x\cos(y^2) = 2$

11. (i) $x^3\cos y + x^2 = 1$
(ii) $e^{2x}\ln(y) = 1$

12.

$xy' = y + 1$	YES	YES	NO	NO	NO
$y'\sin x = e^x - y^2$	NO	NO	NO	NO	NO
$y' = \dfrac{2x-3y}{3x-4y}$	NO	NO	NO	YES	NO
$y' = (x - 3y)^2 - x^2$	NO	NO	YES	NO	NO
$2y' = e^2 - y^2$	NO	NO	NO	NO	YES

13. (i) $y = \tan(x + c) - x$
(ii) $y = \ln(4x - 4 + ce^{-x})$
(iii) $y = \dfrac{x^2}{A - x^2}$
(iv) $x^2 + y^2 - 2xy - 8y + 4x = c$

14. $y = \dfrac{2x}{1 + Ax^2}$

15. (i) linear o.d.e.
(ii) Bernoulli equation
(iii) $y = x + \dfrac{2x}{1 + Cx^2}$
(iv) $y = \dfrac{C + e^x}{C + e^x - xe^x}$

17. $v = \sqrt{\left(u^2 - \dfrac{n^2}{2k^2}\right)e^{-2kx} - \dfrac{n^2x}{k} + \dfrac{n^2}{2k^2}}$

Exercise 3A

1.

	Euler	Taylor series	Runge-Kutta	Modified-Euler
(a)	1.92	2.13962	2.13962	2.13962
(b)	2.7206	2.6734	2.6778	2.6734
(c)	0.63974	0.58584	0.61665	0.59674

2. (i) 2.164472
(ii) 2.155266
(iii) 2.155266

3.

Equation	Better method	Best answer (6D)	Error (3S)
(i)	A	4.539945	4.04×10^{-4}
(ii)	B	0.578153	4.61×10^{-3}
(iii)	B	0.800973	9.73×10^{-4}
(iv)	A	2.320764	3.08×10^{-5}

(the answers are dependent on the calculator/computer/program/ package used)

Exercise 3B

2. (i) L is reduced by a factor of $4^2 = 16$ in (ii), a factor of $4^3 = 64$ in (i), (iii) and (iv), a factor of $4^4 = 256$ in (v) and by a factor of $4^5 = 1024$ in (vi).

(ii) L is reduced by a factor of 10^2 in (ii), 10^3 in (i), (iii) and (iv), 10^4 in (v) and 10^5 in (vi).

3. In (ii), $G \propto h$. In (i), (iii), (iv) $G \propto h^2$. In (v) $G \propto h^3$. In (vi) $G \propto h^4$.

4. (i) G is reduced by factors of 4 in (ii), 4^2 in (i), (iii), (iv), 4^3 in (v) and 4^4 in (vi).

(ii) G is reduced by factors of 10 in (ii), 10^2 in (i), (iii), (iv), 10^3 in (v) and 10^4 in (vi).

5. 0.0032

6.

method	upper bound on h	suitable value of h	no. of steps
Euler	0.0041	0.004	250
Trapezoidal	0.047	0.04	25
Taylor order 2	0.011	0.01	100
Taylor order 4	0.071	0.05	20
Runge-Kutta 2	0.011	0.01	100
Runge-Kutta 4	0.022	0.02	50
Modified Euler	0.00034	0.00025	4000

7. $y(1) = 1.4989$

8. $y(0.5) = 4.826352$

9. (i) $y(0.4) = 4.824021$
   (ii) upper bound on $h = 0.0045$
   $h = 0.004$ and $n = 100$

10. $h = 0.1$, $Y(2) = 2.099476$
    $h = 0.04$, $Y(2) = 2.101025$
    $y(2) = 2.101066$
    $h = 0.004$

11. (ii) (a) $y(3) \approx 2.258340$
    (b) $y(3) \approx 2.258311$

12. $h = 0.3$, $Y(3) = 3.22393$
    $h = 0.15$, $Y(3) = 3.22388$
    $y(3) = 3.22388$

Exercise 3C

1. (i) $u_n = 6(5^n - 5)$
   $u_1 = 25,\ u_2 = 145,\ u_3 = 745,\ u_4 = 3745,\ u_5 = 18745$

   (ii) $u_n = 50 - 5n$
   $u_1 = 45,\ u_2 = 40,\ u_3 = 35,\ u_4 = 30,\ u_5 = 25$

   (iii) $u_n = 2^{n+1} + \dfrac{3}{2^{n-1}}$
   $u_1 = 7,\ u_2 = 9.5,\ u_3 = 16.75,\ u_4 = 32.375,\ u_5 = 64.1875$

2. (i) absolutely ill-conditioned, relatively ill-conditioned
   (ii) absolutely well-conditioned, relatively well-conditioned
   (iii) absolutely well-conditioned, relatively well-conditioned
   (iv) absolutely ill-conditioned, relatively ill-conditioned
   (v) absolutely ill-conditioned, relatively well-conditioned

3. (i) $h > 0$
   (ii) $0 < h < -\dfrac{2}{\lambda}$
   (iii) $0 < h < -\dfrac{2}{\lambda}$

4. (a) (i) $Y_n = A(1.349877)^n + B(-0.905432)^n$
   spurious solution $B(-0.905423)^n \to 0$ as $n \to \infty$

   (ii) $Y_n = A(0.740808)^n + B(-1.10444)^n$
   spurious solution $B(-1.10444)^n \to \pm\infty$ as $n \to \infty$

   (b) (i) $Y_n = A(1.33788)^n + B(0.112117)^n$
   spurious solution $B(0.112117)^n \to 0$ as $n \to \infty$

   (ii) $Y_n = A(0.75)^n + B(-0.2)^n$
   spurious solution $B(-0.2)^n \to 0$ as $n \to \infty$

7. (i) Group *A* methods: Euler, TS2, RK2, Modified Euler.
   Group *B* method: TS4, RK4.

	Group A	Group B	Trapezoidal
(a)	$0 < h < \frac{1}{3}$	$0 < h < 0.463$	all $h > 0$
(b)	$0 < h < 0.2$	$0 < h < 0.278$	all $h > 0$
(c)	$0 < h < 0.211$	$0 < h < 0.293$	all $h > 0$

   (ii) For Group A methods $x = 20$. For Group B, $x = 27.8$.

8. (i) $x(2) = 5.96595$, $y(2) = 9.42054$
   (ii) $x(1) = 3.9524$, $y(1) = 3.6343$, $z(1) = 1.2442$

Exercise 4A

1. (ii) $h=\left(\frac{19.2t}{\pi}\right)^{\frac{1}{3}}$

   (iii) $t = 35.3$ seconds

2. (ii) $h=\left(2-\frac{\sqrt{5}}{4}t\right)^2$

   (iii) $t=\frac{8}{\sqrt{5}}=3.58$ minutes

3. 23.57 minutes

4. (i) 0.777 kg/l, 224 minutes
   (ii) 0.995 kg/l

5. (i) $x=\frac{a}{k}-\frac{b(k\cos\omega t+\omega\sin\omega t)}{\omega^2+k^2}+\left(x_0-\frac{a}{k}+\frac{lk}{\omega^2+k^2}\right)\mathrm{e}^{-kt}$

   where $\omega=\frac{\pi}{12}$

Exercise 4B

1. $y = 13 - 2.78\mathrm{e}^{-0.33t} + 5.51\sin(0.262t + 2.999)$
   3.28 hours

2. (i) $\frac{1}{2T_0^3}\arctan\left(\frac{T}{T_0}\right)-\frac{1}{4T_0^3}\ln\left(\frac{T-T_0}{T+T_0}\right)=at+c$

3. (i) $T=\frac{10}{k}(1-\mathrm{e}^{-kt})$

   $\lim_{t\to\infty} T=\frac{10}{k}$

   (ii) $k = 0.159$

4. (i) $\frac{dy}{dt}=-k(y-T_0)+u$

   (ii) $u = 1.2$

Exercise 4C

1. (i) 4002, 4598, 5282 (millions)
   (ii) Prediction is approximately 0.8 billion short of actual population
   (iii) During 2046

2. (ii) $P \to 10000$ in either case
   (iv) (a) $P = 10000 - 5000e^{-t/20}$
   (b) $P = 10000 + 5000e^{-t/20}$

3. (ii) $h = \dfrac{190.8}{1 + 22.85e^{-0.616t}}$

4. (i) $P \to 5000$

   (ii) (a) $P \to 4208$
   (b) $P \to 3667$

7. $r = 1.5,\ P = \dfrac{9e^{6t}}{(2 + e^{3t})^2}$

   $r = 3,\quad P = \dfrac{\sqrt{3}e^{6t}}{\sqrt{2 + e^{12t}}}$

   $r = 4,\quad P = \dfrac{e^{6t}\,3^{\frac{1}{3}}}{(e^{18t} + 2)^{\frac{1}{3}}}$

8. (a) $P = \exp(b + (\ln p_0 - b)e^{-at})$

9. (i) $x \to \infty, y \to 0$
   (ii) $x \to 0, y \to \infty$
   (iii) $x = 1000, y = 2000$

Exercise 4D

1. 93 metres

2. (a) 42 metres
   (b) 3.3 seconds
   (c) particle model, motion in a straight line, Newton's laws of motion, $g = 9.8\text{ms}^{-2}$

3. (a) $a = 9.8 - 2.83\sqrt{v}$ where a is the acceleration.
(c) 4 seconds

4. (i) 26.03 metres, 2.00 seconds
(ii) 5.6 seconds; speed on impact is 4.95ms^{-1}

5. (a) 2.55 seconds, $x = 478$, $y = 2467$
(b) 4.2 seconds, $x = 3630$, $y = 4983$

6. (b) speed $= \sqrt{4900 - 9800\mathrm{e}^{-0.14t} + 44900\mathrm{e}^{-0.28t}}$

Exercise 5A

1. (i) $y = A\mathrm{e}^{2x} + B\mathrm{e}^{-x}$
(ii) $y = A\mathrm{e}^{-2x} + Bx\mathrm{e}^{-2x}$
(iii) $y = \mathrm{e}^{-x}(A\cos 2x + B\sin 2x)$
(iv) $y = A\mathrm{e}^{-\frac{3}{2}x} + Bx\mathrm{e}^{-\frac{3}{2}x}$
(v) $y = A\mathrm{e}^{\frac{1}{2}x} + B\mathrm{e}^{-\frac{4}{3}x}$
(vi) $y = \mathrm{e}^{x}\left(A\cos\frac{1}{3}x + B\sin\frac{1}{3}x\right)$
(vii) $y = A\mathrm{e}^{2x} + B\mathrm{e}^{-2x}$
(viii) $y = A\cos 2x + B\sin 2x$
(ix) $y = A + B\mathrm{e}^{-4x}$
(x) $y = A\mathrm{e}^{\frac{2}{5}x} + Bx\mathrm{e}^{\frac{2}{5}x}$
(xi) $y = A\mathrm{e}^{\frac{5}{2}x} + B\mathrm{e}^{-\frac{7}{4}x}$
(xii) $y = \mathrm{e}^{\frac{2}{3}x}\left(A\cos\frac{2}{3}x + B\sin\frac{2}{3}x\right)$

2. (i) $y = \sin 2x$
(ii) $y = 3\mathrm{e}^{x}\sin\frac{1}{3}x$
(iii) $y = \mathrm{e}^{2x} + 3\mathrm{e}^{-x}$
(iv) $y = \mathrm{e}^{-x}(\cos 2x + 2\sin 2x)$
(v) $y = 2 - \mathrm{e}^{-4x}$
(vi) $y = 2\mathrm{e}^{-\frac{4}{3}x}$
(vii) $y = 2\mathrm{e}^{2x} + \mathrm{e}^{-2x} = 3\cosh 2x + \sinh 2x$

3. (i) $y = \frac{3}{2}\sin 2x$

(ii) $y = \sqrt{3}e^{-x}\sin\sqrt{3}x$

(iii) $y = 3xe^{-2x}$

(iv) $y = e^{x} - e^{-4x}$

4. (i) $y'' + 4y' - 5y = 0$

(ii) $y'' - 8y' + 25y = 0$

(iii) $y'' - 8y' + 16y = 0$

5. $a = -3$

7. (i) F

(ii) A

(iii) B

(iv) C

(v) A

(vi) E

8. $\theta = A\cos\omega t + B\sin\omega t$

period $= \frac{2\pi}{\omega}$

Exercise 5B

1. (i) -25

(ii) $-2.5x + 0.25$

(iii) $0.2x^2 + 0.04x + 0.104$

(iv) $x^3 + 20x^2 + 125x + 325$

(v) $\frac{1}{5}e^{x}$

(vi) $2e^{2x}$

(vii) $\frac{1}{19}e^{3x}$

(viii) $2e^{x}$

(ix) $6\cos x + 8\sin x = 10\cos(x + 0.927)$

(x) $-\frac{3}{4}\cos 2x + \frac{1}{4}\sin 2x = 0.791\cos(2x + 3.46)$

(xi) $\frac{1}{3}\sin 3x$

(xii) $-2.5\cos 2x - 7\sin 2x = 7.43\cos(2x + 1.91)$

2. (i) $y=-2.5x+0.25-\frac{3}{4}\cos 2x+\frac{1}{4}\sin 2x$

(ii) (a) $-2e^{-x}+3x+2$

(b) $\frac{1}{2}e^{2x}+2\cos x$

(c) $-e^{2x}(5x+4)$

(d) $e^{2x}(\cos x+2\sin x)$

3. (i) $3x$

(ii) $e^{2x}\left(5x-\frac{5}{3}\right)$

(iii) $-\frac{1}{3}e^{-x}(6x+2)$

(iv) $-3x\cos 2x$

(v) $3x^2e^{-2x}$

4. (i) $ax+b$

(ii) $a\cos 3x+b\sin 3x$

(iii) axe^x

(iv) $ae^{-x}+bxe^{2x}$

(v) $e^{2x}(a\cos x+b\sin x)$

(vi) $(ax^2+bx)e^x$

Exercise 5C

1. (i) $y=Ae^{2x}+Be^{-x}-2.5x+0.25$

(ii) $y=Ae^{2x}+Be^{-x}-\frac{3}{4}\cos 2x+\frac{1}{4}\sin 2x$

(iii) $y=e^x\left(A\cos\frac{1}{3}x+B\sin\frac{1}{3}x\right)+2e^{2x}$

(iv) $y=Ae^{2x}+Be^{-2x}-25$

(v) $y=Ae^{\frac{1}{2}x}+Be^{-\frac{4}{3}x}-2.5\cos 2x-7\sin 2x$

(vi) $y=e^{\frac{2}{3}x}\left(A\cos\frac{2}{3}x+B\sin\frac{2}{3}x\right)+2e^x$

(vii) $y=e^{-x}(A\cos 2x+B\sin 2x)+0.2x^2+0.04x+0.104$

(viii) $y=Ae^{-\frac{3}{2}x}+Bxe^{-\frac{3}{2}x}+\frac{1}{3}\sin 3x$

(ix) $y=A+Be^{-4x}+\frac{1}{5}e^x$

(x) $y = A\cos 2x + B\sin 2x + \frac{1}{2}e^{2x} + 2\cos x$

(xi) $y = A\cos 2x + B\sin 2x - 3x\cos 2x$

(xii) $y = e^{-2x}(A + Bx + 3x^2)$

2. (i) $y = 0.75e^{2x} - 3e^{-x} - 2.5x + 0.25$

(ii) $y = 13e^{2x} + 13e^{-2x} - 25$

(iii) $y = e^{2x} - \frac{7}{4}e^{-x} - \frac{3}{4}\cos 2x + \frac{1}{4}\sin 2x$

(iv) $y = e^{\frac{2}{3}x}\left(-\cos\frac{2}{3}x + \sin\frac{2}{3}x\right) + 2e^{x}$

3. (i) $y = 2e^{2x} - 2.5e^{x} + \frac{3}{2}\cos x + \frac{1}{2}\sin x$

(ii) $y = e^{x}(4\sin 2x - 5\cos 2x) + 5e^{-x}$

(iii) $y = e^{\frac{3}{2}x}(6x - 2) + 5\cos x - 12\sin x$

(iv) $y = 4e^{5x} - 14e^{3x} + e^{2x}(6x + 8)$

(v) $y = -e^{3x} + e^{2x}(x - 2) - 2e^{-3x} + 2(x + 2)$

7. $q = \dfrac{E_0}{L(\omega^2 - p^2)}(\cos pt - \cos\omega t)$

$i = \dfrac{E_0}{L(\omega^2 - p^2)}(\omega\sin\omega t - p\sin pt)$

Exercise 5D

1. (i) $y = Ae^{-2x} + x + 2$

(ii) $y = Ae^{-\frac{5}{3}x} + 4e^{5x}$

(iii) $y = A + Be^{4x} + Ce^{-3x}$

(iv) $y = (A + Bx + Cx^2)e^{x} + \sin x - \cos x$

(v) $y = A + e^{x}(B\cos 2x + C\sin 2x) + x + 2e^{x}$

(vi) $y = Ae^{2x} + Be^{-2x} + Ce^{-3x} + Dxe^{-3x}$

(vii) $y = Ae^{2x} + Be^{-2x} + C\cos 3x + D\sin 3x - 2x + 2$

(viii) $y = Ae^{x} + Be^{-x} + C\cos x + D\sin x + \frac{1}{3}\sin 2x$

2. (i) $y = 2e^{x} - 2\cos 2x + 4\sin 2x$

(ii) $y = \frac{3}{2} + \frac{5}{4}e^{2x} + \frac{1}{4}e^{-2x} - \frac{1}{2}x^2$

(iii) $y = \cos 2x + \sin 2x - \cos x - 2\sin x + e^{2x}$

3. (i) $y = 3x^2 + x + A\ln x + B$
(ii) $y = (2x - 1)e^{2x} + Ax^2 + B$
(iii) $y = \cos x + x\sin x + A\sin x + B$
(iv) $y = x + \frac{A}{x} + B$

4. (i) $y = \ln(1 + x^2) + 2\tan^{-1}x + 1$
(ii) $y = (x + 3)e^{-x} + 2x$

5. (i) $y = Ax + Bx^{-7} - 2\ln x - 2$
(ii) $y = Ax^2 + Bx^2\ln x + x^3$
(iii) $y = x^{-2}(A\cos(\ln x) + B\sin(\ln x)) + 10(\ln x)^2 - 16\ln x + 12$

6. $y = \frac{1}{4x^7} + \frac{7}{4}x + \frac{1}{2}x\ln x$

7. $y = -\frac{11}{2}x - 5x\ln x + \frac{1}{2}x^3 - 2\ln x - 5$

8. (i) $y = Ax\cos 2x + Bx\sin 2x + x^2$
(ii) $y = Axe^{3x} + Bxe^{-3x} - 3x\cos x$

9. $y = Ae^{2\sqrt{x}} + Be^{-\sqrt{x}} - 2\sqrt{x} + 0.5$

10. $y = e^{x^2}(A\cos\sqrt{2}x + B\sin\sqrt{2}x + 3)$

11. (i) $x = -3\cos 3t - \sin 3t + 2t + 7$
$y = 2\sin 3t - 4t - 1$

(ii) $x = 3e^t - e^{-t} - 3\cos t$
$y = -e^t + e^{-t} + \sin t + \cos t$

12. (i) $x = 2\cos 2t,\ y = \cos 2t$

(ii) $x = 5\cos t,\ y = \cos t$

(iii) $x = 10\cos 2t - 10\cos t$
$y = 5\cos 2t - 2\cos t$

13. $y(2) \approx 5.617612$, which agrees with the analytical solution, to 6D.

Exercise 6A

1. (i) stiffness $= 8g = 78.4\text{Nm}^{-1}$
   (ii) 1.375 metres

2. (i) $m\ddot{x} + kx = kl$
   (ii) $x = l + \frac{l}{3}\cos\left(\sqrt{\frac{k}{m}}t\right)$
   (iii) amplitude $= \frac{l}{3}$, period $= 2\pi\sqrt{\frac{m}{k}}$, centres is $x = l$

3. (ii) $x = 0.012\sin 8.08t + 0.250\cos 8.08t + l + 0.15$
   (iii) 0.012 seconds
   (iii) 0.395 seconds

4. (ii) $i = A\cos 2.5t + B\sin 2.5t$

5. (i) $x_1 = 0.02\cos 11.07t + l + 0.08$
   $x_2 = 0.02\cos 7t + 0.143\sin 7t + l + 0.2$

   (iii) 0.54 seconds, 0.73 seconds

6. (i) $\theta = \frac{v}{\sqrt{14}}\sin\sqrt{14}t$

   $\theta_{max} = \frac{v}{\sqrt{14}}$ occurs when $t = \frac{n\pi}{\sqrt{14}}$ (n integer)

   (ii) $v < 5\text{ms}^{-1}$

Exercise 6B

1. $m\ddot{x} + r\dot{x} + kx = mg + kl_0$
   (a) $r = 8$ weak damping
   (b) $r = 10$ strong damping

2. (i) $\ddot{x} + \dot{x} + 40x = 38$
   lightly damped oscillations

   (ii) $x = e^{-0.5t}(0.15\sin 6.30t - 0.05\cos 6.30t) + 0.95$

3. Relative response $= \dfrac{200}{3\left|100-\alpha^2\right|}$

Resonance as $|\alpha| \to 10$

4. $r = 1.29$

5. steady state charge $-\dfrac{2}{65}\cos 50t - \dfrac{3}{65}\sin 50t$

transient charge $e^{-10t}\left(\dfrac{2}{65}\cos 30t + \dfrac{17}{195}\sin 30t\right)$

steady state current $\dfrac{20}{13}\sin 50t - \dfrac{30}{13}\cos 50t$

6. (i) $\theta = \dfrac{4\pi}{9}e^{-t} - \dfrac{\pi}{9}e^{-4t}$

(iii) $\theta \to 0$

7. (i) $\ddot{q} + 5\dot{q} + 6.25q = 5 + 5e^{-t}$

$q(0) = 2,\ \dot{q}(0) = 4$

(ii) $q = e^{-\frac{5}{2}t}\left(\dfrac{11}{3}t - \dfrac{46}{45}\right) + \dfrac{20}{9}e^{-t} + \dfrac{4}{5}$

(iv) $q \to 0.8$ as $t \to \infty$

8. (i) $\ddot{x} + 10\dot{x} + 400x = 0.05e^{-2t}\sin 3t$

$x(0) = 0,\ \dot{x}(0) = 0$

(where x is measured from equilibrium)

(ii) $x = e^{-5t}(6.39\times 10^{-6}\cos 5\sqrt{15}t - 1.96\times 10^{-5}\sin 5\sqrt{15}t)$

$-1.33 \times 10^{-4}e^{-2t}\sin(3t + 3.094)$

(iii) $5.9 \times 10^{-5}m$